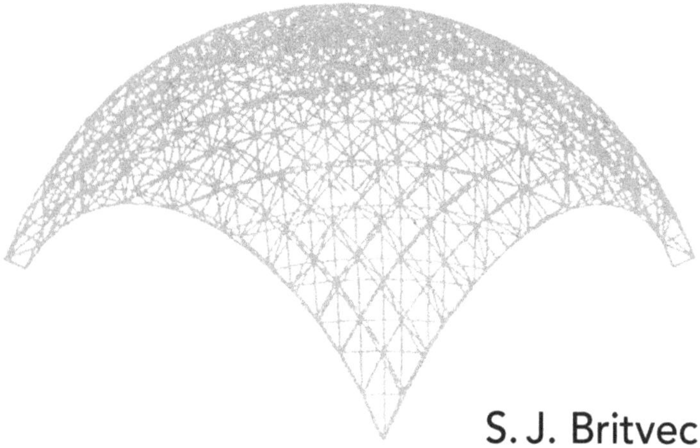

S. J. Britvec

Stability and Optimization of Flexible Space Structures

Birkhäuser Verlag
Basel · Boston · Berlin

Author

Univ. Prof. S.J. Britvec
Universität der Bundeswehr München
Fakultät für Luft- u. Raumfahrt
Institut für Mechanik
Werner Heisenberg Weg 39
D-85577 Neubiberg

Library of Congress Cataloging-in-Publication Data

Britvec, S. J.
Stability and optimization of flexible space structures /
S.J. Britvec
p. cm.
Includes bibliographical references and index.

1. Flexible structures. 2. Structural stability 3. Structural
optimization. I. Title
TA660.F53B75 1995
624. 1'71-dc20

Deutsche Bibliothek Cataloging-in-Publication Data

Britvec, Stanislaus J.:
Stability and optimization of flexible space structures /
S.J. Britvec. - Basel ; Boston ; Berlin : Birkhäuser, 1995

ISBN-13: 978-3-0348-9886-7 e-ISBN-13: 978-3-0348-9053-3
DOI: 10.1007/978-3-0348-9053-3

NE: Britwetz, Stanislaus J.

Softcover reprint of the hardcover 1st edition 1995

Printed on acid-free paper produced of chlorine-free pulp
Cover design: Markus Etterich, Basel

9 8 7 6 5 4 3 2 1

...but God chose what is foolish
in the world to shame the wise,
what is weak in the world to
shame the strong, what is low
and despised in the world, even
things that are not, to bring to
nothing things that are...

1 Cor. 1,2

...Alles reift zu seiner Zeit
und wird Frucht zu seiner Stunde

Divyavadana

Jesus said to them,...If you were blind,
you would have no guilt; but now that you
say, 'We see', your guilt remains.

John 9,41

....

Quantus tremor est futurus,
Quando judex est venturus,
Cuncta stricte discussurus!

....

Mors stupebit, et natura,
Cum resurget creatura,
Judicanti responsura.

Liber scriptus proferetur,
In quo totum continetur
Unde mundus judicetur.

Judex ergo cum sedebit,
Quidquid latet apparebit:
Nil inultum remanebit!

....

About the author

Professor S.J. Britvec (Ph.D. Cambridge University, England 1960, Privatdozent University of Stuttgart, Germany 1972, o. Professor with the University of Zagreb, Croatia 1974) has taught as professor in the United States since 1960 in the areas of the theory and mechanics of structures and materials, at Cornell and Harvard Universities, the University of Pittsburgh, the University of Colorado at Boulder, then the Universities of Maine and Delaware and lately at the University of the Bundeswehr, Fakultät für Luft- und Raumfahrttechnik in Munich, Germany. He has conducted research particularly in the field of elastic stability, dynamics and the nonlinear analysis of flexible light-weight structures in elastic post-buckling and was the first to discover the general characteristic nonlinear forms (paths) of postcritical equilibrium states in imperfection-sensitive elastic systems (trusses, frameworks, space lattices, etc.) and to prove experimentally and theoretically their existence in real structures (Cambridge, England 1957–1960).—The results of his research have been published in many papers and reports and presented at the ASCE Convention in San Francisco in 1963, the International Conferences on Space Structures in England in 1966 and 1984, the International Centre for Mechanical Sciences (CISM) in Udine, Italy in 1975, 1977, 1980 and 1982, the GAMM Conference in 1976 in Graz, Austria, at SECTAM 8 in 1976 in Virginia, USA, the ASME/ASCE International Conference in Albuquerque, N.M., in 1985 and at others and, most notably, in his book *The Stability of Elastic Systems*, Oxford 1973.—He is the inventor of the Brishell System, a new optimizable structural system of light-weight composite and metal space lattices with a variety of applications on earth and outer space.—Prof. Britvec was also associated with the aerospace and related industries in the USA, Britain, France and Germany, e.g. Martin-Marietta Aerospace Corp. in Denver, Colorado, USA, 1978–83, as Principal Investigator with the Office of Naval Research (ONR) in Washington D.C., USA, in a sole source project concerned with the stability and the control of shape of flexible space structures developed by NASA, 1985–87, then with the ONERA in France, W.S. Atkins & Pts. in England, the IABG and others in Germany.

Dr. Britvec is a member of the ASCE, ASME, Hon. Soc. Sigma Xi, the New York Academy of Sciences and is listed in the American Men of Science 1961, 1962 etc., Kürschner's Kalender Deutscher Gelehrter 1982, 1984 etc., Men of Achievement, Cambridge, England 1986, Personalities of America, American Biographical Institute, Raleigh, N.C. 1986, and in others.

Contents

Preface

The purpose in writing this book and in addressing the topics concerning the response and the behavior of structural lattices composed of flexible members was motivated by two main considerations. The first of these is concerned with the development of elastomechanics as it is applicable to elastic systems composed of flexible and slender members of different types, which has progressed very considerably in the last thirty years and which has penetrated into new domains of analysis involving not only various geometrical nonlinearities, which dominate the elastic behavior of flexible structures, but also new physical aspects of this behavior and their mathematical formulations. Particularly important are the changes of physical states associated with the static and dynamic instability in such systems. These states may give rise to the onset of unstable motion, released by influences due to small intrinsic imperfections, which can lead to a catastrophic collapse of the system. The analysis and the design become governed by these considerations in the first place. Allowable stresses in the material are not, therefore, the only, or the main consideration in making this design safe.

The author has made basic contributions to the development of this field by his longstanding research over some thirty years, which he started at Cambridge University in England in 1957 and continued at several institutions in the United States, notably at Cornell and Harvard Universities, the University of Colorado at Boulder and the University of Delaware until 1987. The results of this research were published in many papers and reports which are referenced later. The basic theory was presented systematically for the first time in a book entitled "**The Stability of Elastic Systems**" (**Pergamon Press Oxford, New York, 1973**). It covered the post-buckling behavior of the main categories of such systems in the range of limiting elastic deformations characterized by critical and post-critical states of stress in statical equilibrium, subjected to the influences of small geometrical imperfections. Also, elastic instability and the corresponding modes of failure of elastic systems, manifested by an accelerated motion away from unstable equilibrium, were addressed in this work. However, the theory was confined to the analysis of simpler systems of this kind, involving model structures in which such instabilities are relevant. Most of the theoretical results were verified by tests on these models and these test results agree well with theory.

As the structural systems, to be analyzed by this theory, became more complex, a need arose to develop more general methodologies for their analysis using powerful digital computers. It became clear immediately that, for many reasons, the standard finite element methods could not be suitably adapted to correctly reflect the physical behavior of such systems and this approach had to be replaced by a new methodology, which took into account the branching of equilibrium paths and the axial unloading of the non-buckling members at multiple branching points, which is very different from the classical finite element method.

The analysis of post-critical equilibrium states and of their stability in more complex space lattices and the proposed optimization strategies to improve the global stability limits of imperfection-sensitive structures made of flexible elastic members are based on the characteristic nonlinear theoretical forms of post-buckling equilibrium states (paths) of flexible continuous and discontinuous (quasi pin-jointed) structural elastic systems, such as plane and space lattices and frameworks, first discovered by the author (Cambridge, Selwyn College, England, 1957–1960; **Britvec**, Refs. 6, 7, 8, 9, 10, 11, 13, 15), who was also the first to prove experimentally the existence of these forms in real flexible structures in tests on scaled-down physical model-structures (Refs. 6, 7, 8, 9 et c.). These forms were later confirmed by other authors who are identified in Ref. 15 and later in this book. For a more detailed study of these aspects of the theoretical development the reader is referred to Ref. 15. The results of the basic theories are presented in this book in a form necessary for a further development of the theoretical approaches applicable to the analysis and optimization of the stability limits of more complex space structures and structural lattices intended for various applications in civil and aero-space engineering.

This theory was tested on many model structures of this type and agreement with the experimental observations could be established directly or indirectly on all models which were tested. In fact, the generalized theory for pin-jointed systems yields identical results in simpler cases, which were analyzed and tested before, some of which are reproduced later in this text. Generally speaking, the standard finite element methods are inadequate to analyse flexible structures for several reasons, as we shall see later in the text and throughout this book. Hence the need to present a much more comprehensive theory in the light of some of the most recent research and experimental and theoretical investigations carried out in the last thirty years in this field.

The main thrust of this area of applied mechanics developed as a direct consequence of an increased use for such systems in practical applications, such as the shell-type curved space lattices, suitable for the coverage of large areas or the enclosure of vast volumes, as support structures of large antenna dishes and the like. Lattices of this type may be enveloped by a cover-shell supported centrally at the peripheral nodes of the lattice, if this is required. Such modifications then permit a transmission of hydrostatic and aerodynamic pressures to the lattice and the foundation, where the lattice can be designed and optimized for an exceptionally high elastic stability limit and a load-bearing to weight ratio. Useful applications are today especially in the aerospace field, as space station main body structures are made using such systems. A good example is the first US space station, currently under development by NASA.

The new science emerged under several aspects, e.g. as the modern elastic stability theory, which applied mainly to slender and flexible structural elements, such as flexible bars or beams of different cross-sections, which could be assembled and connected in different ways to form useful structural systems of low weight and a relatively high structural performance. The flexibilities inherent in such systems gave rise to various geometrical and kinematic nonlinearities so that their physical behavior became very different from that of standard linear structures, commonly used in traditional civil engineering practice. Eventually a branch of applied mechanics, known as the "nonlinear mechanics of structures and solids" established itself over the last two decades. In this connection also other effects and characteristics of these systems, such as material nonlinearities and the finite changes in the geometrical configuration under large distortions gained in importance. Also, slight perturbations in the ideal geometry in such systems, the so-called small geometrical and other imperfections, became very important and these play an essential part in this theory, as their effects in the vicinity of unstable physical states are very strong, as we shall see later in Chapters 2 and 3 where this is explained in more detail.

The second consideration, which motivated the author in writing this book, is the present inadequacy of information about the physical behavior and updated analytical formulation, which correctly reflects this behavior and is, at the same time, useful for the analysis, machine computation, design and the optimization of such structures for practical applications in the aerospace and in the modern structural and civil engineering fields. One should point out the variety of elastomechanical response to be found in such systems, especially of interest to modern engineering applications in which passive and active control of the functioning of the system play a part as, for example, in the generic class of problems associated with the response and the active control of elastic nonlinear systems, involving flexible structural beam elements, which are not uncommon in the aerospace field today. Standard control methods used in linear systems cannot be always satisfactorily adapted to deal with problems involving a nonlinear elastic response, especially when minute changes in a parameter in the problem may result in a radical transformation of this response. This may be the case in structural lattices made of flexible members and connected by pinned or quasi-pinned connectors, when these are subjected to influences caused by very small geometrical imperfections. This is so, especially in the proximity of statically and dynamically unstable states, as we shall see later in the text, so that an active control of their response may be necessary. So, not only may the response deviate somewhat from linearity, as in stiffening or softening springs, but entirely new nonlinear forms of load-displacement curves may arise in such systems, which embody abrupt changes concerning the physical state and the performance of such a system and which may be even the cause of its collapse.

We mention one or two other aspects of this behavior which are important for a proper understanding of the performance and the response of the systems under discussion in general. One is the nonrepeatibility of the response in the loading process governed by the somewhat changing intrinsic imperfections which is to be controlled or minimized. This feature results in a kind of structural "degradation" of the system under repeated or pulsating loading, if the loading is dynamic and the magnitude of the load-reversals considerable. This effect should not be underestimated and it should be strictly distinguished from the type of degradation caused by the fatigue phenomenon. It is shown later in the text, that such characteristics are very common in structural systems of this type not only with potential applications in outer space, but also in other "reticulated" structures, particularly if their response embodies statically unstable states under a variety of loading conditions. These imperfections may be sufficient to drastically modify the deformation process under the limiting static or dynamic loading, this depending very much on the response of the perfect system in the post-buckling equilibrium, characterized by the negative gradients of the relevant load-displacement equilibrium paths after the global or local onset of flexural buckling in the lattice.

Another aspect concerns a high sensitivity of such systems to the loading patterns to which they are subjected and to the directions in which static or inertial loads are applied. This raises the possibility of a sudden partial instability of such systems under slightly deflected loads, which may be responsible for an entirely different picture of the response in which the structure reacts to a disturbance and which may result in a total and sudden loss of elastic stability and in a catastrophic event. Such events may be easily brought about in flexible space structures such as reticulated or lattice-type masts currently under development and intended for use in outer space. The characteristics of the collapse of such systems point to new strategies and techniques to be used in future control and structural optimization methods, especially if the flexible parts are large. Traditional methods and control devices that one might be tempted to use to ensure structural stability and to control the structural shape in such nonlinear systems under distortions and displacements which depend strongly nonlinearly on the applied loads, may be totally inadequate for this purpose. Active control of their shape using new types of strong actuators may be necessary and even very effective in the future in eliminating undesirable physical states in such systems which are potentially dangerous.

Flexible members in structural lattices of this type may be manufactured from light composite materials with suitable and reliable mechanical properties. This is certainly the case in lattices to be used in outer space. In such a case piezoelectric actuators and other sensors and devices based on remote control, which could be attached to the individual key members and adapted to produce bending restraints or stiffening effects in the key bars of the lattice, might prove to be very useful in the future to solve control problems in lattices composed of light flexible members, especially if the stiffnesses in key members can be temporarily adapted to suit the need, so that the relatively flexible lattice may be prevented from developing undesirable modes of deformation while undergoing various inertial maneuvres in outer space.

The structural analysis of the post-buckling behavior of complex lattices developed in Chapters 2 and 3 makes it possible to localize such key members in the critical range of static or quasi-static or inertial loading applied to such a lattice. This then provides a means of identifying the members which have to be stiffened in a given case, in order that the performance or the response of the structure may be influenced and, ultimately, optimized. One aspect that emerges from past studies carried out by the author in this regard, points to the fact that new structural optimization strategies may be possible and necessary in the future to relieve or minimize the control effort necessary to produce reliable and safe large space structures, particularly for use in outer space, in order to reduce their overall cost.

The purpose of this work is also to highlight the important aspects in the response of flexible structures of this type of interest to structural and control engineers, particularly in the aerospace field where, especially, statically highly indeterminate or hyperstatic structural lattices are assuming new and important roles and to point out the ways in which their elastomechanical response in the most critical range of loading and elastic deformation may be correctly and adequately formulated. This will reveal that several physical characteristics in such flexible systems are dominant in the type of response which is undesirable and which has to be controlled. We confine our main attention to light structural systems in which flexible members are braced and pin-jointed, or which may be modelled as such, and which have a wide variety of applications as space lattices in large satellites and space stations, space masts and similar large structures, not only in the aerospace field, but also in structural and civil engineering in which the choice of the material may vary from light alloys and high-tensile steels to various composite materials resistant to buckling by creep, as well as to variable humidity effects and temperature changes in the atmosphere, being capable of transmitting high axial loads over long periods of time reliably. The elastomechanical response of such systems is not widely understood and there is a common tendency to treat them as conventional linear structures, which they are not! The flexural or flexural-torsional origin of all deformations in such systems, including those of the nodal displacements which are predominantly governed by the flexural contractions of the compressive members and not by the axial compressions or extensions of the braced members, is most important for the correct understanding and the interpretation of the physical aspects which have been observed both experimentally and theoretically to dominate the response of such systems.

A more comprehensive text that would present the up-to-date theories and methodologies derived from the most recent research in this field has been long overdue. The purpose of this work is also to show that complex systems of this kind can be treated by this theory and that the generalization of the elementary theory, which was published before, is applicable to complex hyperstatic lattices made of flexible members. Both, perfect and geometrically slightly imperfect structures of this type can be treated by this theory as this is shown in Chapters 2 and 3. One important general aspect emerges from this treatment and that is that the gradients, and especially the negative gradients of the equilibrium paths for the equivalent perfect pin-jointed systems reflect the imperfection-sensitivity of the corresponding real system endowed with geometrical imperfections and that the equilibrium paths of the equivalent imperfect system can be computed once these theoretical gradients and a certain parameter are known. This parameter can be evaluated or estimated from tests on a similar scaled-down model structure and it depends on the similitude of the manufacturing process, the fabrication and assembly and the physical characteristics of a particular lattice and its loading. It stands to reason, that the intrinsic geometrical and other imperfections, inherent in such a system, depend on this process. The totality of small influences exerted by inumerable small imperfections is then reflected in such a test from which the mentioned parameter (or parameters) may be estimated. In an actual case, the test should be repeated several times to obtain an average estimate of its value for a given system. This procedure and its justification are given in Chapter 2, where the effects of the initial small geometrical imperfections in pin-jointed flexible structures are discussed. It should be stressed that in rigidly-jointed systems a different parameter plays the key role. This is illustrated and explained in Ref. 47 where these effects are discussed in rigidly-jointed systems. One concludes from this, that the connectors which connect flexible members in a structural lattice, play a very important role in the entire physical behavior of light structures made of flexible slender elastic members. This has been well known since the results of the first investigations conducted by the author at Cambridge University in England were published in 1963 (Refs. 6, 7, 8 and 9). The theoretical formulation of these effects, which confirms this behavior, is presented and explained in Ref. 47. The author very much regrets that, due to the lack of space in the present book, it was not possible to include the treatment of the response of rigidly-jointed systems at this time.

Pin-jointed or quasi pin-jointed or flexibly connected systems of this kind have definite advantages over rigidly-jointed lattices, as the buckled form in lattices connected by pins or hinges is confined to individual critically stressed compressive bars and it does not propagate across the hinges or flexible connectors to other adjointed members, as this is the case in rigidly connected lattices in which the buckling process affects the entire lattice simultaneously, the tensile members included. The buckling process in a pin-jointed lattice may thus be more easily controlled by the stiffening of a limited number of critical compressive bars in the lattice, which are responsible for the generation of undesirable unstable buckling modes. This paves the way to a general optimization strategy, especially in statically highly indeterminate or hyperstatic lattices, which are discussed in Chapter 3.

The internal redundant flexible members play an important part in both pin-jointed and rigidly-jointed systems of this kind. Especially, in pin-jointed lattices they contribute to the increase in the critical external loading and thereby to a higher load-carrying capacity of the system. At the same time, the linear axial displacement-compatibility conditions at the pinned joints no longer apply in the ultimate critical state, from which unrestrained buckling in a lattice may develop. Therefore, such linear relations should not be generally used for the determination of such states. Actually, axially rigid, but otherwise perfectly flexible elastic members may buckle as well in such a state and, as far as their post-buckling is concerned, there is practically no difference between the lattices made from axially highly stiff and axially rigid bars, as we show later in Chapter 1. The ultimate critical states of such structural lattices under centrally applied loads at the joints are determined by considering the compatibility conditions of the flexural contractions of the flexible members attached at each node. These conditions then replace the linear axial strain compatibility conditions used in the conventional linear theory. We show later in Chapters 1 and 3 that flexural contractions in flexible elastic bars beyond that state are much larger than their axial contractions or elongations. The nonlinearities of the equilibrium paths of the real or slightly imperfect structure and its loading are highly pronounced in that range, especially if the equivalent perfect structure is unstable in the ultimate critical state. The equilibrium path of the imperfect structure under the same loading is a slightly different path from the onset of loading and it by-passes the ultimate critical state of the equivalent perfect structure. But, it then converges to the tangent or the gradient of the *relevant perfect* post-buckling equilibrium path, once buckling is fully developed. This behavior of the imperfect structure and the perfect equivalent structure under the same loading is treated and explained in Chapters 1 and 2.

The contents of this work were shaped by a number of lecture series delivered by the author on this topic at various universities and institutions, but especially while the author taught graduate courses on this subject at the Univesity of Colorado at Boulder between 1979 and 1983, the University of Delaware from 1985 to 1987 and the University of the Bundeswehr in Munich between 1988 and to-day. The theory has its origins in the earlier work published by the author in many papers referenced throughout this text and in a first systematic presentation which appeared in a book entitled The Stability of Elastic Systems in 1973 (Pergamon Press, Oxford, New York). **The present work is intended as a text and as a reference book suitable for advanced study, which may prove to be very useful to graduate students and practising engineers in the field of modern structural and aerospace engineering.**

S.J. Britvec

Introduction

Space structures made of flexible light members represent a large category of elastic structural systems forming arbitrary geometrical shapes of braced truss-type three-dimensional arrays or lattices made of pin- or rigidly-jointed members. Such lattices may be flat or curved in space and, although they are composed of flexible straight elastic members, structures of this type are very stiff if supported and loaded by concentrated loads applied centrally at the nodes. The flexible members are then primarily loaded in axial compression or tension. Appreciable transverse loads are usually not applied to such systems, except at the nodes. External loads are normally applied centrally at the joints to which the flexible members may be attached by different types of connectors. Thus, the structural members of such a lattice are loaded mainly axially, so that only relatively small end shearing-forces and end-moments may be induced in the bars, arising from various excentricities of loading at the nodes or from very small geometrical initial imperfections inherent in the structure. However, in a perfect system in which the members are perfectly straight and in which no geometrical imperfections occur, the flexible members may, after the onset of flexural buckling, be subjected to internal end-moments and shearing forces induced by the external concentrated loads applied centrally at the joints, this depending very much on the intensities of the axial loads in the bars and on the manner in which the members are interconnected. Under rigid connectors, end-moments and shearing forces are set up at the member ends, which cause the members to buckle flexurally in two and three-dimensional shapes involving torsional-flexural deformations. The buckled form propagates in this case across the rigid nodes, as the continuity of the geometric form is conserved by the rigid connector elements.

On the other hand, if the bars are connected by means of pins or hinges, no end-moments or shearing forces are set up in the members after the onset of flexural buckling and no such propagation of flexural deformations takes place across the nodes, as the buckled form is confined only to the critically stressed bars which buckle and shorten flexurally in the post-buckling process. Therefore, in pin-jointed lattices buckling is constrained within the space lattice, so that the buckled form can be eliminated by a corresponding stiffening of the critically stressed bars. This is particularly true in statically indeterminate or hyperstatic pin-jointed systems, in which pin-connected members may be axially constrained by other adjacent bars, so that they remain practically straight while supporting more or less constant critical loads until the next intermediate critical state is reached, as the external loads are being increased in a constrained buckling process. Such states, in which certain bars are critically stressed, yet prevented from buckling freely, are called the *intermediate critical states* of a hyperstatic pin-jointed system. Only when a sufficient number of redundant compressive members become critically stressed, and then in a required sequence, can such a critical state of the system be followed by the unconstrained buckling of a set of critical bars and by the development of a post-buckling mechanism. This distinction between pin-jointed or quasi-hinged and rigidly-jointed lattices is very important, as the buckled postcritical states in pin-jointed systems may be more easily eliminated by the appropriate stiffening of the bars, than this is the case in rigidly-jointed structures of this kind.

In practical systems of this kind, the hinge-action has to be ensured, at least when the critical state in a bar is reached, so that sufficient end-rotations of the member may be developed in a post-buckling mechanism. Perfectly pinned connectors in conventional structures are practically non-existent today. In a first approximation, in an axially statically determinate structure axial loads in the not-so-slender bars of a conventional truss loaded centrally at the joints can be evaluated from statics, even if the bars are rigidly connected. In such trusses only secondary bending moments may be induced on account of various eccentricities.

In lightweight structures or lattices of this type made of flexible members, special connectors are necessary to ensure this hinge-action to make sure that several advantages of such systems over rigidly-jointed lattices may be exploited in potential practical applications. One type of such connectors, usually referred to as the Brishell or quasi-pin-jointed connectors, is discussed and described in Chapter 3. This connector develops a local hinge-like flexibility which provides the desired hinge-action precisely when it is required. The analysis of such quasi pin-jointed systems and their optimization for global stability and an increased load-carrying capacity is the same as for ideally hinged or pin-jointed structures, as we show later in the text. Quasi-pinned connectors of this type may be imagined to develop hinges at the ends of the attached members just before the axial loads in these members attain their critical (near-Eulerian) values, so that an unrestrained post-buckling mode or mechanism of deformation may develop beyond the ultimate critical state, as we show later, similarly as in the corresponding case in which these connectors are replaced by ideal hinges. Under the reduced working subcritical loads such connectors may be rigid, as their hinge-action is then not required. In practical design, based on this type of structural optimization, only the potential of the connectors to develop this hinge-like action in a *predetermined* state of stress in the member is actually required, so that the structural lattice may be optimized for the global stability in the critical range of loading under an increased load-bearing capacity. This is so, because the reduced working loads are limited by a factor of safety, so that under these loads the connectors are rigid and thereby contribute to the global stiffness of the lattice. The design under these circumstances results in a much more efficient and safe structural system, compared to the conventional design of a geometrically similar structure subjected to the same loading. Thus, an improvement, especially in regard to the global stability and a substantially increased global load-carrying capacity to weight ratio, becomes possible, as we show later in Chapter 3.

The statical analysis of a perfect lattice before the onset of buckling in the members is straightforward and it does not differ from the conventional linear theory. In pin-jointed systems, this analysis is confined to the determination of internal axial forces and the calculation of stresses and strains. At higher axial loads, when the critical states of certain members may be approached, the linear theory no longer holds and recourse must be made to the nonlinear theory and the post-buckling analysis, developed later in this text, which takes into account the effects of small geometrical imperfections.

In rigidly-jointed lattices secondary bending stresses may be induced in rigidly interconnected members on account of the axial extensibility of the members and the application of rigid connectors at the nodes, so that the geometrical continuity conditions may be conserved at the rigid joints and these are then responsible for the bending stresses induced in the members. However, in slender flexible members the axial extensibility is small compared to the flexural contractions induced by buckling, as the subcritical axial loads are not so high as in the stocky bars found in heavy conventional trusses. Thus, in the so called subcritical range of loading the dominant stresses in the flexible bars of a lightweight lattice are caused mainly by the compressive and tensile axial loads. Here, the elementary methods of statics may be used for their evaluation.

"Flexible space structures" or more precisely, space structures or space lattices *made from flexible members*, may be manufactured from different materials. Light metal alloys and temperature resistant composite materials may be used. Geometrically, such systems may be formed into any structural shape, such as flat or curved lattices in the form of beams, plates or shells, assembled from three-dimensional arrays of different braced polyhedral space elements made of light flexible tubular or solid members, commonly referred to as "reticulated" beams, plates or reticulated shells. Particularly important applications of such structural systems are nowadays in outer space, where three-dimensional structural arrays of this type may be part of the body of a space station (an example is the international space station currently under development in the U.S. space program). Other sophisticated systems of this type are support structures for large antennas, radio telescopes or radiometers, for example, which may be used on earth or in outer space. Reticulated structures of this kind and, particularly reticulated shells, exhibit a high global stiffness once their global stability is under control. The ways in which this stability may be optimized and controlled are discussed later in this book.

Shell-type structures of this kind may be used on earth for the coverage of relatively large areas without intermediate supports, capable of transmitting appreciable pressures and distributed loads to the foundations. It has been shown, Refs. 16, 20, that such reticulated shells, when enveloped by a continuous cover shell which is supported at the peripheral nodes of the lattice, can carry relatively high useful loads compared to the dead weight or mass of the lattice. In fact, if structures of this kind are properly optimized in the limiting range of loading, their load-bearing capacity to weight ratios may be by an order of magnitude higher (say 10 or 100 times higher) than those of comparable conventional shell structures made of conventional materials such as steel or reinforced and prestressed concrete, as this is shown and explained in more detail later in Chapter 3. Optimization strategies suitable for the structural optimization of hyperstatic lattices of this kind are discussed later in the text.

The analysis of all types of pin- or rigidly-jointed frames and lattices made of flexible, light members are in the first place governed by relatively large elastic flexural deformations of flexible members compared to conventional elastic structures. This is particularly reflected in the displacements of the nodes to which such members are connected, which are dominated by the flexural contractions of all the members in the system. These flexural shortenings become so large, that they by far overshadow the axial elastic elongations or axial contractions of the flexible bars caused by relatively low axial loads. Since, in a first approximation, the flexural shortening of such a member is proportional to the square of its lateral deflection, characterized by a suitable lateral coordinate of deformation, then the nodal displacements of such a system stand in a nonlinear relationship to the lateral deflections of the members. This is true after the onset of buckling in perfectly straight members, or before this onset, if certain small excentricities of loading and other intrinsic small geometrical imperfections are present in the system.—In rigidly-jointed frames or lattices also tensile members shorten flexurally. However, the flexural contractions of tensile members are much smaller than those encountered in compressive buckling bars.—In contrast to this, in pin-jointed or quasi pin-connected lattices or frames, only the compressive members can contract flexurally. But, these flexural contractions are very large, so that axial elongations or axial contractions of flexible members are negligibly small compared to their flexural shortenings.

The analysis of flexible compressive pin-jointed members sufficient to analyse their post-buckling behavior is presented in Chapter 1, Section 1.1. In this section also a new general relationship expressing the change in the axial load in relation to the flexural contraction is derived. This constitutive relationship is then generalized, so that it applies also to tensile and to subcritically stressed compressive straight pin-jointed members in the lattice after the onset of buckling regardless of whether or not this lattice is a statically determinate or an internally or externally statically indeterminate system and whether a critical bar actually buckles and shortens flexurally or whether it unloads and remains straight in the post-buckling process. In this generalized form this type of relationship is referred to in the text as the "general law" of pin-jointed members. It can be used in the analysis of any post-buckling process of a system of this kind.

One should point out that at the infinitesimal scale at any point and at any stage of deformation in the elastic system the elastic deformations are governed by Hooke's law, which is a linear relationship between the local stress and strain. But, globally the relationships between an applied load and a resulting elastic coordinate of deformation in such a system is far from being governed by a linear law. This is so, because of the geometrical nonlinearities and the nonlinear relations between different types of coordinates of deformation in the highly flexible system. We shall see later that qualitatively and quantitatively the deviations from linearity in the load-deformation patterns in such systems can be considerable. In fact, we show later in Chapter 2 and Chapter 3 that, under the influence of the initial geometrical imperfections, these relationships become highly nonlinear.

The analysis of plane pin-jointed statically determinate frames, loaded centrally at the nodes is developed in Section 1.2 in the first Chapter, using a simplified theory and the results obtained in Section 1.1. It is intended thereby to provide the reader with a clear understanding of a very different physical behavior in elastic systems composed from flexible slender bars from that encountered in conventional structures of this type in which the structural members are non-flexible or stocky. The differences are especially noticeable in the limiting range of loading before the onset of elastic instabilities in the system takes place. There, the elastic load-deformation relationships are not necessarily uniquely defined, so that several equilibrium load deformation paths may be possible beyond the first critical state of the system, this depending on the combinations of the buckling members and the critically stressed bars which may unload and remain straight in the post-buckling process. Such a behavior is then radically different from that encountered in conventional structures where equilibrium is always uniquely defined. This behavior is analysed on several model structures to give the reader a straightforward demonstration of the application of this theory. The same model structures were also tested experimentally and the experimental results are presented. These results show that in all cases they are in good agreement with the theory. Thus, the general validity of this theory is confirmed also experimentally. The theory may be used to analyse routinely the post-buckling processes in any type of pin-jointed structural lattices made from flexible elastic members. For more experimental evidence of this kind and a more elaborate presentation of certain elementary parts of this theory the reader is referred to Ref. 15, especially to Chapters 3, 6 and 7.

The statical equilibrium of pin-jointed or quasi pin-jointed complex three-dimensional elastic structural systems, connected by connectors which may develop hinge-like flexibilities under critical or near-critical axial compressive and tensile loads, is treated in detail in Chapters 2 and 3 in this book to cover also complex systems of this kind. The theory applies to such complex statically determinate and indeterminate three-dimensional systems of arbitrary geometrical shapes. The reader will notice that the theory differs radically from that used for the analysis of conventional linear structures. These differences are a consequence of a very different physical behavior of such systems on account of the high elastic flexibilities in the members and the relationships in which these affect the global response of the system, which is very different from the linear response of a conventional stable structure.

Also, a new concept of the total potential energy hypersurface for such an elastic system and the applied loading is developed in Chapter 3 to be used in the computational procedure for a direct evaluation of the so called most degrading buckling modes, which govern the static and the dynamic response of the structure after the onset of buckling. These modes are found from the criterion that the largest conversion of the total potential into the kinetic energy of an unstable elastic system of this kind takes place from its ultimate critical state in a dynamic collapse under those elastic deformations, which characterize the most negative global regions of the total potential energy hypersurface of the given structure and its loading. This assumption is found to be in agreement with experimental observations, as long as there are no appreciable energy losses in the process. The minimization process which makes it possible to evaluate directly the most degrading post-buckling modes in equilibrium and in motion is discussed in detail in Chapter 3. This theory, the evaluation of the most adverse modes of failure and their successive elimination are illustrated on a practical model reticulated spherical shell suitable for the transmission of high hydrostatic pressures. Further references to the literature are given in the text, where such and similar problems are treated by this theory.

Failure of structural systems composed of flexible members generally occurs in the elastic range of deformation on account of unstable equilibrium states on the primary loading path. The first such state is referred to as the first *ultimate critical state* in which one or several compressive flexible members are critically stressed and may be on the point of buckling. However, such members do not necessarily buckle as some of these may unload and remain straight, while others contract flexurally in the post-buckling process, which may be characterized by many equilibrium and non-equilibrium modes of failure. This instability is characterized by a sudden onset of unstable motion in the structure usually under constant loading and a dynamic collapse which follows. This collapse is essentially triggered off by various geometrical and other imperfections which are activated in an unstable equilibrium state. The mode in which the system usually collapses is referred to as the most degrading mode in which theoretically the largest amount of the total potential energy can be converted into kinetic energy.

Thus, the first event which takes place in the failure mechanism of such a system is this motion, which develops under the elastic deformation in the members. This is followed by a growth of elastic distortions while the system is in motion and, eventually, by a material yielding and a break-down at some point in the structure. It is important to note that the sequence in this failure mechanism is precisely opposite to that found in conventional structures, where, first, a material break-down takes place, which is then usually followed by a catastrophic collapse in motion.

Buckled states of flexible structures are also possible in statical equilibrium, provided the applied loads are appropriately adjusted. However, even then such states are mostly *statically unstable*, so that equilibrium is only possible under laboratory conditions, if the system is artificially stabilized. Buckling then takes place in one of many possible equilibrium modes of elastic deformation of the members, which may be based on very different elastic distortions in the buckling and the non-buckling members from those encountered under a *dynamic* collapse by buckling, say, in a *most degrading dynamic mode* under constant critical loads, mentioned earlier. The stability theorem developed in Chapter 1, Section 1.2, may be extended to a general critical state of a quasi pin-connected system of this type, to establish from an equilibrium analysis or test, whether the system is statically unstable in a critical state, from which a bifurcation of loading paths is possible in equilibrium. At that, the instability of the system is easier to prove than its stability. The concept of the total potential energy hypersurface, developed in Chapter 3 and applied to the global stability and instability analysis of such a system, is much more powerful and it leads to very useful analytical and numerical methods suitable for the direct computation of the most unstable or the most degrading post-buckling modes in statical equilibrium or in dynamic collapse, which are generally different. The equilibrium modes are found to be in very good agreement with experimental observations, as the reader may verify from this work and also from the tests reproduced in Ref. 15, in Chapters 3, 5, 6 and 7.

So, in flexible systems of this kind the dynamic collapse comes first, which is followed by a material breakdown in the structure. Therefore, the criterion for a safe design becomes not merely to prevent the stresses from exceeding the allowable limits, but *primarily to eliminate the onset of unstable motion in the first critical state* on the loading path, which is the cause of such a collapse. Thus, the design strategy becomes not only to limit the allowable stresses, as in conventional structures, but to prevent a change of physical states to occur, followed by a dynamic failure of the system. As we shall see later in the text, this tendency changes the entire scope and the strategy in the procedure where an improvement or an optimization of such structural systems is desired.

There have been several attempts, especially in the recent American technical literature, e.g. Refs. 2, 35, 36, 39 etc., to use the classical finite element and similar methods in the analysis of flexible structures of this kind. In this regard the question is posed, whether the post-buckling behavior of pin- or quasi pin-jointed lattices and of other similar elastic systems, assembled from flexible elastic bar-elements, can be analysed by standard finite element methods, so well known in the linear structural analysis. The answer to this is that the classical finite element method in its present form, *cannot* be meaningfully adapted to correctly reflect the physical behavior of such a system in the post-buckling process, when the members are connected by hinges or hinge-like connectors, or even when they are elastically or rigidly-jointed and that it generally leads to *erroneous* results which disagree with the experiments, as evidenced by the mentioned sources and the theoretical and experimental verifications presented in this and the related works referenced herein. The reasons are fundamental, as we shall see in a moment and throughout this text, as flexural buckling is a very different elastomechanical process from the linear elastic deformation process treated by the classical theory. This process embodies many bifurcations of equilibrium states which may result in different modes of nonlinear deformations in which each critically stressed compressive member has a choice to either contract flexurally, or to unload axially remaining straight beyond a bifurcation point or critical state, when unrestrained buckling develops from that state. This choice is automatically satisfied, if a new constitutive relation is introduced, referred to as the *general law of pin-jointed members*, developed in Section 1.1, combined with an energy or other relevant criterion, such as that of the largest conversion or loss of total potential energy discussed in Chapter 3. It is important to state in this connection that the linear displacement-compatibility conditions, usually employed to analyse linear hyperstatic lattices, do not apply generally to flexible members which contribute to the displacements of the nodes principally by flexural contractions. The procedure, how the statically indeterminate axial loads in the redundant members of a hyperstatic pin-jointed lattice are determined using the compatibility conditions of flexural contractions and an energy method, is described in Chapter 3 and in Ref. 15 in Chapter 7, pp. 329 to 389. In the behavior of rigidly-jointed hyperstatic braced frames and simple lattices these aspects are treated briefly in Ref. 15, pp. 269 to 311.

This behavior is qualitatively different from that of linear elastic structures. For example, not all critically stressed compressive members necessarily buckle and contract flexurally, as this is sometimes assumed in the technical literature, in a process essentially governed by nonlinear kinematic relations between these flexural contractions and the lateral deflections. Some of the members may do so, but others may choose to remain straight and unload axially if the kinematic and other conditions, which we discuss later in the text, permit this. Straight bars may then suffer only body rotations in the post-buckling mechanism of which they become a part. Which critically stressed bars buckle flexurally and contract and which do not and in which combinations the buckling and non-buckling axially unloading members form the physically most degrading modes that can be observed experimentally to occur in practical model structural systems of this kind and how this is decided and by what physical laws, are questions which cannot be resolved by the standard finite element theory. All these questions reflect new aspects of a different physical behavior and a different theory of much more complex systems, than those routinely covered by the classical theories. This type of behavior is not even addressed by the classical elasticity theory, nor can it be modeled by the classical finite element methods.

Hence, the necessity to present a new theory which takes into account all these features and correctly reflects the physical laws which govern the buckling and the post-buckling processes under finite elastic deformations of space structures and lattices made from flexible members subjected to the influence of small geometrical initial imperfections. Such a theory must be evidently very different from the standard finite element theory. The geometrical nonlinearities, inherent in these systems, may lead to a change of a physical state of the whole system and to a loss of statical stability, such that the system may relinquish all states of equilibrium and collapse dynamically in unstable motion. All this is reflected in such a theory, as we shall see later in the text.

Therefore, the answer to the original question posed at the beginning of this discussion is that, in order to describe or to reflect all these characteristics correctly, a theory must embody much more than this is the case in the standard finite element method used mainly to analyse linear or piece-wise linear structures, which are not subject to multiple bifurcations of equilibrium states on the loading path and to other influences caused by small geometrical imperfections and other factors, nor to changes of physical states which lead to a loss of stable equilibrium.

The theory presented in this work addresses mainly the analysis of pin-jointed or quasi pin-connected systems made from flexible members. Rigidly-jointed lattices under similar loading and other conditions show a radically different behavior and these are characterized by the nonlinear parabolic equilibrium paths. Such systems are considered briefly in the light of a more accurate theory in Ref. 47 and by an approximate theory in Ref. 15. Also, the effects of initial or intrinsic geometrical imperfections, in this case, are different from those encountered in pin-jointed systems of this kind which are treated in Chapter 2, Section 2.14 in this book.

In this connection it is relevant to point out that the deformations of elastically restrained members are strongly influenced by the continuity conditions at the nodes, which are non-existent in pin-jointed structures. An updated and more complete theory of the elastica of flexible members, which are elastically restrained at their ends, is also developed in Ref. 47, where this theory is applied in the equilibrium analysis of buckled rigidly-jointed plane braced frames. Experimental results obtained earlier on such model structures are also presented and these match well the theoretical results obtained by this updated theory. Further, the modified equilibrium paths in the presence of small geometrical imperfections are also analysed in some generality in the same reference by means of a perturbation theory under small finite changes of the geometry of the structure under loading in the post-critical equilibrium. Some experimental evidence of the existence of these modified equilibrium paths caused by small imperfections and modeled by this theory is also presented in Ref. 47. It was not possible to include these developments in the present book, because of the space limitations imposed on the length of the text. The author is planning to present these in a subsequent volume to make them accessible to a wider readership, as the Ref. 47 appeared only in a limited number of copies.

1 The Post-Buckling Analysis of Pin-Connected Slender Prismatic Members

1.1 The Postbuckling Behavior of Single Pin-Ended Elastic Members—General Law of Pin-Jointed Members

Many types of structural elastic systems are composed of pin-ended or hinged bars or members. The most common types are plane and three-dimensional space trusses commonly used in structural engineering applications. More recently new types of structural trusses composed of light, slender and very flexible but axially very stiff members have been increasingly used especially in the aerospace applications and in applications in outer-space as masts, or as supporting structures for space stations, large telescopes and radiometers and in other similar light structures of large dimensions. It turns out that such structural systems are very stiff, if connected by precise member fasteners and loaded or supported by concentrated loads centrally at the nodes which connect the members. Such structures can be formed into arbitrary global geometrical shapes, if the members are given different lengths. First, space elements such as tetrahedrons, octahedrons, cubes combined with tetrahedrons etc. are formed from such flexible members and connected. Then the space elements are assembled into a three-dimensional array forming such a space structure in which any two adjacent nodes are connected by a simple bar. As, in general, global geometry of such a structure may be curved as a shell, we speak of shelltype lattices or of *reticulated* shells[1]. Such structures may be enveloped by a membrane or cover-shell if necessary and they may be used effectively for the coverage of large areas, on the earth, for example.

The primary or basic element of such a structure is a light pin-jointed slender member, which is usually characterized by a fairly high slenderness ratio (S). Here ($S = L/r$) where (L) is the original length of the member and ($r = (I/A)^{1/2}$) is the radius of gyration of the cross-sectional area (A) of the member. (I) represents its moment of inertia about the principal bending axis and (E) its elastic modulus. If the members in a truss-type structure are slender, failure of the structure normally occurs by elastic instability (and not by the failure of the

[1] Derived from the Latin word reticulum ≡ net or network.

material!) as the critical stresses are exceeded when one or several members may buckle simultaneously. However, not all compressive members, which are initially on the point of buckling, necessarily develop unrestrained buckling, even if the truss is statically determinate, as some of the critically stressed members may unload remaining straight as they rotate bodily in a buckling mechanism during the collapse. How this happens and why, we shall see later, when we derive the general constitutive relations, called the general law for pin-ended bars.

First, we consider very accurately the elastic deformations of a single prismatic pin-jointed bar subjected to two equal and opposite compressive axial loads (P) applied without any eccentricities and increased to the critical value, when elastic flexural buckling of the bar becomes possible. We then analyse the system beyond that stage, when the buckled member continues to deform under an increasing post-critical axial load in what is known as the post-buckling range of deformation, or simply, in post-buckling. This behavior is most important for the response of the entire structure composed of such bars and it has to be analysed very accurately considering also the nonlinear effects.

Elasto-statically there is no difference between a pin-ended member and an elastic column, although, usually, elastic columns may be less slender. It is, therefore, of interest to examine the effects of the *axial extensibility* as well as that of the *elastic flexibility* on the overall buckling behavior of a column. This is accomplished by using the *energy method.*

We show first that in the post-buckling range the overall contraction of the column is followed by an increase in the end-load. Further, we demonstrate that the elasticity of the centroidal axis of the column has a negligible effect on the buckling behavior and that the load-contraction relation for an elastic column may be approximated by that corresponding to an elastic column with an *inextensional* centroidal axis. Such members are sometimes referred to as *axially-rigid.* Following this analysis, we consider the generalized load-displacement relations embodying the buckling behavior of *tensile* and *compressive* pin-ended members within a system.

a) Consider a uniform, perfectly elastic, straight column (Fig. 1.1.1) initially of length (L). Buckling of the column is assumed to take place in one plane only;

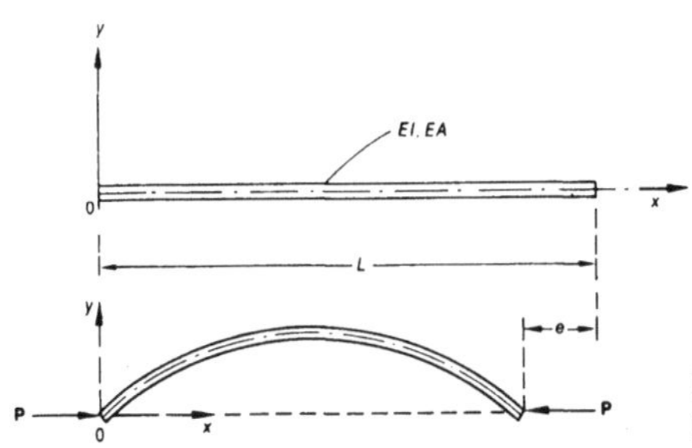

Figure 1.1.1
Initially straight member under axial compression

this plane is to be xy-plane, this implies that the cross-sectional form of the column is one for which buckling is restricted to the xy-plane. In Fig. 1.1.1 Ox is coincident with the initial centroidal longitudinal axis of the column. Oy is perpendicular to Ox. Flexural buckling of the column in the xy plane is considered under an axial compressive load (P), the column contracting an amount (e) between the loaded ends.

b) Suppose the point (a), on the centroidal axis of the straight column, is displaced to (a') in the buckled mode (Fig. 1.1.2). The point undergoes displacements (u), (v) parallel to Ox, Oy, respectively. Suppose that the normal to Ox at (a) is rotated by an angle (θ) due to buckling. The point (b) is adjacent to (a), at a small distance (δx). In the buckled mode, this distance from a' to b' along the deformed centroidal axis is (δs). Then

$$(\delta s)^2 = (\delta x + \delta u)^2 + (\delta v)^2 \tag{1.1.1}$$

so that, in the limit as (δx) becomes very small but remains different from zero, i.e. as $\delta x \to 0$

$$\left(\frac{ds}{dx}\right)^2 = \left(1 + \frac{du}{dx}\right)^2 + \left(\frac{dv}{dx}\right)^2 \tag{1.1.2}$$

The angle of rotation (θ) may be expressed in terms of (u) and (v) in the convenient form

$$\cos\theta = \left(1 + \frac{du}{dx}\right)\frac{dx}{ds} \tag{1.1.3}$$

The extensional linear strain of the centroidal axis is

$$\varepsilon = \frac{ds}{dx} - 1 \tag{1.1.4}$$

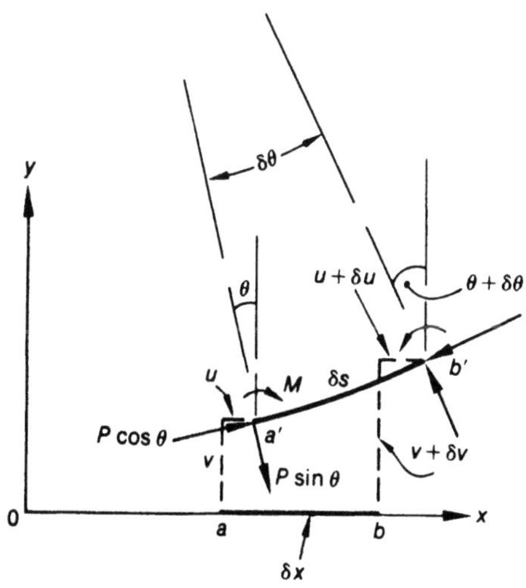

Figure 1.1.2
Displacements and stress resultants on an element length of the deflected member

But from Eq. (1.1.3)

$$\frac{ds}{dx} = \left(1 + \frac{du}{dx}\right)\frac{1}{\cos\theta} \tag{1.1.5}$$

so that the linear strain becomes

$$\varepsilon = \left(1 + \frac{du}{dx}\right)\frac{1}{\cos\theta} - 1 \tag{1.1.6}$$

c) It is evident there can be *no lateral* forces at the ends in the column in equilibrium. The only external forces are the two equal and opposite forces (P). The axial compressive force along the deformed centroidal axis is ($P\cos\theta$) and the bending moment (m) at any point of the buckled centroidal axis is

$$m = -Pv \tag{1.1.7}$$

d) We next derive the total potential energy change (V) of the buckled elastic member. There are *two forms of elastic energy*, i.e. that of *contraction due to axial loads, and that of bending*. The strain energy (U_c) of contraction may be written

$$U_c = \int_0^L \frac{P^2}{2EA}\cos^2\theta\, dx \tag{1.1.8}$$

since the axial thrust normal to the cross-section of the column is ($P\cos\theta$). It is convenient to express the strain energy, (U_B), of bending in terms of (θ), as

$$U_B = \int_0^L \frac{1}{2}EI\left(\frac{d\theta}{dx}\right)^2 dx \tag{1.1.9}$$

The total potential energy change of the column in the buckled form is then

$$V = \frac{P^2}{2EA}\int_0^L \cos^2\theta\, dx + \frac{1}{2}EI\int_0^L \left(\frac{d\theta}{dx}\right)^2 dx + P(u_L - u_0) \tag{1.1.10}$$

where (u_0), (u_L) are the end-displacements at ($x = 0, x = L$).

e) The linear extensional strain (ε), of the centroidal axis of the column is given by Eq. (1.1.6). This strain can be related to the axial thrust ($P\cos\theta$) by Hooke's law

$$\varepsilon = \left(1 + \frac{du}{dx}\right)\frac{1}{\cos\theta} - 1 = \frac{P\cos\theta}{EA} \tag{1.1.11}$$

so that

$$\frac{du}{dx} = (\cos\theta - 1) - \frac{P}{EA}\cos^2\theta \tag{1.1.12}$$

The contraction (e) of the column is

$$e = -(u_L - u_0) = \int_0^L \left(1 - \cos\theta + \frac{P}{EA}\cos^2\theta\right) dx \tag{1.1.13}$$

and the total potential energy change may be written

$$V = \frac{1}{2} EI \int_0^L \left(\frac{d\theta}{dx}\right)^2 dx - \frac{P^2}{2EA} \int_0^L \cos^2\theta \, dx - P \int_0^L (1 - \cos\theta) \, dx \tag{1.1.14}$$

Hence the total potential energy is a *functional* of (θ). For small values of (θ), this form of (V) may be suitably expanded into a power series to study approximate solutions.

f) When (θ) is small, approximately

$$\begin{aligned} \cos\theta &= 1 - \tfrac{1}{2}\theta^2 + \tfrac{1}{24}\theta^4 \\ \cos^2\theta &= 1 - \theta^2 + \tfrac{1}{3}\theta^4 \end{aligned} \tag{1.1.15}$$

Making use of the first few terms of each series, V becomes

$$\begin{aligned} V = &-\frac{P^2 L}{2EA} + \frac{1}{2} EI \int_0^L \left(\frac{d\theta}{dx}\right)^2 dx - \frac{1}{2} P\left(1 - \frac{P}{EA}\right) \int_0^L \theta^2 \, dx \\ &+ \frac{1}{24} P\left(1 - \frac{4P}{EA}\right) \int_0^L \theta^4 \, dx + \cdots \end{aligned} \tag{1.1.16}$$

Suppose the buckled shape of the column is given by the equation

$$\theta = \theta_0 \cos \frac{\pi x}{L} \tag{1.1.17}$$

where $(+\theta_0)$, $(-\theta_0)$ are the end-rotations at $(x = 0)$ and $(x = L)$ respectively. Therefore, the buckled form is symmetrical about the mid-length of the column. Then

$$V = \frac{P^2 L}{2EA} + \frac{1}{4}\left[P_E - P\left(1 - \frac{P}{EA}\right)\right] L\theta_0^2 + \frac{1}{64} PL\left(1 - \frac{4P}{EA}\right)\theta_0^4 + \cdots \tag{1.1.18}$$

where (P_E) is the Euler load, defined by

$$P_E = \frac{\pi^2 EI}{L^2} \tag{1.1.19}$$

Since the buckled mode is in statical equilibrium, $(\partial V/\partial\theta_0 = 0)$, which gives the non-trivial solution for $(\theta_0 \neq 0)$,

$$P_E - P\left(1 - \frac{P}{EA}\right) + \frac{P}{8}\left(1 - \frac{4P}{EA}\right)\theta_0^2 + \cdots = 0 \tag{1.1.20}$$

At $(\theta_0 = 0)$ this identity becomes

$$P_E - P\left(1 - \frac{P}{EA}\right) = 0 \tag{1.1.21}$$

Denoting the solution[2] of this equation by (P^0), we get

$$P_E - P^0\left(1 - \frac{P}{EA}\right) = 0 \tag{1.1.22}$$

so that

$$P^0 = \frac{EA}{2}\left(1 - \sqrt{1 - \frac{4P_E}{EA}}\right) \tag{1.1.23}$$

The load ($P = P^0$) defines the initial buckling of the flexible member *when its axial extensibility is considered.* But,

$$\frac{P_E}{EA} = \frac{\pi^2}{(L/r)^2} \tag{1.1.24}$$

where ($r^2 = I/A$), and (r) is the relevant radius of gyration of the cross-section. In practice, the smallest values or (L/r) are of the order of 30, so that P_E/EA is not likely to be greater than about 10^{-2}, which is small compared with unity. Equation (1.1.22) gives approximately,

$$P^0 = P_E\left(1 + \frac{P_E}{EA}\right) \tag{1.1.25}$$

The elastic buckling load of an extensible column is somewhat greater than the Euler load (P_E). It is important to remember that (P_E/EA) is not likely to be greater than approximately 1%.

g) An expression for the total axial contraction (e) of the column has already been derived, Eq. (1.1.13). When (θ) is small, (e) approximates to

$$e = \frac{PL}{EA} + \frac{1}{2}\left(1 - \frac{2P}{EA}\right)\int_0^L \theta^2\,dx - \frac{1}{24}\left(1 - \frac{8P}{EA}\right)\int_0^L \theta^4\,dx \tag{1.1.26}$$

With $\theta = \theta_0 \cos(\pi x/L)$, and only the second order powers of (θ_0) this contraction is

$$e = \frac{PL}{EA} + \frac{L}{4}\left(1 - \frac{2p}{EA}\right)\theta_0^2 \tag{1.1.27}$$

Ignoring higher powers of θ_0 Eq. (1.1.20) reduces to

$$P_E - P\left(1 - \frac{P}{EA}\right) + \frac{P}{2}\left(1 - \frac{4P}{EA}\right)\left(\frac{(e/L) - (P/EA)}{1 - (2P/EA)}\right) = 0 \tag{1.1.28}$$

Thus, the axial-flexural contraction parameter (λ) becomes,

$$\lambda = \frac{e}{L} = \frac{P}{EA} + 2\left(\frac{1 - (2P/EA)}{1 - (4P/EA)}\right)\left(1 - \frac{P}{EA} - \frac{P_E}{P}\right) \tag{1.1.29}$$

When (EA) is very large,

$$\lambda = \frac{e}{L} \to 2\left(1 - \frac{P_E}{P}\right) \tag{1.1.30}$$

[2] Britvec, S.J., 1963 "Elastic Buckling of Pin-jointed Frames", Int. J. Mech. Sci. Vol 5, pp. 447–461.

and for this particular case the slope of the post-buckling curve at $(P = P_E)$ is

$$\left(\frac{dP}{de}\right)_{P=P_E} = \frac{1}{2}\frac{P_E}{L} \tag{1.1.31}$$

More generally[3] then, at the buckling load $(P = P^0)$ from Eq. (1.1.29)

$$\left(\frac{dP}{de}\right)_{P=P^0} = \frac{1}{\dfrac{L}{EA} + \dfrac{2L}{P^0}\dfrac{(1-2P^0/EA)^2}{1-4P^0/EA}} \tag{1.1.32}$$

When $(P^0/EA) \ll 1$, this becomes

$$\left(\frac{dP}{de}\right)_{P=P^0} = \frac{P_E}{2L}\left(1 + \frac{P_E}{2EA}\right) \tag{1.1.33}$$

so that the P_E/EA term has only a very small effect on the slope of the post-buckling curve. The effect of P_E/EA on the post-buckling behavior is shown in Fig. 1.1.3. Since this effect is so small, it may be safely assumed that

$$\left(\frac{dP}{de}\right)_{P=P^0} = \frac{1}{2}\frac{P_E}{L} \tag{1.1.34}$$

The slope to the load displacement characteristic after the onset of flexural buckling is positive and the post-buckling of an elastic column takes place in *stable equilibrium.*

h) The relation between the contraction of the column and its lateral deflection can easily be found. The bending moment at any section is given by

$$m = EI\frac{d\theta}{dx} = -Pv \tag{1.1.35}$$

---

[3] Op. cit.

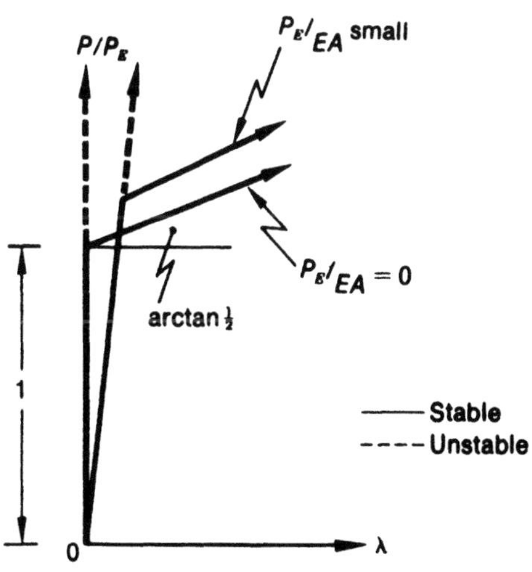

Figure 1.1.3
Load-contraction relations for compressed elastic members

so that

$$v = -\frac{EI}{P}\frac{d\theta}{dx} \tag{1.1.36}$$

If $\theta = \theta_0 \cos(\pi x/L)$, then

$$v = \frac{\pi EI}{PL}\theta_0 \sin\frac{\pi x}{L} \tag{1.1.37}$$

The maximum deflection occurs at the mid-length, and has the value

$$v_{\max} = \frac{\pi EI}{PL}\theta_0 = \frac{L}{\pi}\theta_0 \tag{1.1.38}$$

If (EA) is very large, then from Eq. (1.1.27) the flexural contraction is proportional to the square of the end-rotation (θ_0), i.e.

$$e \to \tfrac{1}{4}L\theta_0^2 \tag{1.1.39}$$

and, because of the limiting value of Eq. (1.1.39)

$$\frac{v_{\max}}{L} = \frac{2}{\pi}\sqrt{\lambda} \tag{1.1.40}$$

Therefore, in a first approximation, the lateral deflection is proportional to the square root of the flexural shortening parameter (λ).

From this discussion we conclude that the effect of the extensibility of the centroidal axis on the load-displacement relationship of a pin-ended flexible elastic member plays an *insignificant* part and that this effect may be neglected in the post-buckling behavior of flexible members.

In general, therefore, an elastic member may be replaced in the post-buckling range of deformation by an *inextensional* flexible member. When a *pin-jointed axially rigid or inextensional* member is on the point of buckling within a system the increase (δP) in the axial force gives rise to the purely flexural contraction (e) which may be expressed according to Eq. (1.1.34)‡

$$\frac{\delta P}{P^0} = \frac{1}{2}\frac{e}{L} = \frac{1}{2}\lambda \tag{1.1.41}$$

where $(P^0 = P_E)$ is the Euler load and (L) the initial length of the member. (λ is the non-dimensional form of e, i.e. $\lambda = e/L$). Generally, (P^0) represents the axial force in a particular member corresponding to a critical state of that member or of the whole system. This means, *if a particular member within the system is in a critical state*, then if the system is statically *determinate*, also the *whole system*

‡ Note that $\delta e = e - e_0 = e$, as $e_0 = 0$ in the case of an initially straight inextensible member.

is in a critical state of equilibrium. Therefore, the axial forces in other members depending on (P^0) are also critical. This state may be followed by the flexural contraction of a compressive member and an *increase* in its axial force, or by a decrease in its axial force, in which case the member does not contract flexurally but remains *straight.* (In Fig. 1.1.4 the unloading path of the member is indicated along a gradient whose value approaches ($-\infty$).) In reality this gradient is proportional to the negative value of the elastic *axial* stiffness (EA/L) of the bar which is very large so that in practice the unloading path becomes ($\delta P/\delta e \to -EA/L \simeq -\infty$), if ($EA$) is assumed to become infinitely large. In a tensile member this state may be followed either by a *decrease or by an increase*, but never by a *flexural extension* or a *flexural contraction* of the pin-jointed member, Fig. 1.1.5.

The buckling behavior of pin-ended members within a discrete system may, therefore, be expressed by the *general formula*

$$\frac{\delta P}{P^0} = K\frac{e}{L} = K\lambda \tag{1.1.42}$$

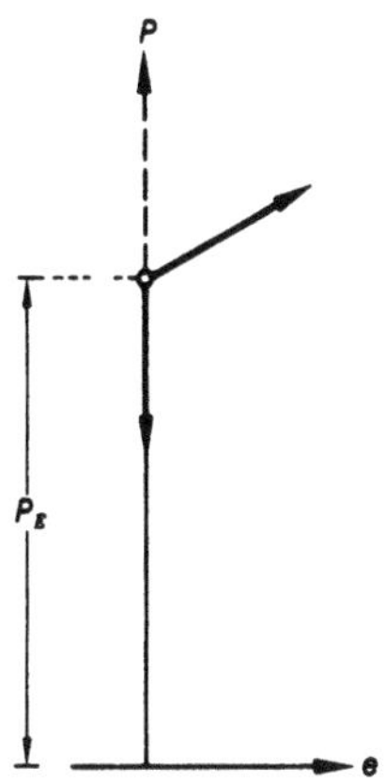

Figure 1.1.4
Buckling characteristics of compressive members

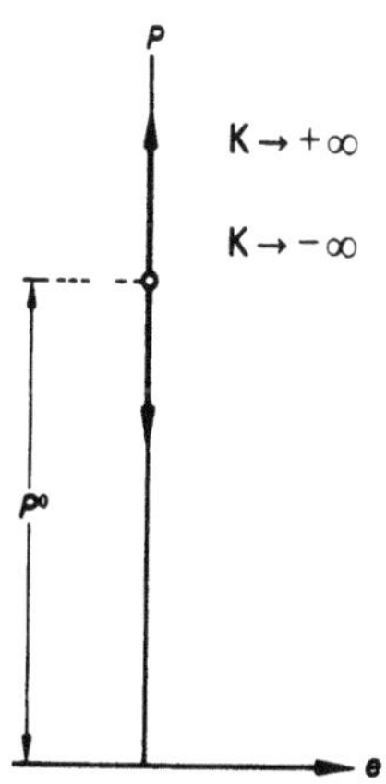

Figure 1.1.5
Buckling characteristics of tensile members

which constitutes the *general law* for pinned or hinged members, originally introduced by the author[4]. Here (e) is a small but finite change. (Note that $\delta\lambda = \lambda - \lambda^0 = \lambda$, since $\lambda^0 = e^0 = 0$.) The buckling constant (K) may take any values between ($+\infty$) and ($-\infty$). (P^0) is the axial force in the member corresponding to the critical state. Any decrease of the compressive load can occur without flexural shortening, so that ($e = \lambda = 0$). Where post-buckling may take place *with the decreasing load*—and this may be true of a non-prismatic member with a particular cross-sectional profile or property or of the plastic buckling of columns, see Ref. (15, 26)—a small change (δP) of compressive force may, or may not, be accompanied by a small contraction (e). This is indicated in Fig. 1.1.6.

Therefore, in the case of a uniform bar, Fig. 1.1.4, $K = 1/2$ or $-\infty$.

In the case of tensile members shown in Fig. 1.1.5 $K \to +\infty$ or $-\infty$.

Finally, in the case shown in Fig. 1.1.6 $K =$ constant or $-\infty$.

We note here that also non-prismatic members have critical axial loads which differ from the Euler load depending on the geometry of the particular bar. The post-buckling behavior of non-prismatic bars may take place in one plane, if buckling is not accompanied by twist. In such cases the constant (K) may have also positive values which are different from 1/2. [At present, the evaluation of the constants (K) for different non-prismatic elastic members is not pursued. The author wishes to point out that such data are important and useful and their evaluation should be pursued at a later stage.] Therefore in the case in Fig. 1.1.6 (K) may be generally a positive or a negative constant. (K) may be also negative if the member is not ideally hinged, but plastic hinges develop in the connectors, as explained later in Section 3.12. The load-displacement characteristics in Figs. 1.1.4 to 1.1.6 actually represent equilibrium paths in the coordinate planes (P, e) or (p, λ), where (p) and (λ) are the equivalent non-dimensional quantities, i.e. ($p = PL^2/EI$) and ($\lambda = e/L$).

[4] Britvec, refs. (7, 8) etc.

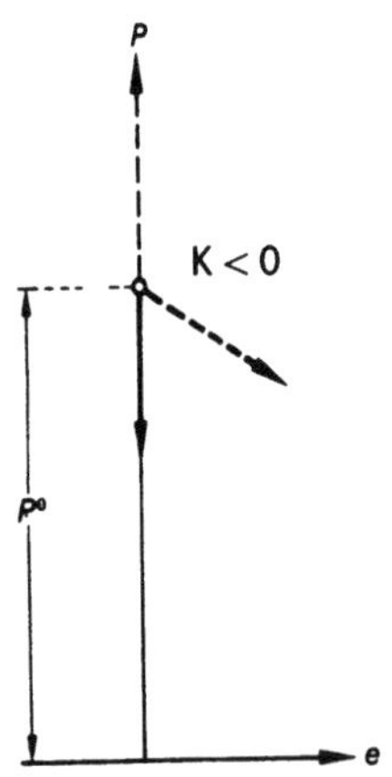

Figure 1.1.6
Buckling characteristics of unstable compressive members

1.2 Elastic Buckling of Pin-Jointed Plane Isostatic Trusses Composed of Flexible Bars

The buckling or post-buckling behavior of trusses connected by hinges or pins, or those connected by very flexible connectors, usually referred to as pin-jointed frames, is similar to the buckling behavior of a simple pin-ended member or column under axial thrust. While the simple column is always in stable equilibrium after buckling, pin-jointed frames may buckle under *stable* or *unstable* equilibrium conditions, depending on their properties and on the system of the applied loads. It must be emphasized here that in contrast to this, continuous or rigidly-jointed braced frames or trusses are *generally unstable* in the initial phase of buckling and the form of their equilibrium paths in this phase is highly non-linear and radically different from that of pin-jointed frames (Refs. 8, 9, 15).

The post-buckling behavior in this section is investigated theoretically on the assumption that the members in the frame are perfect and initially straight and their distortions after buckling small but not negligible compared to their lengths. Distortions of the buckled members are assumed to be due to flexure. It is confirmed by tests on model frames that statical equilibrium paths in post-buckling may be stable or unstable.

Elastic Stability in Statical Equilibrium

Consider the configuration of a conservative elastic system consisting of a lattice of pin-jointed flexible, initially straight, members under the action of external loads (W_i) acting centrally at the joints (i), when the system is in equilibrium in the critical state under the loads (W_i^0) and in an adjacent configuration after buckling, characterised by small nodal displacements (δs_i). The external loads (W_i) and the nodal displacements (δs_i) are referred to a global Cartesian coordinate system, so that (W_i) and (δs_i) actually represent the Cartesian components. If buckling of the structure comes about as a result of the gravitational field in which the masses concentrated at the nodes exert the external loads (W_i), a change (δH) in the potential energy (excluding the change in the elastic or strain energy) of the system from the critical state is then negative and given by

$$\delta H = -W_i^0\, \delta s_i; \qquad \text{summation on } i; \quad i = 1, 2, \ldots, J \tag{1.2.1}$$

where, usually, (δs_i) are positive, but may also be negative displacements. The external loads (W_i) or (W_i^0) need not be applied at all the joints and they may be zero in some coordinate directions of the global coordinate system. Suppose now the system is maintained in statical equilibrium in the adjacent configuration by a new and slightly different system of external forces ($W_i^0 + \delta W_i'$), where the increments ($\delta W_i'$) are essentially arbitrary and they may be applied at all the joints (i), in all coordinate directions, or only at some joints, the joints loaded by the external loads (W_i^0) included.

The load changes $(\delta W_i')$ are finite and it is assumed that the ratios $(\delta W_i'/\delta s_k)$, where (i) and (k) denote generally different joints, are also finite in the limit, when (δs_k) are very small quantities. Then, a change (ΔU) in the elastic or strain energy of the system may be written along the equilibrium path originating in the critical state as

$$\Delta U = W_i^0\,\delta s_i + \tfrac{1}{2}\,\delta W_i'\,\delta s_i \qquad \text{summation on } i; \quad i = 1, 2, \ldots, J \tag{1.2.2}$$

where (J) is the total number of loaded joints, and where (δs_i) generally embody the displacements of the joints loaded by the (W_i^0). The total mechanical energy (E) of an elastic conservative system is composed of the elastic energy (U), the potential energy (H) and the kinetic energy (T). The elastic and the potential energies at an instant constitute the total potential energy (V) of the system. If there are no losses on a path from the critical state, the change in the total energy (E) is zero, i.e.

$$\Delta E = \Delta U + \Delta H + \Delta T = 0 \tag{1.2.3}$$

or

$$\Delta E = \Delta V + \Delta T = 0$$

Thus

$$\Delta T = -\Delta V \tag{1.2.4}$$

The last result means that a conversion of energy from the total potential into kinetic energy in a motion between the critical and an adjacent state (not necessarily an equilibrium state) characterized, in this case, by the nodal displacements (δs_i) is possible only then if (ΔV) is *negative* in the adjacent configuration of the system. This, in fact, as we shall see later, is often the case in pin-jointed trusses or lattices loaded centrally at the joints.

It is now easy to calculate this change in the total potential energy in terms of the applied loads and the additional (hypothetical) load-changes $(\delta W')$. Putting $(\Delta V = \Delta U + \Delta H)$ and substituting from (1.2.1) and (1.2.2), we get

$$\Delta V = W_i^0\,\delta s_i + \tfrac{1}{2}\,\delta W_i'\,\delta s_i - W_i^0\,\delta s_i \tag{1.2.5}$$

or

$$\Delta V = \tfrac{1}{2}\,\delta W_i'\,\delta s_i; \qquad \text{summation on } i = 1, 2, \ldots, J \tag{1.2.6}$$

Thus, the change (ΔV) in the total potential energy of the system in the neighbourhood of an equilibrium position is in the first approximation given by half the sum of the small changes $(\delta W_i')$ in the externally applied loads at the nodes (i) and the small displacements (δs_i) of these nodes, such that, in the adjacent position, defined by the displacements (δs_i), the elastic system can be maintained in statical equilibrium by the loads $(W_i^0 + \delta W_i')$. From this we deduce the following

Theorem 1†

(i) If at least one physically possible adjacent statical equilibrium position can be found for which

$$\Delta V = \tfrac{1}{2}\,\delta W_i'\,\delta s_i < 0; \qquad \text{sum on } i = 1, 2, \ldots, J$$

is *negative*, then in the original position the statical equilibrium of the system under the loads (W^0) is *unstable*. This means that, under a small disturbance an onset of unstable motion under a kinetic energy conversion ($\Delta T = -\Delta V$) becomes possible.

(ii) If in *every* physically possible and kinematically admissible adjacent statical equilibrium position the total potential energy change given by

$$\Delta V = \tfrac{1}{2}\,\delta W_i'\,\delta s_i > 0; \qquad \text{sum on } i = 1, 2, \ldots, J$$

is *positive*, then in the original position the statical equilibrium of the system under the loads (W_i^0) is *stable*. This means that, under a small disturbance, the system returns to its original position of equilibrium.

(iii) If, the total potential energy change

$$\Delta V = \tfrac{1}{2}\,\delta W_i'\,\delta s_i = 0; \qquad \text{sum on } i = 1, 2, \ldots, J$$

is zero *in all* statically and kinematically admissible adjacent positions, the equilibrium of the system in the original position is regarded as neutral to the accuracy considered.

If, however in some adjacent positions ($\Delta V > 0$) and in others ($\Delta V = 0$), the stability of the system in the original position under the loads (W_i^0) is not uniquely defined in the sense explained under (i) and (ii), then higher order terms in the expression for (ΔV) must be considered.

Since the incremental components ($\delta W_i'$) of the applied loads and the corresponding displacements (δs_i) can be measured experimentally in an equilibrium test of a model structure, then, provided the frictional and other losses incurred in the test are small, Eq. (1.2.6) may be used to predict the stability of an elastic system experimentally. This is shown later on examples of model pin-jointed trusses.

Statical Equilibrium Conditions of the Buckled Structure

Having investigated the post-buckling of single members and the question of elastic stability in statical equilibrium, we proceed to examine the equilibrium conditions of a buckled plane truss-type structure which is statically internally and externally determinate.

Consider, therefore, a pin-jointed frame consisting of initially straight members. A typical frame is shown in Fig. 1.2.1. If (N) is the total number of members of

† See also Theorem 5, Ref. 15, page 74.

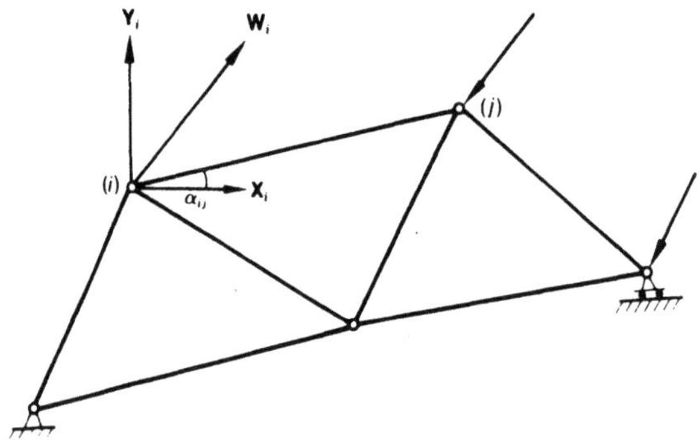

Figure 1.2.1
A typical pin-jointed frame

the frame and (J) the total number of joints, then for the frame to be internally isostatic

$$N = 2J - 3$$

The frame is subjected to a system of external loads (W) applied centrally at the joints. It is assumed that before buckling the components $(X_i), (Y_i)$ of the applied loads W_i are proportional to a load parameter (W). Thus

$$X_i = \xi_i W, \qquad Y = \eta_i W; \qquad i = 1, 2, \ldots \tag{1.2.7}$$

Buckling of the frame occurs when the thrusts (P_{ij}) in one or more of the compressed members reach their critical values (P_{ij}^0). The external loads are then defined by the equilibrium relations at the joints

$$X_i^0 = \xi_i W^0 = \sum_j (P_{ij}^0 \cos \alpha_{ij}) \tag{1.2.8a}$$

$$Y_i^0 = \eta_i W^0 = \sum_j (P_{ij}^0 \sin \alpha_{ij}) \tag{1.2.8b}$$

where (W^0) is the critical value of the load parameter and (P_{ij}^0) is positive when compressive. (ξ_i) and (η_i) are load-proportionality constants and (α_{ij}) is the angular co-ordinate of member (ij).

For the equilibrium of the system to be possible after buckling the applied loads are assumed to change by a small amount (δW_i). Moreover, the components of the load changes (δW_i) are supposed to be proportional to a reference increment of load (δW). Thus

$$\delta X_i = \mu_i \delta W \tag{1.2.9a}$$

$$\delta Y_i = v_j \delta W \tag{1.2.9b}$$

where (μ_i) and (v_i) are load-proportionality constants and may be different from (ξ_i) and (η_i).

Immediately after buckling the members of the frame suffer small bodily rotations (ψ) and the buckling members contract flexurally by finite amounts (e),

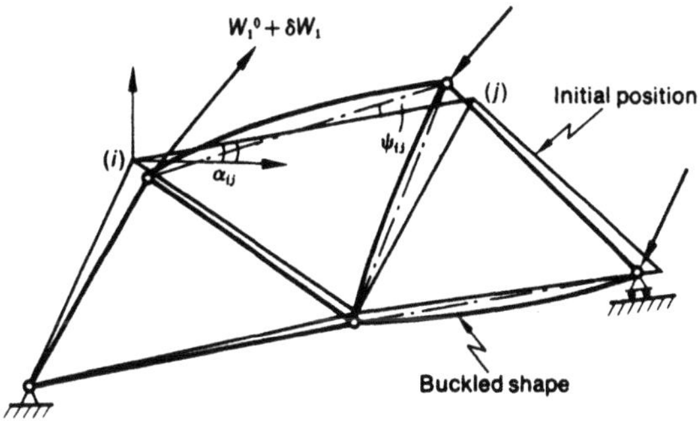

Figure 1.2.2
Buckled form of a pin-jointed frame

Fig. 1.2.2. The geometrical configuration of the buckled frame may be completely defined in terms of the flexural contractions (e). The relationship between the (ψ) and the (e) may be established on considering the geometrical compatibility of closed independent loops of members in the frame, Fig. 1.2.3. Each member of initial length (L_i) shortens by an amount (e_i) and rotates bodily[5] by a small angle (ψ). (e_i) is positive when it represents a shortening and (ψ) is assumed to be positive when the rotation is counter-clockwise. (α_i) is also positive counter-clockwise. Then if all the members shorten and rotate counter-clockwise, projecting horizontally,

$$\sum_{i=1}^{m} (L_i - e_i) \cos(\alpha_i + \psi_i) = 0 \tag{1.2.10a}$$

projecting vertically,

$$\sum_{i=1}^{m} (L_i - e_i) \sin(\alpha_i + \psi_i) = 0 \tag{1.2.10b}$$

The summation is taken around a closed loop of (m) members as indicated in Fig. (1.2.3).

[5] The dummy index (i) now denotes a member in the numbering sequence in a closed loop of bars.

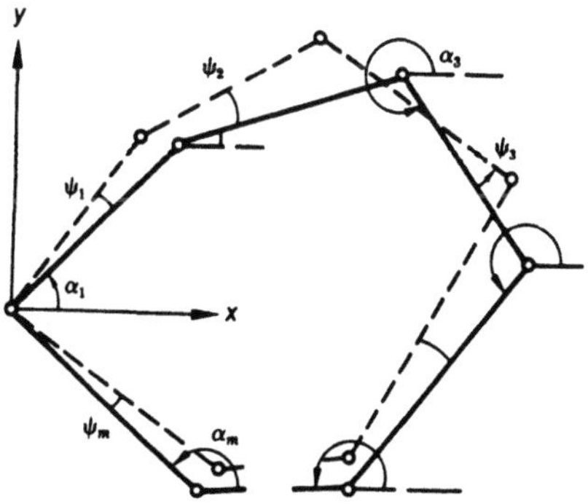

Figure 1.2.3
A closed loop of members in a pin-jointed frame

When $(e_i = 0)$ and $(\psi_i = 0)$ these equations reduce to

$$\sum_{i=1}^{m} (L_i \cos \alpha_i) = 0 \tag{1.2.11a}$$

$$\sum_{i=1}^{m} (L_i \sin \alpha_i) = 0 \tag{1.2.11b}$$

On linearizing the relations (1.2.10) and allowing for (1.2.11), the following geometrical compatibility equations are obtained

$$\sum_{i=1}^{m} (e_i \cos \alpha_i + L_i \psi_i \sin \alpha_i) = 0 \tag{1.2.12a}$$

$$\sum_{i=1}^{m} (e_i \sin \alpha_i - L_i \psi_i \cos \alpha_i) = 0 \tag{1.2.12b}$$

Equations (1.2.12) may be written for any closed loop members in the frame.

An isostatic triangulated frame of (N) members contains (T) independent loops (for proof see Ref. 15, p. 201) where

$$2T = N - 1. \tag{1.2.13}$$

Another compatibility condition is derived from the fact that the projection of the distance between the simple supports of the frame onto the direction in which the roller is prevented to move is constant. Then the $(2T)$ independent relations (1.2.12) together with the condition of support give $(2T + 1 = N)$ equations required to determine the (ψ) in terms of the (e) in the form

$$\psi_i = \sum_j c_{ij} \lambda_j; \qquad i, j = 1, 2, \ldots, N \tag{1.2.14}$$

where $(\lambda_j = e_j / L_j)$ and (c_{ij}) are known geometrical constants.

The equations of statical equilibrium at joint (i) of the *buckled* frame are

$$X_i^0 + \delta X_i = \xi_i W^0 + \mu_i \delta W = \sum_j (P_{ij}^0 + \delta P_{ij}) \cos(\alpha_{ij} + \psi_{ij}) \tag{1.2.15a}$$

$$Y_i^0 + \delta Y_i = \eta_i W^0 + v_1 \delta W = \sum_j (P_{ij}^0 + \delta P_{ij}) \sin(\alpha_{ij} + \psi_{ij}) \tag{1.2.15b}$$

Here the (δP_{ij}) represent small but finite changes in the axial loads from their critical values. On linearizing these equations and taking into account, Eqs. (1.2.8),

$$\delta X_i = \mu_i \delta W = \sum_j (\delta P_{ij} \cos \alpha_{ij} - P_{ij}^0 \psi_{ij} \sin \alpha_{ij}) \tag{1.2.16a}$$

$$\delta Y_i = v_i \delta W = \sum_j (\delta P_{ij} \sin \alpha_{ij} + P_{ij}^0 \psi_{ij} \cos \alpha_{ij}) \tag{1.2.16b}$$

Making use now of the relation (1.1.42) and of the compatibility condition (1.2.14) the equilibrium equations (1.2.16) can be written in the form

$$\mu_i \, \delta W = \sum_j P_{ij}^0 \left[K_{ij} \lambda_{ij} \cos \alpha_{ij} - \left(\sum_{k=1}^{N} c_{(ij)k} \lambda_k \right) \sin \alpha_{ij} \right] \tag{1.2.17a}$$

$$v_i \, \delta W = \sum_j P_{ij}^0 \left[K_{ij} \lambda_{ij} \sin \alpha_{ij} + \left(\sum_{k=1}^{N} c_{(ij)k} \lambda_k \right) \cos \alpha_{ij} \right] \tag{1.2.17b}$$

where (K_{ij}) is the buckling constant in the general law which, for prismatic members is equal to $(1/2)$ or $(\pm\infty)$. There are $(2J - 3 = N)$ equations of this type. Hence, the small flexural contractions (λ) of the buckled members can be found in terms of the change (δW) in the load parameter. If member (i, j) shortens $(K_{ij} = 1/2)$; if it remains straight, $(K_{ij} \to \pm\infty$ and $\lambda_{ij} \to 0$, such that $(\lim[K_{ij} P_{ij}^0 \lambda_{ij}] = \delta P_{ij})$ which is finite and may be positive, negative or zero. Then, also the changes (δP_{ij}) in the axial loads can be found in terms of the load parameter (δW). If a member is on the point of buckling but remains straight in post-buckling, (δP_{ij}) is always negative or zero. Finally, the Cartesian component joint displacements (δs_{xi}) and (δs_{yi}) can also be expressed in terms of (δW).

Applications of the Theory—Tests[6] on Simple Model Pin-Jointed Frames

Tests on simple model pin-jointed frames under the action of one and two external loads are presented to verify the post-buckling behavior of pin-jointed trusses in statical equilibrium. The compressive members in these frames were made from high-tensile spring steel. The cross-sectional areas (nominally 1 in. wide by 1/16 in. = 0.15875 cm thick) and the initial lengths of the compressive members (usually 15 in. = 38.10 cm) were chosen to obtain high slenderness ratios in order that buckling of the members may take place in the elastic range. Thus the elastic axial shortenings of the members could be reduced to a negligible minimum. However, some initial curvature of the members was always present.

The pinned attachment of the adjoining compressive and tensile members was achieved by introducing knife-edges at both ends of the struts and flat notches at the adjoining points on the tensile members, thus ensuring the freedom of rotation at the joints in the plane of the frame.

A hydraulic device was used to study the unstable post-buckling paths of the model frames. The method consisted in loading the frame by means of dead weight (lead shot) in a cylindrical container which was allowed to sink into a matching water vessel. The diameters of the container and the water vessel were arranged so that the rate of fall-off of the applied load was much greater than the

[6] Britvec, S.J., 1963 "Elastic Buckling of Pin-jointed Frames", Int. J. Mech. Sci. Vol 5, pp. 447–460.

rate of fall-off of the load along the equilibrium path in the frame. The applied load was measured directly by a dynamometer placed between the frame and the lead-shot container. The whole assembly, including the frame and the loading device, represents a stable system, so that an *unstable* equilibrium path of the loaded frame could be studied without the motion of the frame developing under decreasing loads. Measurements of vertical displacements of the frame were carried out by optical means. The evaluation of the incipient critical loads for the pin-ended members was carried out separately using the Southwell plot (Ref. 15, p. 231).

Test 1

The simple frame shown in Fig. 1.2.4 consists of members 1 and 2 which are initially of the same length ($L = 15$ in. $= 38.10$ cm). The members are made from high-tensile steel strip and are reasonably straight initially. The critical axial load

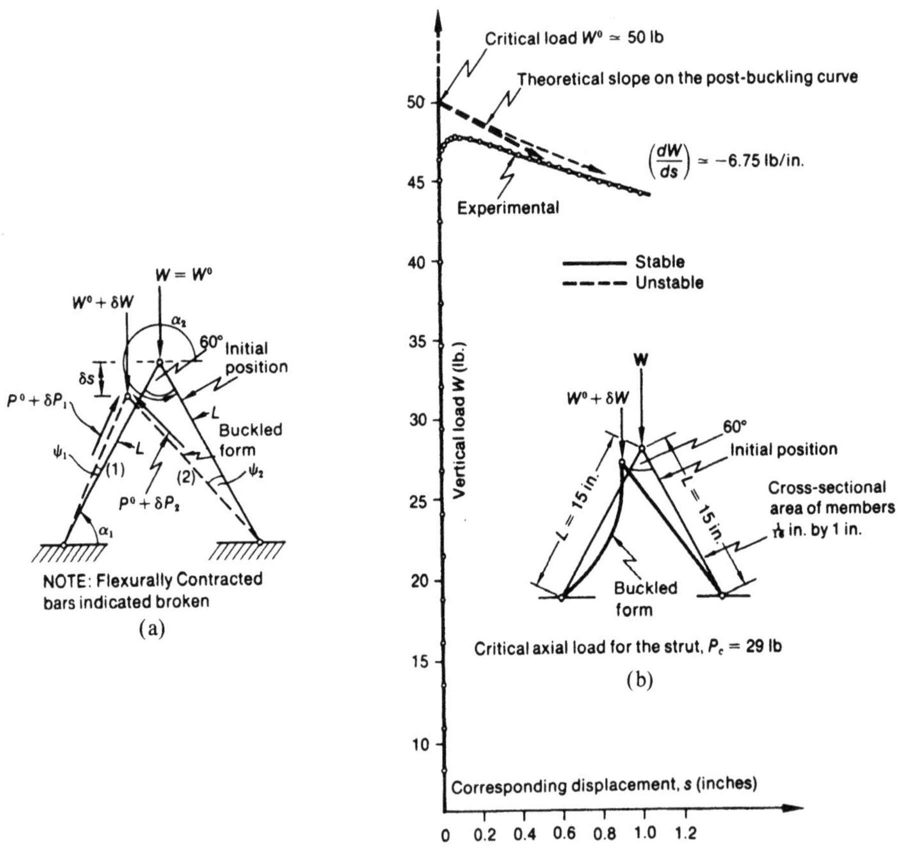

Figure 1.2.4
Elastic buckling of pin-jointed frames. Test 1

(P^0), for the strut equals (29 lb. = 13.15 kg*[7]). The corresponding applied load (W^0) equals (50 lb. = 22.68 kg*).

A vertical load (W) is applied at the free joint. It is assumed that both members are on the point of buckling and may shorten by small amounts (e_1) and (e_2) respectively. The Eqs. (1.2.1) may be applied to the closed loop consisting of members 1 and 2, and the distance between the supports, Fig. 1.2.4a.

Then, the geometry after buckling requires that,

$$e_1 \cos 60^\circ + L\psi_1 \sin 60^\circ + e_2 \cos 300^\circ + L\psi_2 \sin 300^\circ = 0 \tag{1.2.18a}$$

$$e_1 \sin 60^\circ - L\psi_1 \cos 60^\circ + e_2 \sin 300^\circ - L\psi_2 \cos 300^\circ = 0 \tag{1.2.18b}$$

Hence

$$\sqrt{3}\psi_1 L = e_1 - 2e_2 \tag{1.2.18c}$$

$$\sqrt{3}\psi_2 L = 2e_1 - e_2 \tag{1.2.18d}$$

In this case it is sufficient to consider the equilibrium of forces at the loaded joint.

Resolving horizontally

$$(P_1^0 + \delta P_1)\cos(60 + \psi_1) - (P_2^0 + \delta P_2)\cos(60 - \psi_2) = 0 \tag{1.2.19a}$$

Resolving vertically

$$(P_1^0 + \delta P_1)\sin(60 + \psi_1) + (P_2^0 + \delta P_2)\sin(60 - \psi_2) - (W^0 + \delta W) = 0 \tag{1.2.19b}$$

At the incipient critical condition

$$P_1^0 = P_2^0 = P^0 \tag{1.2.20a}$$

$$2P^0 \sin 60 = W^0 \tag{1.2.20b}$$

On linearizing the equations (1.2.19) and substituting for (ψ_1) and (ψ_2) from (1.2.18),

$$\frac{\delta P_1}{P^0} - \frac{\delta P_2}{P^0} - 3\frac{e_1}{L} + 3\frac{e_2}{L} = 0 \tag{1.2.21a}$$

$$3\frac{\delta P_1}{P^0} + 3\frac{\delta P_2}{P^0} - \frac{e_1}{L} - \frac{e_2}{L} = 6\frac{\delta W}{W^0} \tag{1.2.21b}$$

Using the general law in Eq. (1.1.42), these equations may be written in the form

$$\begin{pmatrix} K_1 - 3 & -(K_2 - 3) \\ 3K_1 - 1 & 3(K_2 - 1) \end{pmatrix} \begin{pmatrix} \lambda_1 \\ \lambda_2 \end{pmatrix} = 6\frac{\delta W}{W^0} \begin{pmatrix} 0 \\ 1 \end{pmatrix} \tag{1.2.22}$$

[7] Here, kg* denotes Kg. weight.

where $(\lambda_i = e_i/L)$, $i = 1, 2$. Denoting the small corresponding displacement of (W) by (δs),

$$\delta s = L \sin 60 - (L - e_1)\sin(60 + \psi_1)$$

On linearizing and substituting for (ψ_1) and (ψ_2) from (1.2.18) the expression becomes

$$\sqrt{3}\,\delta s = e_1 + e_2 \tag{1.2.23}$$

(i) *One Member Buckling*

If only member 1 shortens, then $K_1 = 1/2$ and $K_2 \to \pm\infty$.

$$\lambda_1 = 6\frac{\delta W}{W^0} = \frac{\begin{vmatrix} 0 & -(K_2 - 3) \\ 1 & 3K_2 - 1 \end{vmatrix}}{\begin{vmatrix} K_1 - 3 & -(K_2 - 3) \\ 3K_1 - 1 & 3K_2 - 1 \end{vmatrix}}$$

Dividing both determinants by K_2 and putting $K_1 = 1/2$.

$$\lambda_1 = 6\frac{\delta W}{W^0} \quad \frac{\begin{vmatrix} 0 & -1 + \dfrac{3}{K_2} \\ 1 & 3 - \dfrac{1}{K_2} \end{vmatrix}}{\begin{vmatrix} -\dfrac{5}{2} & -1 + \dfrac{3}{K_2} \\ \dfrac{1}{2} & 3 - \dfrac{1}{K_2} \end{vmatrix}}$$

Putting further $K_2 \to \pm\infty$ and evaluating the ratio of determinants,

$$\lambda_1 = -\frac{6}{7}\frac{\delta W}{W^0} \quad \text{and} \quad \lambda_2 = 0 \tag{1.2.24}$$

From (1.2.23), if $(\lambda_2 = 0)$,

$$\delta s\sqrt{3} = e_1 \tag{1.2.25}$$

and from (1.2.24) and (1.2.25)

$$\frac{\delta W}{\delta s} = -\frac{7\sqrt{3}}{6}\frac{W^0}{L} = -6.75 \text{ lb./in.} = -1.20 \text{ kg*/cm}^{(8} \tag{1.2.26}$$

Then the slope of the post-buckling load-displacement characteristic at the incipient critical condition is negative, Fig. 1.2.4.

If member 2 remains straight, the axial force in that member must decrease by a small amount (δP_2).

---

8 *Note:* 1 kg* denotes the force exerted by a mass of 1 kg in the field of gravity on earth.

From (1.2.21a), since ($\lambda_2 = 0$)

$$\frac{\delta P_2}{P^0} = -3\frac{e_1}{L} + \frac{\delta P_1}{P^0}$$

Putting

$$\frac{\delta P_1}{P^0} = \frac{1}{2}\frac{e_1}{L} \quad \text{gives} \quad \frac{\delta P_2}{P^0} = -\frac{5}{2}\frac{e_1}{L} = -\frac{5}{2}\lambda_1 \tag{1.2.27}$$

Since (λ_1) is positive, the small change (δP_2) in the axial load of member 2 is negative.

(ii) *Both Members Buckling*

If both members shorten ($K_1 = K_2 = 1/2$). The solution of (1.2.22) gives in this case

$$\lambda_1 = \lambda_2 = 6\frac{\delta W}{W^0} \tag{1.2.28}$$

Using equations (1.2.23) and (1.2.28) the slope

$$\frac{\delta W}{\delta s} = \frac{\sqrt{3}}{12}\frac{W^0}{L} = 4.81 \text{ lb./in.} = 0.86 \text{ kg*/cm} \tag{1.2.29}$$

Then, if both members shorten the slope is positive; if only one member shortens it is negative. It follows that two equilibrium paths in post-buckling are possible.

The experimental load-displacement characteristic after buckling is shown in Fig. 1.2.4b. Experimentally the frame is observed to remain practically undistorted until a certain critical range of loading is reached. In this region large distortions begin to take place. At first both members show the tendency to buckle; however, only one of the members continues to deform appreciably while the other remains nearly straight.

The stability of the system at the onset of buckling may be determined according to Theorem 1. In the case of only one member buckling

$$\Delta V = -\frac{6.75}{2}(\delta s)^2 < 0$$

i.e. the perfect system is initially unstable.

The change in the applied load may be expressed in terms of the maximum lateral deflection of the buckling member using relation (1.1.40). Then

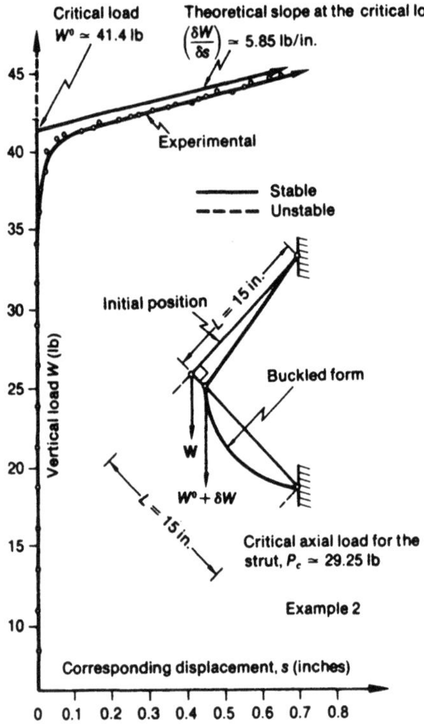

Figure 1.2.5
Elastic buckling of pin jointed frames. Test 2

approximately,

$$\delta W = -\frac{7\pi^2 W^0}{24}\left(\frac{v_{m1}}{L}\right)^2 = -0.64 v_{m1}^2 \tag{1.2.30}$$

where (v_{m1}) is the maximum lateral deflection of member (1). It is evident from this, that for the same decrease in the applied load, the lateral deflection is much larger than the corresponding displacement of the loaded joint.

Test 2

The frame shown in Fig. 1.2.5 consists of two equal members initially at right-angles. The members have roughly the same properties as in the previous example. The analysis is carried out similarly as in the first example. The critical compressive load ($P^0 = 29.25$ lb. $= 13.27$ kg*) and the corresponding applied load ($W^0 = 41.40$ lb. $= 18.78$ kg*). In this case there is only one post-buckling mode. The post-buckling equilibrium path is given by

$$\frac{\delta W}{\delta s} = \frac{3\sqrt{2}}{2}\frac{W^0}{L} = 5.85 \text{ lb/in.} = 1.04 \text{ kg*/cm}$$

which is in good agreement with the experimental result. The post-buckling takes place in stable equilibrium.

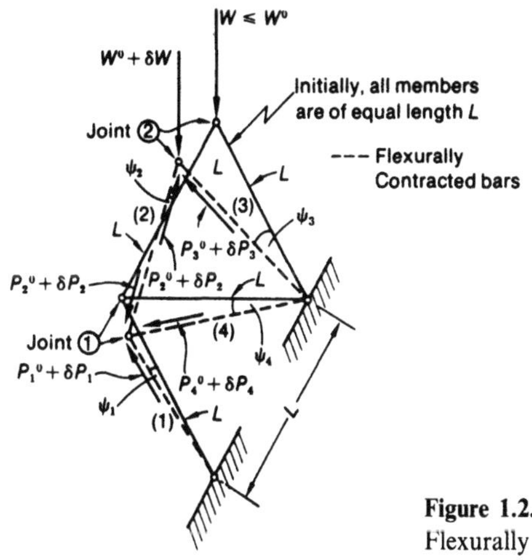

Figure 1.2.6
Flexurally contracted bars indicated broken

Test 3

Consider the frame shown in Fig. 1.2.6. It consists of four elastic members of initial length ($L = 15$ in. $= 38.10$ cm), ($P_0 = 29$ lb. $= 13.15$ kg*). The frame is loaded with a vertical load (W) applied at the top joint. Members 1, 2 and 3 are compressive and member 4 is tensile. If members 1, 2 and 3 are of the same cross-sectional area, then all the three members are on the point of buckling simultaneously when the applied load (W) reaches its critical value (W^0).

Geometry After Buckling

If all the members are assumed to shorten by the respective small amounts (e), and also to rotate bodily counterclockwise by small amounts (ψ), then, applying Eqs. (1.2.12) to the closed loop consisting of members 2, 3 and 4, and also to the loop consisting of members 1 and 4, and the distance (L) between the supports, the following relations may be derived,

$$\begin{aligned}
\psi_1\sqrt{3} &= \lambda_1 - 2\lambda_4 \\
\psi_2\sqrt{3} &= 2\lambda_1 + \lambda_2 - 2\lambda_3 \\
\psi_3\sqrt{3} &= 2\lambda_1 + 2\lambda_2 - \lambda_3 - 2\lambda_4 \\
\psi_4\sqrt{3} &= 2\lambda_1 - \lambda_4
\end{aligned} \tag{1.2.31}$$

where the small rotations (ψ) are assumed positive counterclockwise and ($e_i/L = \lambda_i$); $i = 1, 2, 3, 4$.

Equilibrium of Forces

Before the onset of buckling in the bars,

$$P_1^0 = P_2^0 = P_3^0 = \frac{W^0}{\sqrt{3}} \tag{1.2.32}$$

and

$$P_4^0 = -\frac{W^0}{\sqrt{3}} \tag{1.2.33}$$

Resolving the forces after buckling and retaining the linear terms in (ψ); at joint (1), parallel and normal to the distance between the supports,

$$\delta P_1 \cos 60 - P_1^0 \psi_1 \sin 60 - \delta P_4 \cos 60 - P_4^0 \psi_4 \sin 60 - \delta P_2 = 0 \tag{1.2.34}$$

and

$$\delta P_1 \sin 60 + P_1^0 \psi_1 \cos 60 + \delta P_4 \sin 60 - P_4^0 \psi_4 \cos 60 - P_2^0 \psi_2 = 0 \tag{1.2.35}$$

At joint (2), horizontally and vertically

$$\delta P_2 \sin 30 - P_2^0 \psi_2 \cos 30 - \delta P_3 \sin 30 - P_3^0 \psi_3 \cos 30 = 0 \tag{1.2.36}$$

and

$$\delta P_2 \cos 30 + P_2^0 \psi_2 \sin 30 + \delta P_3 \cos 30 - P_3^0 \psi_3 \sin 30 = \delta W \tag{1.2.37}$$

Introducing the relations between the changes in the axial forces and the corresponding small flexural shortenings, i.e. the general law (1.1.42),

$$\frac{\delta P_1}{P_1^0} = K_1 \lambda_1, \qquad \frac{\delta P_3}{P_3^0} = K_3 \lambda_3$$

$$\frac{\delta P_2}{P_2^0} = K_2 \lambda_2, \qquad \frac{\delta P_4}{P_4^0} = K_4 \lambda_4$$

On substititing for (ψ_1, ψ_2, ψ_3) and (ψ_4) from Eq. (1.2.31) and satisfying Eqs. (1.2.32) and (1.2.33), the Eqs. (1.2.34) to (1.2.37) may be written in matrix form as follows,

$$\begin{bmatrix} 1+K_1 & -2K_2 & 0 & 1+K_4 \\ 3K_1 - 1 & -2 & 4 & -3(1+K_4) \\ 4 & 3-K_2 & K_3 - 3 & -2 \\ 0 & 3K_2 - 1 & 3K_3 - 1 & 2 \end{bmatrix} \begin{bmatrix} \lambda_1 \\ \lambda_2 \\ \lambda_3 \\ \lambda_4 \end{bmatrix} = 6 \begin{bmatrix} 0 \\ 0 \\ 0 \\ 1 \end{bmatrix} \frac{\delta W}{W^0} \tag{1.2.38}$$

Small Corresponding Displacement

The small corresponding displacement (δs) of the applied load (W) equals,

$$\delta s = L \cos 30 - (L - e_3)\cos(30 + \psi_3) \tag{1.2.39}$$

Expanding, and retaining the linear terms

$$\delta s = L\psi_3 \sin 30 + e_3 \cos 30$$

Substituting for ψ_3 from Eq. (1.2.31)

$$\delta s = \frac{\sqrt{3}}{3}(e_1 + e_2 + e_3 - e_4) \tag{1.2.40}$$

Members (1) and (2), Buckling Simultaneously

From Eq. (1.2.38)

$$\lambda_1 = 6\,\frac{\begin{vmatrix} 0 & -2K_2 & 0 & 1+K_4 \\ 0 & -2 & 4 & -3(1+K_4) \\ 0 & 3-K_2 & K_3-3 & -2 \\ 1 & 3K_2-1 & 3K_3-1 & 2 \end{vmatrix}}{\begin{vmatrix} 1+K_1 & -2K_2 & 0 & 1+K_4 \\ 3K_1-1 & -2 & 4 & -3(1+K_4) \\ 4 & 3-K_2 & K_3-3 & -2 \\ 0 & 3K_2-1 & 3K_3-1 & 2 \end{vmatrix}}\,\frac{\delta W}{W^0} \tag{1.2.38a}$$

Dividing the determinants by (K_3, K_4) and putting

$$K_3 \to \pm\infty, \qquad K_4 \to \pm\infty$$

$$K_1 = K_2 = 1/2$$

the contraction parameter λ_1 becomes

$$\lambda_1 = \frac{e_1}{L} = 6\,\frac{\begin{vmatrix} 0 & -1 & 0 & 1 \\ 0 & -2 & 0 & -3 \\ 0 & 5/2 & 1 & 0 \\ 1 & 1/2 & 3 & 0 \end{vmatrix}}{\begin{vmatrix} 3/2 & -1 & 0 & 1 \\ 1/2 & -2 & 0 & -3 \\ 4 & 5/2 & 1 & 0 \\ 0 & 1/2 & 3 & 0 \end{vmatrix}}\,\frac{\delta W}{W^0} = -\frac{6}{19}\frac{\delta W}{W^0} \tag{1.2.41}$$

It is easily shown that in this case $(\lambda_2 = \lambda_1)$ and that

$$\sqrt{3}\,\delta s = e_1 + e_2 \tag{1.2.42}$$

The post-buckling slope is

$$\frac{\delta W}{\delta s} = -\frac{19\sqrt{3}}{12}\frac{W^0}{L} \tag{1.2.43}$$

From the last of Eqs. (1.2.38) it follows that

$$\frac{\delta P_3}{P_3^0} = -\frac{39}{6}\frac{e_2}{L} \tag{1.2.44}$$

As (e_2) is positive, the small change (δP_3) in the axial force of member 3 is negative, as it should be, as the straight member unloads.

Therefore, member 3 remains straight as buckling of members 1 and 2 develops.

The experimental curve and the theoretical slope confirm this as shown in Fig. 1.2.7.

Test 4

The frame considered in Test 3 is now loaded with two vertical loads W_1 and W_2 such that $(W_1 = 3W_2)$. Member 4 is here prevented from buckling. Figure 1.2.8 shows a mirrored image of this frame.

The analysis is generalized so that all members are assumed to contract by an amount (e). The case under consideration is then deduced from the general result. The geometrical relations are given again by Eq. (1.2.31) in Test 3 in which (ψ) is positive counterclockwise according to Fig. 1.2.6.

Equilibrium of Forces Before the Onset of Buckling

$$\begin{aligned} &P_1^0 = \frac{\sqrt{3}}{3}(W_1^0 - W_2^0) = \frac{2\sqrt{3}}{3}W_2^0; \qquad P_3^0 = \frac{2\sqrt{3}}{3}W_2^0 \\ &P_2^0 = -\frac{\sqrt{3}}{3}W_2^0; \qquad P_4^0 = \frac{\sqrt{3}}{3}(W_1^0 + W_2^0) = \frac{4\sqrt{3}}{3}W_2^0 \end{aligned} \tag{1.2.45}$$

Small Changes in the Axial Forces After Buckling

Since, by the generalized Eq. (1.1.42),

$$\begin{aligned} &\frac{\delta P_1}{P_1^0} = K_1\frac{e_1}{L}; \qquad \frac{\delta P_3}{P_3^0} = K_3\frac{e_3}{L} \\ &\frac{\delta P_2}{P_2^0} = K_2\frac{e_2}{L}; \qquad \frac{\delta P_4}{P_4^0} = K_4\frac{e_4}{L} \end{aligned} \tag{1.2.46}$$

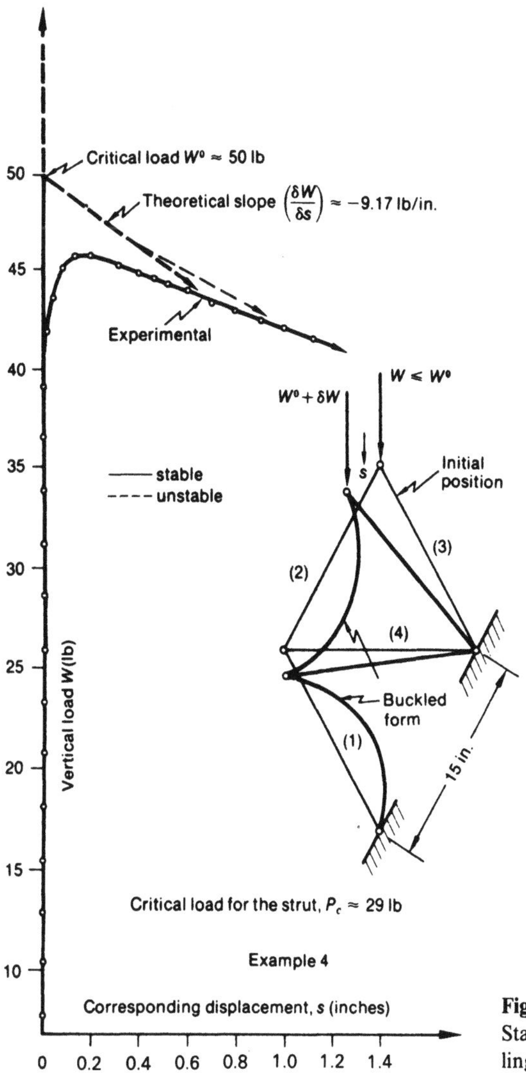

Figure 1.2.7
Statical equilibrium in the elastic post-buckling of simple isostatic frames. Test 3

substituting for (P_1^0, P_2^0, P_3^0) and (P_4^0) from Eq. (1.2.45) the respective small changes (δP) in the axial forces are

$$
\begin{aligned}
\delta P_1 &= \frac{\sqrt{3}}{3}(W_1^0 - W_2^0)K_1\lambda_1 \\
\delta P_2 &= -\frac{\sqrt{3}}{3} W_2^0 K_2 \lambda_2 \\
\delta P_3 &= \frac{2\sqrt{3}}{3} W_2^0 K_3 \lambda_3 \\
\delta P_4 &= \frac{\sqrt{3}}{3}(W_1^0 + W_2^0)K_4\lambda_4
\end{aligned}
\tag{1.2.47}
$$

where $(e_i/L = \lambda_i;\ i = 1, 2, 3, 4)$.

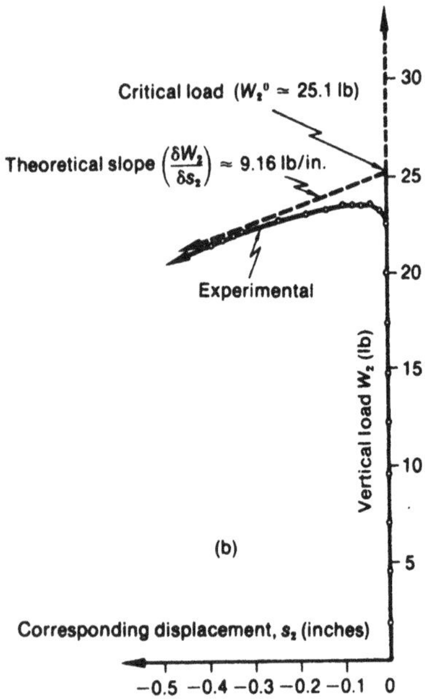

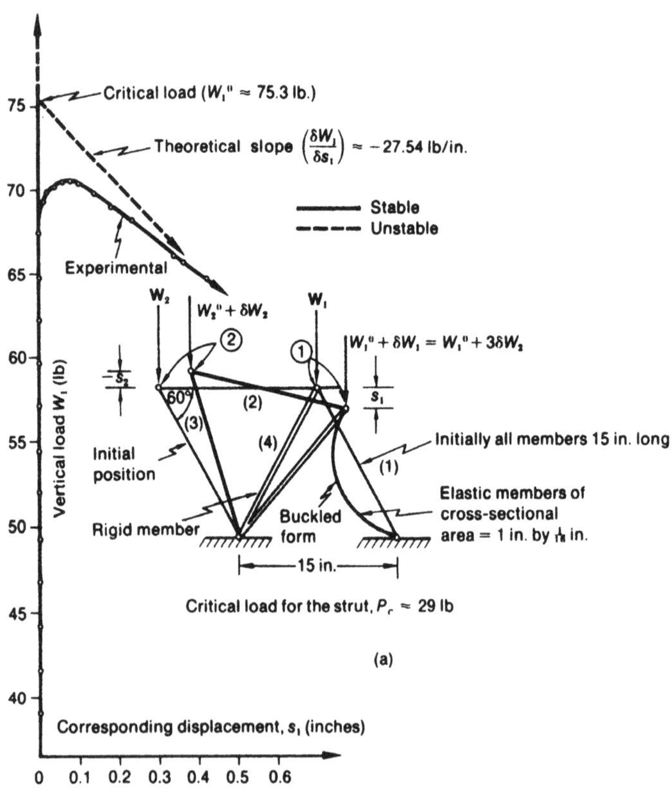

Figure 1.2.8
Statical equilibrium in the elastic post-buckling of simple isostatic frames. Test 4

Equilibrium of Forces After the Onset of Buckling

Resolving the forces in the horizontal and vertical directions at joint (1),

$$\begin{aligned}(P_1^0+\delta P_1)\cos(60+\psi_1)-(P_4^0+\delta P_4)\cos(60-\psi_4)\\ -(P_2^0+\delta P_2)\cos\psi_2=0\end{aligned}\tag{1.2.48}$$

and

$$\begin{aligned}(P_1^0+\delta P_1)\sin(60+\psi_1)+(P_4^0+\delta P_4)\sin(60-\psi_4)\\ -(P_2^0+\delta P_2)\sin\psi_2-(W_1^0+\delta W_1)=0\end{aligned}\tag{1.2.49}$$

At joint (2),

$$(P_2^0+\delta P_2)\cos\psi_2+(P_3^0+\delta P_3)\cos(60+\psi_3)=0\tag{1.2.50}$$

and

$$(P_2^0+\delta P_2)\sin\psi_2+(P_3^0+\delta P)\sin(60+\psi_3)-(W_2^0+\delta W_2)=0\tag{1.2.51}$$

On expanding to the first order of small quantities, further, substituting for (ψ_1, ψ_2, ψ_3) and (ψ_4) from Eq. (1.2.31) in Test 1, for $(\delta P_1, \delta P_2, \delta P_3)$ and (δP_4) from Eq. (1.2.47) above, and satisfying Eq. (1.2.45), the Eqs. (1.2.48) to (1.2.51) may be written in matrix form

$$\begin{bmatrix} -\left(3\frac{W_1^0}{W_2^0}+1\right)+K_1\left(\frac{W_1^0}{W_2^0}-1\right) & 2K_2 & 0 & 3\frac{W_1^0}{W_2^0}-1-K_4\left(\frac{W_1^0}{W_2^0}+1\right)\\ 1-\frac{W_1^0}{W_2^0}+3K_1\left(\frac{W_1^0}{W_2^0}-1\right) & 2 & -4 & 3-\frac{W_1^0}{W_2^0}+3K_4\left(\frac{W_1^0}{W_2^0}+1\right)\\ 2 & 2+K_2 & -(1+K_3) & -2\\ 0 & 1 & 1+3K_3 & -2 \end{bmatrix}$$

$$\cdot\begin{bmatrix}\lambda_1\\ \lambda_2\\ \lambda_3\\ \lambda_4\end{bmatrix}=\begin{bmatrix}0\\ 6\frac{\delta W_1}{W_2^0}\\ 0\\ 3\frac{\delta W_2}{W_2^0}\end{bmatrix}\tag{1.2.52}$$

Member 1 Shortening Alone

Then,

$$e_2=e_3=e_4=0\tag{1.2.53}$$

From Eq. (1.2.52), on allowing for Eq. (1.2.45)

$$\lambda_1 = 3 \frac{\begin{vmatrix} 0 & 2K_2 & 0 & 8-4K_4 \\ 6 & 2 & -4 & 12K_4 \\ 0 & 2+K_2 & -(1+K_3) & -2 \\ 1 & 1 & 1+3K_3 & -2 \end{vmatrix}}{\begin{vmatrix} -10+2K_1 & 2K_2 & 0 & 8-4K_4 \\ -2+6K_1 & 2 & -4 & 12K_4 \\ 2 & 2+K_2 & -(1+K_3) & -2 \\ 0 & 1 & 1+3K_3 & -2 \end{vmatrix}} \frac{\delta W_2}{W_2^0} \tag{1.2.54}$$

On dividing both determinants by (K_2, K_3, K_4), putting

$$K_2 \to \pm\infty, \qquad K_3 \to \pm\infty, \qquad K_4 \to \pm\infty$$

and also

$$K_1 = 1/2$$

the Eq. (1.2.54) reduces to

$$\lambda_1 = 3 \frac{\begin{vmatrix} 0 & 2 & 0 & -4 \\ 6 & 0 & 0 & 12 \\ 0 & 1 & -1 & 0 \\ 1 & 0 & 3 & 0 \end{vmatrix}}{\begin{vmatrix} -9 & 2 & 0 & -4 \\ 1 & 0 & 0 & 12 \\ 2 & 1 & -1 & 0 \\ 0 & 0 & 3 & 0 \end{vmatrix}} \frac{\delta W_2}{W_2^0} = -\frac{6}{19}\frac{\delta W_2}{W_2^0} \tag{1.2.55}$$

Denoting the small corresponding displacements of (W_1) and (W_2) by (δs_1) and (δs_2) respectively (Fig. 1.2.8)

$$\delta s_1 = L \sin 60 - (L - e_1)\sin(60 + \psi_1) \tag{1.2.56}$$

and

$$\delta s_2 = L \sin 60 - (L - e_3)\sin(60 + \psi_2) \tag{1.2.57}$$

On expanding to the first order of small quantities,

$$\delta s_1 = e_1 \sin 60 - L\psi_1 \cos 60$$

$$\delta s_2 = e_3 \sin 60 - L\psi_3 \cos 60$$

Substituting for (ψ_1) and (ψ_3), one gets

$$\sqrt{3}\,\delta s_1 = e_1 + e_4 \tag{1.2.58}$$

and

$$\sqrt{3}\,\delta s_2 = -e_1 - e_2 + 2e_3 + e_4 \tag{1.2.59}$$

Satisfying now Eq. (1.2.53), we get finally,

$$\sqrt{3}\,\delta s_1 = e_1$$

and

$$\sqrt{3}\,\delta s_2 = -e_1 \tag{1.2.60}$$

From Eqs. (1.2.55) and (1.2.60), the post-buckling slopes are

$$\frac{\delta W_1}{\delta s_1} = -\frac{19\sqrt{3}}{2}\frac{W_2^0}{L} \tag{1.2.61}$$

and

$$\frac{\delta W_2}{\delta s_2} = \frac{19\sqrt{3}}{6}\frac{W_2^0}{L} \tag{1.2.62}$$

These two results are correlated with the data from the test on the model frame in Fig. 1.2.8a and b. It may be shown using the last of the Eqs. (1.2.52) that the axial force in the critically stressed member 3 decreases by the amount

$$\frac{\delta P_3}{P_3^0} = \frac{\delta W}{W^0} = -\frac{19}{6}\frac{e_1}{L} \tag{1.2.63}$$

which, therefore, remains straight in post-buckling. This is confirmed by the experiment.

Stability of the system may be evaluated according to Theorem (1). In this case

$$\Delta V = \tfrac{1}{2}(\delta W_1\,\delta s_1 + \delta W_2\,\delta s_2) = -\frac{\delta W_2}{\delta s_2}(\delta s_2)^2$$
$$= -\frac{19\sqrt{3}}{6}\frac{W_2^0}{L}(\delta s_2)^2 < 0 \tag{1.2.64}$$

Therefore, clearly, the system is *unstable*.

Conclusions

The simple theory of elastic buckling of pin-jointed frames developed in this chapter is thus confirmed by a number of tests on model frames. A more elaborate treatment and additional tests may be found in Ref. 15, pp. 196 to 234 and 312 to 389. A pin-jointed frame or truss may develop several modes in post-buckling equilibrium depending on how many members in the frame are on the point of buckling.

It is most important to realize that *not* all critically stressed members necessarily buckle, as we saw this in the first test, but that some members *unload and remain straight* in post-buckling. We shall see later that in complex lattices this unloading of the critical members plays a very important part and it can be accounted for by the constitutive relation (1.1.42), called the "general law of pin-jointed members".

Moreover, the post-buckling equilibrium paths may be *stable* or *unstable* in the dynamic sense, depending on the buckling properties of the members, the geometrical configuration of the truss and the system of loading.

It is observed from tests on model frames that the theoretical equilibrium paths are never reached experimentally. This phenomenon is attributed to the presence of *imperfections* such as the slight initial curvature in the members, eccentricities of loading etc. Equilibrium paths under the influence of initial imperfections are discussed in the subsequent sections.

1.3 Thermal Buckling of Axially Constrained Compressive Elastic Pin-Jointed Slender Members

Consider an axially elastically constrained straight elastic slender pin-jointed member (Fig. 1.3.1) at an ambient temperature (T). If (T) is increased by (ΔT) the member tends to expand axially, but it is resisted in this expansion by elastic axial restraints, so that an axial increasing compressive load is induced in the

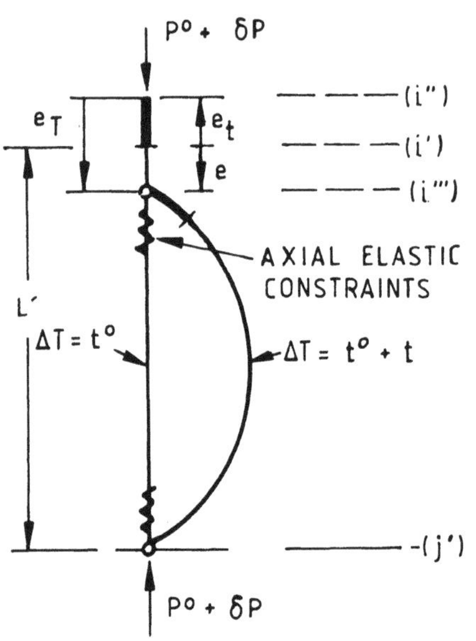

Figure 1.3.1
Compressive member under axial elastic end-restraints in thermal buckling

member. When ($\Delta T = t^0$), this axial force assumes a critical value (P^0), so that under a further increase of the temperature the member becomes unstable in its straight configuration. The member is then on the point of buckling, the axial load having the magnitude (P^0). But, it does not buckle, unless the temperature is increased further. If the temperature is decreased at this point, the axial load in the member also decreases and the member remains straight, the danger of its buckling flexurally having been eliminated.

Now, suppose the critically stressed axially elastically restrained straight pin-jointed member is in the position (i'), (j') at the critical temperature ($\Delta T = t^0$). If at this point the member were not further axially restrained, it would extend freely, under the constant axial load (P^0) and the temperature increase (t), by the amount

$$e_t = \alpha_T L' t \tag{1.3.1}$$

where (α_T) is the positive thermal expansion coefficient for the material of which it is made, (L') its length in the position (i'), (j') Fig. 1.3.1. Its total expanded length would then be located between the points (i'') and (j'). If, on the other hand, the axial restraint against extension persists, the axial load increases from its critical value (P^0) by an amount (δP), as the temperature is increased by a further small amount (t). As the straight configuration of the member under the axial load ($P^0 + \delta P$) is statically *unstable*, the member buckles *thermally*. The amount of its flexural shortening in this process depends on the increase (δP). The total flexural shortening (e_T) (thermoelastic shortening) of the member, in the location (i''') of its end (i), in this case, must be reckoned from its total expanded length, i.e.

$$e_T = e_t + e \tag{1.3.2}$$

where (e) is the *additional* or *visible* flexural shortening, measured from the original critical position in which the member is on the point of thermal buckling and its length is (L'). This additional flexural shortening may now be negative, but its negative value cannot exceed the amount ($-e_t$), as then its total flexural contraction is zero and the member is straight in the position (i'', j'), having been heated to the temperature ($\Delta T = t^0 + t$). Its axial load remains to be critical and equal to (P^0). From (1.3.2),

$$e_T = e_t + e = 0$$

and the maximum negative *visible* flexural contraction is,

$$e = -e_t \tag{1.3.3}$$

At this stage the straightened member may unload further by expanding *axially* and remaining straight.

Before continuing this discussion, let us examine how the critical thermal axial loads (P^0) may be determined from the consideration of the thermoelasticity of the entire structural system.

The critical thermal axial load (P^0) in the compressive member depends on its location (i'), (j') between two displaced joints (i') and (j') of a space lattice or truss, to which it is connected. The displaced locations of these joints, on the other hand, depend to a large extent on the linear thermal strain compatibility conditions between the adjoining members at each node and on the deformation process in which these members remain straight, *without* buckling flexurally. This is usually true under perfect geometrical conditions. Therefore, the determination of the axial loads in an internally statically indeterminate pin-jointed system, such as a truss or a lattice, is a standard problem in linear thermoelasticity and it is not pursued here in any generality. To illustrate this point, we examine only two cases.

(i) Suppose a prismatic elastic member of initial length (L) and temperature (T) is rigidly held at its ends (i), and (j) by hinges attached to two opposite rigid walls Fig. 1.3.2.

If the temperature is increased by an amount ($\Delta T = t^0$) and the member were not axially restrained, it would expand freely by the amount

$$e_t^0 = \alpha_T L t^0 \tag{1.3.4}$$

where (α_T) is the thermal expansion coefficient and (e_t^0) an axial extension. As the rigid walls do not permit this expansion, an axial force (P^0) is induced in the member, which cancels the full amount of the hypothetical free extension. According to Hooke's law

$$e_t^0 = \alpha_T L t^0 = \frac{P^0 L}{EA} \tag{1.3.5}$$

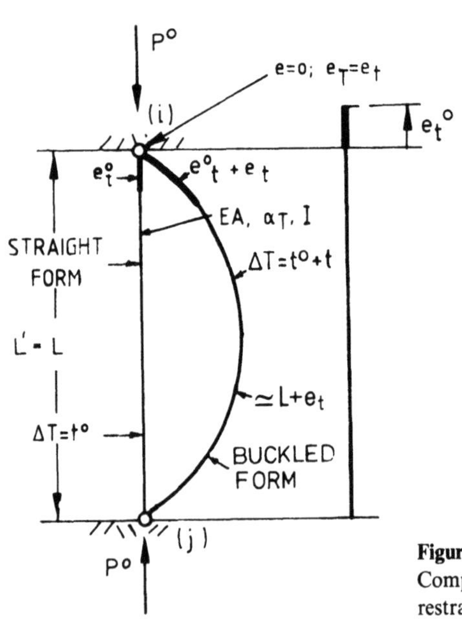

Figure 1.3.2
Compressive critical member under rigid axial end-restraints in thermal post-buckling

Hence

$$P^0 = \alpha_T E A t^0 \tag{1.3.5a}$$

Assuming that the critical load, at which the member is on the point of thermal buckling, is equal to the Euler load (P_E) for the member and that ($P^0 = P_E$), we deduce that

$$\alpha_T E A t^0 = \frac{\pi^2 EI}{L^2} = P_E \tag{1.3.6}$$

where

E: Young's modulus
I: Moment of inertia of cross-section
A: Cross-sectional area of member

The critical temperature (t^0) in the critical state, such that any further increase (t) in the temperature may cause it to buckle flexurally, is given by

$$t^0 = \frac{P_e}{\alpha_T EA} = \frac{\pi^2}{\alpha_T}\frac{I}{AL^2} = \frac{\pi^2}{\alpha_T}\left(\frac{r}{L}\right)^2 = \frac{\pi^2}{\alpha_T S^2} \tag{1.3.7}$$

where ($S = L/r$) is the slenderness ratio and (r) the radius of gyration of its cross-sectional area (A).

(ii) Consider the plane braced truss-type elastic system shown in Fig. 1.3.3a. It consists of three bars of lengths (L) and ($\sqrt{3}L$) respectively, and it is once hyperstatic. Suppose that after assembly and at no prestress it is subjected to a temperature rise ($\Delta T = t^0$).

The thermal expansion coefficient for all bars is (α_T), and the elastic modulus (E), moment of inertia of cross-section (I) and the cross-sectional area (A).

The axial loads are induced in the members (1), (2) and (3), by the elastic axial restraints of the structure. These loads may be determined using any analysis compatible with the linear thermoelasticity theory. We choose the method of *virtual work*, which is perhaps the most effective in this case. To this end we define *two independent* groups of external forces or agencies (i) and (k). The group (i) is applied before the group (k), but the deformations caused by the group (i) are of no interest here. We exploit the physical principle that the virtual work of the external forces of group (i) on the displacements of the group (k) is equal to the internal elastic virtual work of the internal equilibrated forces of the group (i) on the corresponding internal elastic deformations caused by the group (k). The external agency of the group (k) may be identified with the temperature rise (t^0) in this case, which causes the hyperstatic system to deform thermally. Suppose we cut member (3), which may be considered as redundant, close to node (1) and apply two equal and opposite unity forces in the direction of member (3) belonging to the group (i), as shown in Fig. 1.3.3b and 1.3.3c.

The disconnected system represents now the *basic isostatic system* to which the unity loads are applied. The internal axial loads caused by group (i) are indicated

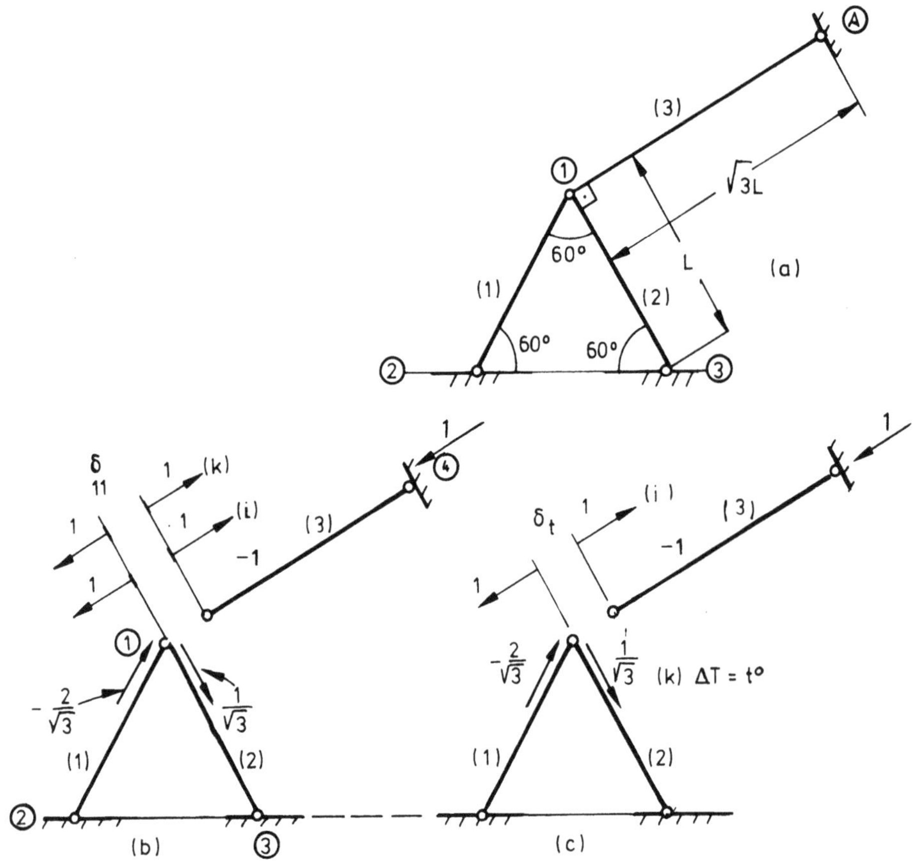

Figure 1.3.3

in the figures, where *now the minus sign* implies that the axial load is *compressive*. The linear strain compatibility condition in this case is very simple. If (δ_t) is the gap caused by the temperature difference ($\Delta T = t^0$) in the direction of the unknown axial force (X) in member (3) on the initial geometry, Fig. 1.3.3c, and (δ_{11}) the relative displacement in the same direction between the cut-off ends caused by two equal and opposite unity forces of the group (i), as shown in Fig. 1.3.3b, the strain-compatibility condition can be written

$$X \cdot \delta_{11} + \delta_t = 0 \tag{1.3.8}$$

where (X) is the unknown axial force in the redundant bar (3). To compute (δ_t) we equate the external virtual work of the group (i) on the external elastic displacements, caused by the temperature increase (t^0), Fig. 1.3.3c, to the equivalent internal work. Thus

$$1 \cdot \delta_t = \sum_{\alpha=1}^{3} N_\alpha^{(i)}(\alpha_T L_\alpha t^0) = \alpha_T t^0 \sum_{\alpha=1}^{3} N_\alpha^{(i)} L_\alpha \tag{1.3.9}$$

where ($N_\alpha^{(i)}$) represent the axial loads in the structure induced by the group (i) and ($\alpha_T L_\alpha t^0$) the axial extensions of the bars caused by the temperature increase

($\Delta T = t^0$). (Note that the dummy index (α) is unrelated to (α_T).) There are no bending or shearing forces induced in the pin-jointed bars in this case. Hence

$$\delta_t = \alpha_T t^0 \left(-\frac{2}{\sqrt{3}} L + \frac{1}{\sqrt{3}} L - 1 \cdot \sqrt{3} L \right) = -\frac{4}{3} \sqrt{3} \alpha_T L t^0 \qquad (1.3.10)$$

Similarly, (δ_{11}) is calculated from two identical groups of forces (i) and (k), shown in Fig. 1.3.3b, as follows

$$1 \cdot \delta_{11} = \sum_{\alpha=1}^{3} N_\alpha^{(i)} \frac{N_\alpha^{(k)} L_\alpha}{(EA)_\alpha} = \sum_{\alpha=1}^{3} \frac{(N_\alpha^{(i)})^2 L_\alpha}{(EA)_\alpha} \qquad (1.3.11)$$

where $N_\alpha^{(k)} \equiv N_\alpha^{(i)}$. Evaluating

$$\delta_{11} = \frac{L}{EA} \left[\left(-\frac{2}{\sqrt{3}} \right) \left(-\frac{2}{\sqrt{3}} \right) + \frac{1}{\sqrt{3}} \cdot \frac{1}{\sqrt{3}} + (-1)(-1)\sqrt{3} \right]$$
$$= \frac{5 + 3\sqrt{3}}{3} \frac{L}{EA} \qquad (1.3.12)$$

Substituting in Eq. (1.3.8)

$$X = -\frac{\delta_t}{\delta_{11}} = +\frac{4\sqrt{3}}{5 + 3\sqrt{3}} \alpha_T E A t^0 = +2(9 - 5\sqrt{3}) \alpha_T E A t^0 \qquad (1.3.13)$$

The critical axial loads in the original hyperstatic system, caused by the temperature increase ($\Delta T = t^0$), are now readily found. Introducing again the *original positive sign* conventions for compressive critical loads, we find

$$P_1^0 = \frac{2}{\sqrt{3}} X = \frac{8}{5 + 3\sqrt{3}} \alpha_T E A t^0 \le P_{1E} = \frac{\pi^2 EI}{L^2} \qquad (1.3.14a)$$

$$P_2^0 = -\frac{1}{\sqrt{3}} X = -\frac{4}{5 + 3\sqrt{3}} \alpha_T E A t^0 \le P_{2E} = \frac{\pi^2 EI}{L^2} \qquad (1.3.14b)$$

$$P_3^0 = X = \frac{4\sqrt{3}}{5 + 3\sqrt{3}} \alpha_T E A t^0 \le P_{3E} = \frac{\pi^2 EI}{(\sqrt{3}L)^2} = \frac{1}{3} \frac{\pi^2 EI}{L^2} \qquad (1.3.14c)$$

where P_{1E}, P_{2E}, P_{3E} are the Euler loads for the bars. The critical temperature (t^0) is the highest positive value which satisfies the three inequalities in the Eqs. (1.3.14a to c). Thus

$$t^0 \le \frac{5 + 3\sqrt{3}}{8} \frac{P_{1E}}{\alpha_T EA} = \frac{5 + 3\sqrt{3}}{8} \frac{\pi^2}{\alpha_T S^2} \qquad (1.3.15a)$$

$$t^0 \ge -\frac{5 + 3\sqrt{3}}{4} \frac{P_{2E}}{\alpha_T EA} = -\frac{5 + 3\sqrt{3}}{4} \frac{\pi^2}{\alpha_T S^2} \qquad (1.3.15b)$$

$$t^0 \le \frac{5 + 3\sqrt{3}}{4\sqrt{3}} \frac{P_{3E}}{\alpha_T EA} = \frac{5 + 3\sqrt{3}}{12\sqrt{3}} \frac{\pi^2}{\alpha_T S^2} \qquad (1.3.15c)$$

where

$S = L/r$: slenderness ratio for bar of length (L)
r: radius of gyration of cross-sectional area (A).

Clearly, the lowest critical temperature (t^0) is equal to

$$t^0 = \frac{5+3\sqrt{3}}{12\sqrt{3}}\frac{\pi^2}{\alpha_T S^2} = \frac{5\sqrt{3}+9}{36}\frac{\pi^2}{\alpha_T S^2} \tag{1.3.16}$$

and the critical axial loads in the bars of the hyperstatic system are

$$P_1^0 = \frac{2}{\sqrt{3}}X = \frac{8}{5+3\sqrt{3}}\alpha_T EAt^0 = \frac{2\sqrt{3}}{9}\pi^2\frac{EA}{S^2} = \frac{2\sqrt{3}}{9}\frac{\pi^2 EI}{L^2} \tag{1.3.17a}$$

$$P_2^0 = -\frac{1}{\sqrt{3}}X = -\frac{4}{5+3\sqrt{3}}\alpha_T EAt^0 = -\frac{\sqrt{3}}{9}\pi^2\frac{EA}{S^2} = -\frac{\sqrt{3}}{9}\frac{\pi^2 EI}{L^2} \tag{1.3.17b}$$

$$P_3^0 = X = \frac{4\sqrt{3}}{5+3\sqrt{3}}\alpha_T EAt^0 = \frac{1}{3}\pi^2\frac{EA}{S^2} = \frac{1}{3}\frac{\pi^2 EI}{L^2} \tag{1.3.17c}$$

If one compares this result with that in Eq. (1.3.7), in the first example, one concludes that in this case the critical temperature (t^0) is lower if (S) is the same and that thermal buckling is more critical than in the previous case. At the temperature change (t^0) member (1) is more highly stressed than member (3), since ($2\sqrt{3} > 3$), but member (3) buckles first, because its critical axial load is considerably lower. Buckling would occur only if the temperature were increased by an amount (t) from the critical value (t^0). This, then fully determines the thermal critical state of the system. The thermally extended or compressed lengths of the bars, which were previously denoted by (L'), can now be computed by adding to the original lengths the axial elongations or contractions. However, these are so small that the original lengths of the bars can be readily used in the continuation of this analysis. Similarly, the thermal displacements of the nodes of the hyperstatic structure can be readily evaluated using the linear thermoelasticity theory. The most effective method in this case may be again that of virtual work, already employed. However, we do not pursue this point further, since this method of analysis is a standard methodology, familiar to the educated reader, and proceed with our discussion concerning *thermal post-buckling analysis*, when the temperature is increased by an amount (t), beyond the critical value ($\Delta T = t^0$).

We now return to the thermally critically stressed but axially elastically restrained compressive bar in Fig. 1.3.1 and the Eq. (1.3.2). To calculate the axial load increase (δP) in the buckled member caused by the temperature increase (t), we use the result of Eq. (1.3.2) and the general law for compressive members in the form

$$\frac{\delta P}{P^0} = K\frac{e_T}{L} = \frac{K}{L}(\alpha_T Lt + e) = K\lambda_T \tag{1.3.18}$$

In prismatic members ($K = 1/2$) and, generally,

$$e_T = e_t + e = \alpha_T Lt + e \qquad \text{(1.3.2)bis}$$

is the flexural thermoelastic contraction, caused by the increase (δP) in the axial restraining force, which includes the free thermal extension ($e_t = \alpha_T Lt$). Also, in the nondimensional form

$$\lambda_T = \frac{e_T}{L} = \alpha_T t + \lambda \tag{1.3.19}$$

where (α_T) is a *positive* thermal expansion coefficient for the material and $(e = \lambda L)$ the actual (visible) flexural shortening of the critically stressed bar of length (L'), reckoned from the critical state in which the bar is held at the critical temperature $(\Delta T = t^0)$. Note that (e_T) may be greater than (e), so that in a thermally buckled member an additional length (e_t) is fed into the bowed out bar which is produced by the temperature increase (t). Numerically, (L') may be replaced by the initial bar length (L) at $(\Delta T = 0)$, as the difference in length is very small. In calculating the geometrical changes of the flexurally distorted structure, such as the bar body rotations (ψ), for example, measured from the thermally critical configurations of the structure, only the visible flexural shortenings of the chords $(e = \lambda L)$ should be introduced into the geometrical and kinematic compatibility conditions 1.2.12(a) and (b), as we show later.

If a thermally critically stressed compressive member in a system does not buckle, but it remains straight under a temperature increase (t), the axial load (δP) is either negative or zero. In that case the bar extends slightly. If (δP) is negative, i.e. a tensile increment, the bar must extend linearly. In that case the axial thermoelastic contraction $(\bar{e}_T)$ under the load change (δP) is negative and it represents a thermoelastic *extension* of the axially restrained bar. Then, from Hooke's law,

$$\bar{e}_T = \frac{\delta P L'}{EA} \simeq \frac{\delta P L}{EA} \tag{1.3.20}$$

where, Fig. 1.3.4,

$$-\bar{e}_T + e_t = -\bar{e}$$

or

$$\bar{e}_T = e_t + \bar{e} = \alpha_T L t + \bar{e} \tag{1.3.21}$$

Here, (e_t) represents the free thermal extension of the bar under the constant load (P^0) as before. $(-\bar{e})$ is now the linear visible extension of the bar from its critical state length (L') and $(\bar{e}_T)$ is the axial contraction of the bar under a positive incremental *compressive* force (δP); (vice versa, $(-\bar{e}_T)$ is an axial extension under a *tensile* incremental axial load.) It follows from the Eqs. (1.3.20 and 21) that for (δP) to be negative, the axial contraction $(\bar{e})$ must also be negative—and therefore, an extension—and greater than the free thermal extension $(e_t = \alpha_T L t)$, i.e., in that case, $(\bar{e} < -e_t = -\alpha_T L t)$ and

$$\delta P = \frac{EA}{L}\bar{e}_T = \frac{EA}{L}(e_t + \bar{e}) = \frac{EA}{L}(\alpha_T L t + \bar{e}) \tag{1.3.22}$$

is negative. When the visible extension of the bar $(\bar{e} = -\alpha_T L t)$, the bar extends only under a temperature increase (t) and the increase in the axial load (δP) is zero.

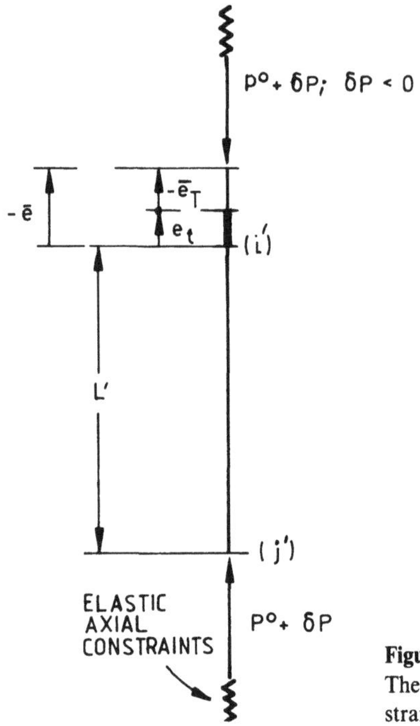

Figure 1.3.4
The unloading of a thermally critically stressed compressive straight bar under elastic axial end-restraints

Dividing the last result by (P^0), where ($P^0 = P_E(\text{Euler}) = \pi^2 EI/L^2$), in this case, we find

$$\frac{\delta P}{P^0} = \frac{S^2}{\pi^2}(\alpha_T t + \bar{\lambda}) = \frac{S^2}{\pi^2}\frac{\bar{e}_T}{L} = \frac{S^2}{\pi^2}\bar{\lambda}_T \tag{1.3.23}$$

where

$$\bar{\lambda}_T = \frac{\bar{e}_T}{L} = \alpha_T t + \frac{\bar{e}}{L} = \alpha_T t + \bar{\lambda} \tag{1.3.24}$$

and (S) is the slenderness ratio of the bar. We see that Eq. (1.3.23) has the form of the general law of buckling members, introduced in the last section (also Ref. 15, Chapter 2) except that now the constant (K) takes the value

$$K = \bar{K} = \frac{S^2}{\pi^2} \tag{1.3.25}$$

and the flexural contraction (e) is replaced by the thermoelastic axial contraction ($\bar{e}_T$). The axial loading and unloading process of a thermally critically stressed compressive bar may be summarized graphically as shown in Fig. 1.3.5. ($\bar{K}$) is usually a large constant compared to unity. The slenderness ratio of an elastic bar which buckles thermally is not likely to be lower than about 30 in practice. In that case ($\bar{K}$) is of the order of 100. If (S) is in the region of (100) or higher, ($\bar{K}$) is of the order of 1000 or higher.

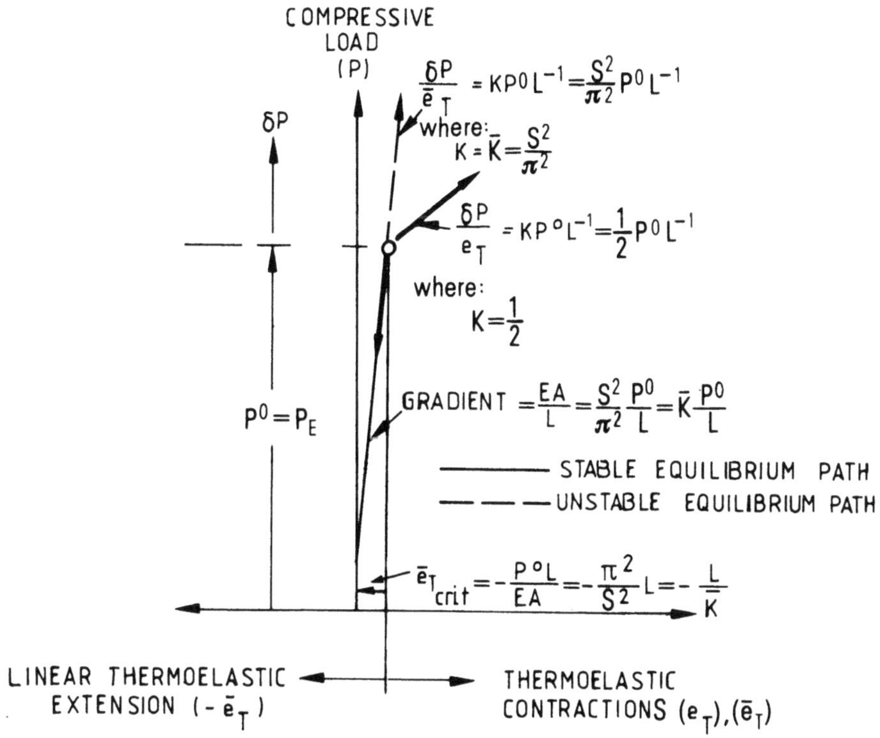

Figure 1.3.5
Load-displacement equilibrium paths of an axially compressed critical member of slenderness (S) in thermal post-buckling

Consider next a *subcritically* stressed axially elastically constrained compressive bar (i'), (j') in the thermally critical state of the system (Fig. 1.3.6), such as the bar (1) in the second example, in which the axial load at this state is only ($P_1^0 = (2\sqrt{3}/3)P_3^0$) where ($P_3^0$), in this case, is the critical load for the system, Eq. (1.3.17c). As the temperature is increased by the amount (t) from (t^0), where now ($\Delta T = t^0 + t$), the bar extends by the amount ($e_t = \alpha_T L't \simeq \alpha_T Lt$) to the point ($i''$) if it is allowed to expand freely. Since it is axially restrained an increase (δP) in the axial load (P^0) is induced, according to Hooke's law. This causes the bar to contract *axially* by the amount

$$\bar{e}_T = e_t + \bar{e} = \frac{\delta PL'}{EA} \simeq \frac{\delta PL}{EA} \tag{1.3.26}$$

where ($\bar{e}_T$) is the axial thermoelastic contraction caused by the compressive axial load increase (δP) and ($\bar{e}$) the positive *visible axial* contraction of the bar-length (L') from the critical state. (A horizontal bar above (e) indicates that the contraction is axial or linear as opposed to flexural.) The end (i') now moves to its final position (i'''), Fig. 1.3.6. Obviously ($\bar{e}$) can also be negative. If ($\bar{e} = -e_t$), the bar extends at no decrease in the axial load (P^0). On the other hand, if ($\bar{e} < -e_t = -\alpha_T Lt$) the thermoelastic contraction ($\bar{e}_T$) is negative and the axial

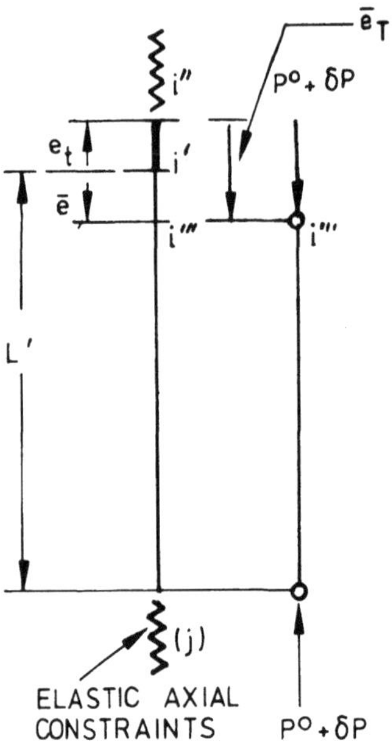

Figure 1.3.6
A subcritically compressed straight bar under elastic axial end-restraints

load decreases by the amount (δP) according to Eq. (1.3.26) under an increase (t) in the ambient temperature. In that case, from Hooke's law,

$$\delta P = \frac{EA}{L}(e_t + \bar{e}) = \frac{EA}{L}\bar{e}_T$$

If ($P_E = \pi^2 EI/L^2$) represents the critical (Euler) load of the subcritically stressed bar under the load (P^0), where ($P^0 < P_E$), the last result may be written

$$\frac{\delta P}{P^0} = \frac{P_E}{P^0}\frac{S^2}{\pi^2}\left(\frac{e_t}{L} + \frac{\bar{e}}{L}\right) = \frac{P_E}{P^0}\frac{S^2}{\pi^2}(\alpha_T t + \bar{\lambda}) = K\frac{\bar{e}_T}{L} = K\bar{\lambda}_T \tag{1.3.27}$$

where now

$$K = \frac{P_E}{P^0}\frac{S^2}{\pi^2} = \frac{P_E}{P^0}\bar{K} = \frac{EA}{P^0} \tag{1.3.28}$$

and ($\bar{e}_T = e_t + \bar{e}$) is the thermoelastic linear axial contraction, with ($\bar{e}$) representing the visible amount of the axial contraction of the bar length (L') from the critical state. ($e_t = \alpha_T Lt$) is the free thermal expansion of the bar, as before.

If a subcritically stressed member unloads from the thermal critical state of the system of which it is a part, ($\bar{e}_T$) is negative, as the member extends under a negative axial load change (δP), Fig. 1.3.7. Then, from Eq. (1.3.26)

$$\delta P = \frac{EA}{L}\bar{e}_T = \frac{EA}{L}(e_t + \bar{e}) \tag{1.3.29}$$

and, in this case, ($\bar{e}_T = e_t + \bar{e} < 0$), or ($\bar{e} < -e_t = -\alpha_T Lt$).

Putting ($e_t = \alpha_T L t$), and writing

$$\frac{\delta P}{P^0} = \frac{P_E}{P^0}\frac{S^2}{\pi^2}\frac{\bar{e}_T}{L} = \frac{P_E}{P^0}\frac{S^2}{\pi^2}(\alpha_T t + \bar{\lambda}) = \frac{P_E}{P^0}\frac{S^2}{\pi^2}\bar{\lambda}_T \tag{1.3.30}$$

we obtain again the *form of the general law*

$$\frac{\delta P}{P^0} = K\bar{\lambda}_T \tag{1.3.31}$$

where again

$$K = \frac{P_E}{P^0}\frac{S^2}{\pi^2} = \frac{P_E}{P^0}\bar{K} = \frac{EA}{P^0} \tag{1.3.32}$$

but, where ($\bar{e}_T = \bar{\lambda}_T L$) is now a negative axial thermoelastic contraction or an axial thermoelastic extension, as indicated in Fig. 1.3.7.

A similar situation occurs if in the thermally critical state of the system a tensile member within the system is considered. Let us now assume that a tensile axial load (P^0) is induced in the member at the temperature change ($\Delta T = t^0$), which is critical for the system in the sense that some other critically stressed member may buckle thermally under a further temperature increase (t). If now (t) is an increase and the tensile bar under consideration is axially elastically restrained, its axial tensile load may suffer a positive or a negative load change (δP), this depending on the total axial extension of the bar, which is goverened by other

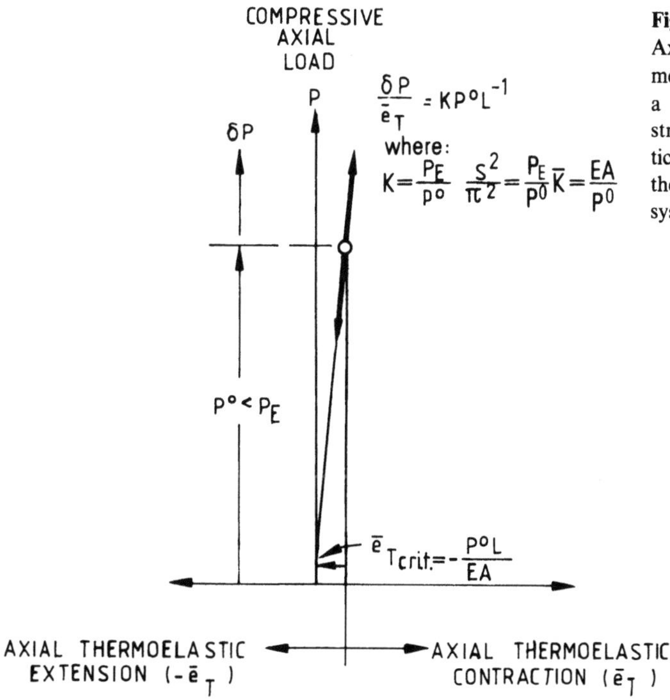

Figure 1.3.7
Axial load - axial displacement equilibrium paths of a compressive subcritically stressed member under elastic axial end-restraints at a thermal critical state of the system

members of the structure in a thermal buckling process. A tensile load may be relieved by a temperature rise (t) if the bar is allowed to extend axially. If (t) is a decrease, the opposite may be true.

If we introduce the same symbols as before, where ($\bar{e}$) is the axial contraction of the bar from its critical state length (L'), (or ($-\bar{e}$) its axial extension from that length), and ($\bar{e}_T$) the total thermoelastic axial contraction under a compressive load increase (δP), or ($-\bar{e}_T$) the total thermoelastic axial *extension* under an increase (δP) in the tensile load (P^0), then, from Fig. 1.3.8

$$-\bar{e}_T + e_t = -\bar{e} \tag{1.3.34}$$

or

$$-\bar{e}_T = -(e_t + \bar{e}) \quad \text{viz. } \bar{e}_T = e_t + \bar{e} \tag{1.3.35}$$

If (δP) is now an increase in the *tensile* axial load, we deduce from Hooke's law, Fig. 1.3.8,

$$-\bar{e}_T = \frac{\delta P L'}{EA} \simeq \frac{\delta P L}{EA} \tag{1.3.36}$$

where ($-\bar{e}_T$) is a positive thermoelastic extension and (δP) an increase in the tensile load. Or

$$\delta P = \frac{EA}{L}(-\bar{e}_T) = -\frac{EA}{L}(e_t + \bar{e}) \tag{1.3.37}$$

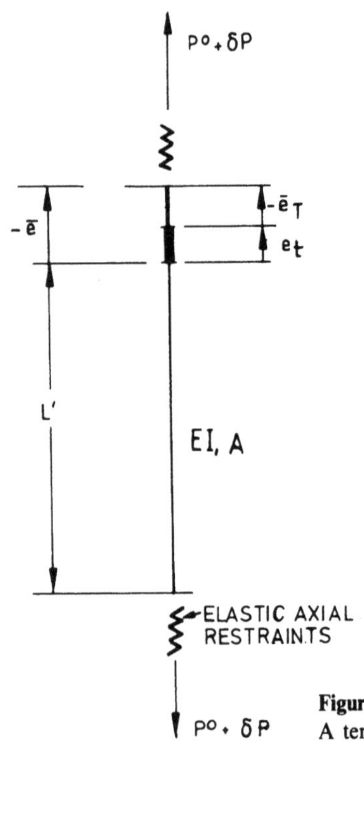

Figure 1.3.8
A tensile member under elastic axial end-restraints

where ($\bar{e}$) is the axial contraction of the critical length (L'). If the geometrical compatibility conditions, concerning the changes in the lengths of the members in a thermal buckling mechanism of a pin-jointed structure require that ($\bar{e}$) is *negative* and larger in magnitude than ($e_t = \alpha_T Lt$), then (δP) is a positive tensile increment, as this requires a further lengthening of the bar beyond the amount (e_t) of the free thermal expansion under the constant tensile load (P^0), as shown in Fig. 1.3.8. If, on the other hand, ($\bar{e} \geq -e_t = -\alpha_T Lt$) or, if ($\bar{e}$) is positive, the change (δP) in the axial tensile load is negative, i.e. a compressive increment, and it relieves the tension in the bar. If ($\bar{e} = -e_t$) there is no change in the axial tensile load under the temperature increase (t). From (1.3.37)

$$\frac{\delta P}{P^0} = -\frac{P_E}{P^0}\frac{S^2}{\pi^2}\frac{\bar{e}_T}{L} = -\frac{P_E}{P^0}\frac{S^2}{\pi^2}\frac{\alpha_T Lt + \bar{e}}{L}$$
$$= -\frac{P_E}{P^0}\frac{S^2}{\pi^2}(\alpha_T t + \bar{\lambda}) = -\frac{P_E}{P^0}\frac{S^2}{\pi^2}\bar{\lambda}_T \qquad (1.3.38)$$

where ($\bar{\lambda}_T = \bar{e}_T/L = \alpha_T t + \bar{\lambda}$) is the axial thermoelastic contraction, as before. Thus

$$\frac{\delta P}{P^0} = K\frac{\bar{e}_T}{L} = K\bar{\lambda}_T \qquad (1.3.39)$$

where

$$K = -\frac{P_E}{P^0}\frac{S^2}{\pi^2} = -\frac{P_E}{P^0}\bar{K} = -\frac{EA}{P^0} \qquad (1.3.39a)$$

and (S) is the slenderness ratio of the tensile bar. (K) is now a large negative constant compared to unity. The incremental axial load-displacement equilibrium paths for this case are indicated in Fig. 1.3.9.

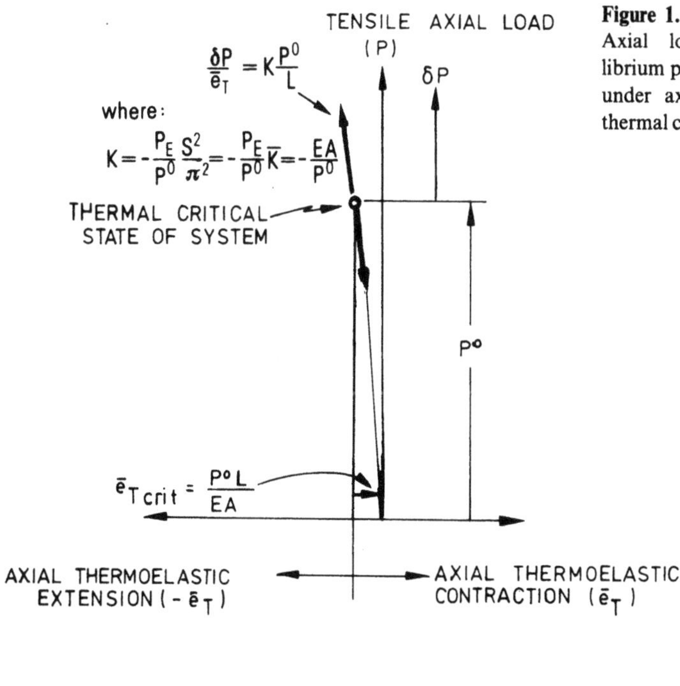

Figure 1.3.9
Axial load-displacement equilibrium paths of a tensile member under axial end-restraints at a thermal critical state of the system

All cases of thermal stressing and deformation under a temperature increase (t) beyond the critical temperature (t^0) in a member can now be summarized. These are: the thermal buckling of compressive critical bars under a load increase (δP) and a positive thermoelastic flexural contraction (e_T), the axial unloading of compressive critically stressed bars suffering an axial or linear extension ($-\bar{e}_T$) under a load decrease (δP), the post-critical thermal stressing and unstressing of a straight subcritical bar suffering only an axial contraction ($\bar{e}_T$) or extension ($-\bar{e}_T$) under a respective positive or negative load change (δP) and the stressing or unstressing of a tensile member under those circumstances. These cases may be summarized in the form of a generalized constitutive relation or law, governing the *thermal buckling of all axially elastically constrained pin-connected bars*, similarly as this was the case in pin-jointed systems subjected to external applied loading.

Thus an increase in the axial load (δP) due to the thermoelastic stress under the temperature increase (t) may be expressed in the form

$$\frac{\delta P}{P^0} = K\frac{e'_T}{L'} \simeq K\frac{e'_T}{L} \tag{1.3.40}$$

where

P^0:	axial load in the bar under the critical temperature rise ($\Delta T = t^0$),
L':	the length of the bar in the thermally critical state at ($\Delta T = t^0$),
L:	the original length of the bar at ($\Delta T = 0$),
K:	a buckling constant depending on whether the member buckles thermally or not,
e'_T:	the generalized thermoelastic contraction of the bar under the temperature rise (t) from its length (L') in the thermally critical state and an axial load increase (δP).
$e_t = \alpha_T L t$:	the free thermal expansion of the bar under the constant axial load (P^0),
e:	the visible flexural contraction of a buckling bar from its critical length (L'),
$\bar{e}$:	the visible axial or linear contraction of a straight bar from its critical length (L').

The load-deformation patterns in compressive axially restrained bars subjected to a total temperature increase ($t^0 + t$) are summarized graphically in Fig. 1.3.10.

In the case of compressive prismatic members, which buckle thermally

$$e'_T = e_T = e_t + e = \alpha_T L t + e \tag{1.3.41}$$

and

$$K = 1/2 \quad \text{and} \quad P^0 = P_E.$$

In *compressive* critical members which unload and remain straight

$$e'_T = \bar{e}_T = e_t + \bar{e} = \alpha_T L t + \bar{e} \quad \text{and} \quad \bar{e} = -e_t + \bar{e}_T \tag{1.3.42}$$

where ($\bar{e}_T$) is *negative* and it represents an extension under a negative change

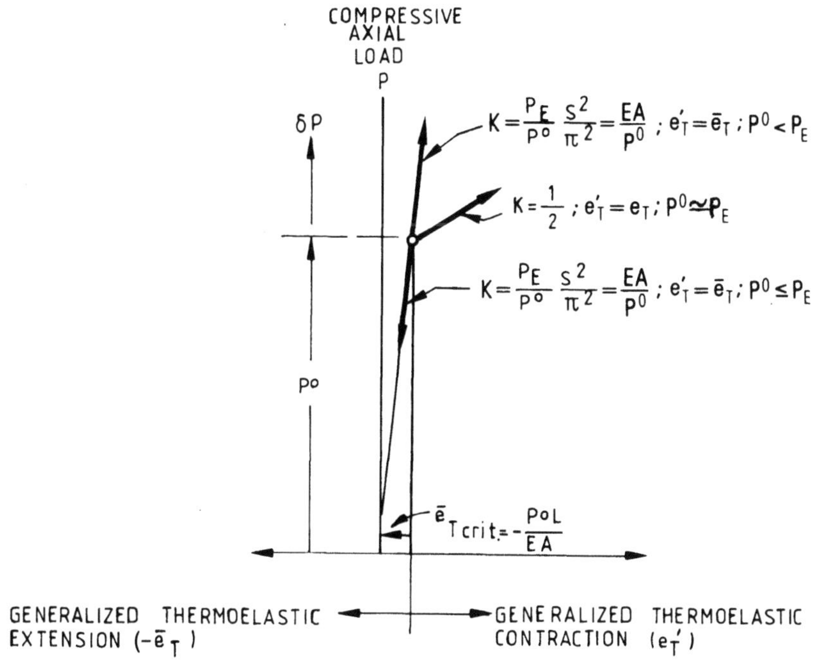

Figure 1.3.10
Axial load-displacement relationships for compressive thermally stressed elastic members at the critical state

(δP) in the axial load, which relieves the critically compressed bar for which ($P^0 \simeq P_E$). In that case

$$K = \frac{P_E}{P^0}\frac{S^2}{\pi^2} = \frac{P_E}{P^0}\bar{K} = \frac{EA}{P^0} \quad \text{and} \quad e'_T = \bar{e}_T$$

where (S) is the slenderness ratio of the bar. In subcritically stressed compressive bars (K) is given by the same formula, only ($P^0 < P_E$).

Tensile critical members are characterized by Eq. (1.3.39), where ($\bar{e}_T$) is again given by Eq. (1.3.42), but now (δP) is reckoned positive in tension and

$$K = -\frac{P_E}{P^0}\frac{S^2}{\pi^2} = -\frac{P_E}{P^0}\bar{K} = -\frac{EA}{P^0} \quad \text{and} \quad e'_T = \bar{e}_T$$

as shown in Fig. (1.3.9).

1.4 Thermal Post-Buckling of Flexible Elastic Members in Statically Indeterminate Pin-Jointed Lattices— An Illustration of the Basic Theory

The beam elements in statically indeterminate or hyperstatic space structures are attached at their ends by means of partly pinned or flexible (sometimes referred to as quasi-pinned) connectors. When such lattices are subjected to a temperature increase (ΔT), high thermal axial stresses may be induced in the bars of the lattice. This is true of lattices exposed to temperature changes in the atmosphere as well as in those intended for use in outer space. In that case, temperature differences of up to a few hundred degrees (°F) may arise between opposite sides due to solar irradiation on one side and shadows on the other, causing high axial forces to be induced in members constrained within the highly hyperstatic lattice. Under certain irradiation intensities high axial forces in the compressive members may cause buckling and bending of the constrained part of the structure out of the plane of the initial lattice geometry. In this way a form of thermal buckling of the irradiated parts of the lattice becomes possible. Several axially constrained members, in the affected lattice region, may thus be axially compressed simultaneously and sudden thermal buckling of several members or a whole region or of the lattice may ensue. The sequence in which this can happen is *not* accidental, but the buckling behavior under finite deformations, sometimes referred to as thermal postbuckling, is governed by definite laws, with due regard to the existence of unstable and decreasing equilibrium paths, marking the relationship between the temperature changes on one hand and the finite flexural deformations of the lattice on the other. This process of buckling and bending of the flexible bars is at first gradual, as the critical levels of compressive axial stresses in the lattice are approached. Thermal buckling of axially constrained members causes a weakening of the lattice stiffness and its effectiveness in reliably transmitting external payloads. Thermal buckling may also cause unwanted distortions and a change of shape of the lattice. Shape control is particularly important, if the lattice supports sensitive instruments, such as a radio-telescope. Therefore, it is essential to interpret correctly the response of the lattice to thermal excitation, in order that a proper control of the shape may be implemented. The process becomes much more pronounced and rapid as the critical state is surpassed, and it may result in a sudden onset of instability by motion in the affected part of the lattice. This instability may, under many circumstances, be very sudden if not prevented, causing considerable changes and perturbations in the original structure and in its physical state. Such perturbations may, in turn, upset the orbital motion of a LSS (large space structure), such as a large solar platform or a space station. The intrinsic geometrical imperfections, which may be very small, are responsible for the occurrence of these large distortions before the critical region of an essentially unstable system is reached under thermal stress. These imperfections arise from different causes, such as eccentricities, curvatures of beams and various deviations from straightness of the members in the lattice, eccentricities at joints and the like.

In addition to these effects, the lattice being of a relatively large size, has low natural frequencies of vibration. Vibrations at low frequencies in such structures may be readily induced by thermal variations in temperature levels between day and night and by the shifting intensities of solar irradiation depending on the position of the platform relative to the sun, as well as on the geometry and the lay-out of the space lattice.

The thermal bending or buckling process in such statically indeterminate space lattices can be predicted by the thermal postbuckling analysis of quasi pin-jointed space frames. Evaluation of only critical states, without due regard for the whole prebuckling and postbuckling processes, is not sufficient, because only the postbuckling equilibrium paths give an indication as to how, under the influence of initial imperfections, the thermal buckling process develops and proceeds and which of the critically stressed beams or members buckle, why and in which sequence, and which remain straight and unload axially, once the critical state is exceeded.

Buckling of such a lattice may occur under a combination of situations due to direct thermal compressive stresses in the lattice and partly due to the stresses induced by low frequency deformations which may occur concurrently with the thermal effects. In large trusses, supporting a space station, for example, low frequency oscillations may also arise by the veering motion of boosters and other periodic inertial effects related to the steering of the platform in orbit.

To demonstrate how the thermal buckling effect on the stability of the whole structure can be estimated or evaluated by means of an elementary analysis, one can consider a *much simpler* situation. The behavior can be detected already in a plane hyperstatic structure, such as the plane lattice typified in Fig. 1.4.1. Imagine that the hexagon shown by thick lines is constrained within the lattice and that it is irradiated by solar energy, this irradiation being responsible for the rise in temperature in its members.

If the lattice is heated to a critical temperature ($\Delta T = t^0$) all the members within this hexagon become simultaneously critically stressed in compression provided they are of equal material, equal cross-sectional areas, and equal flexural stiffness.

The question arises, which of these members develop thermal buckling if the temperature is increased beyond the critical value (t^0) *by an amount* (t).

This can be only explained by the postbuckling analysis and by the stability or instability of postbuckling equilibrium states in the lattice. Namely, the post-

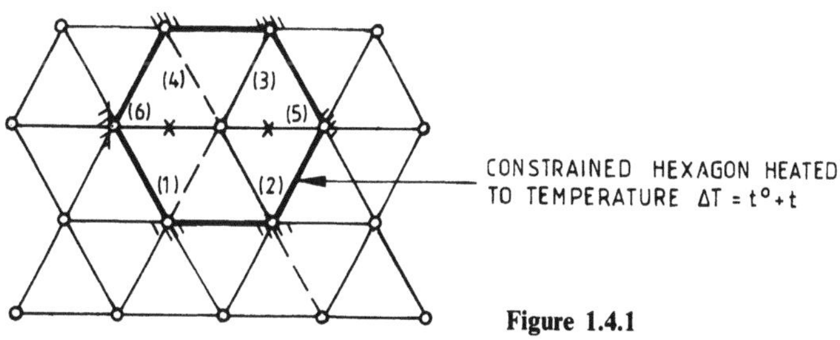

Figure 1.4.1

buckling analysis reveals the imperfection-sensitivity of the constrained structure, in this case of the constrained hexagon.

To make the calculation somewhat simpler and more easily understandable, let us assume that the bars (5) and (6), marked by a cross, are disconnected and consider only the bars (1), (2), (3) and (4) as being affected by this irradiation.

Then, using the elementary theory in Reference (15) and in Section 1.2, as demonstrated in the continuation of this text, it can be shown, that either members (1) and (4) develop buckling beyond the critical temperature, or members (2) and (3). Members (1) and (2) or (3) and (4) *cannot* buckle simultaneously under the imperfection-sensitivity and finite thermal distortions of the lattice. This pattern then determines the thermal buckling mode. The part of the frame showing members (1), (2), (3) and (4) is shown in detail in Fig. 1.4.2. It is indicated there, that while members (1) and (4), say, buckle, those marked (2) and (3) are prevented from buckling simultaneously in this case.

Once the problem is understood in a simple case such as this one, it is easily seen that this behavior is generally very important for a better understanding of the general performance and of the thermal stability of a hyperstatic large space structure and in regard to the development of its thermal post- or near-critical deformations. Large space structures are relatively highly imperfection-sensitive on account of the somewhat not-so-precise or remotely-controlled assembly techniques, so that an assessment of the control of their deformations, as carriers of solar energy panels and other instrumentation is of high relevance.

The cut-out structure from the hexagon in Fig. 1.4.1 is reproduced in Fig. 1.4.2. The initially straight members (1), (2), (3), (4) are of equal length, cross-sectional areas and stiffness. If they are heated to a temperature ($\Delta T = t^0$), their

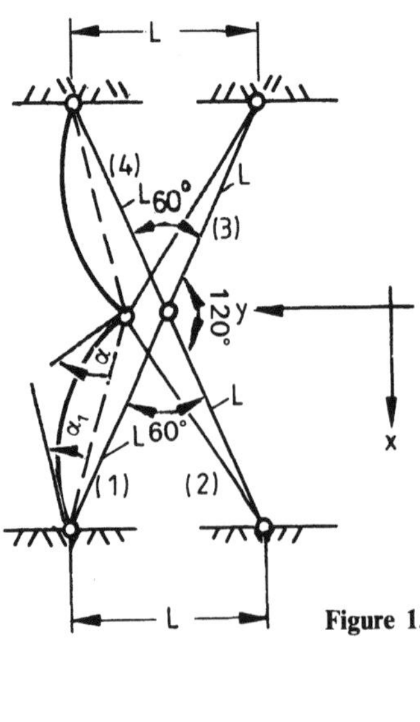

Figure 1.4.2

expansion is prevented by the restraints along the boundaries. As the temperature is increased, the axial forces are induced in these straight members, until the axially constrained system gets onto the point of buckling at the temperature increase ($\Delta T = t^0$). In this case the four critical axial forces are equal and all four members are on the point of buckling *simultaneously*.

The question arises which members buckle first? The joint can move horizontally or vertically. If it moves horizontally to the left say, members (1) and (2) buckle. It is therefore of interest to find the correct answer, which, as it will be shown, depends on the *loading path* which the members must undergo in these two alternative modes of thermal stressing.

If an axially fully constrained straight bar is heated, its axial force increases until $P = P^0_{\text{crit}} \simeq P_{\text{Euler}}$. For values of $P < P_E$, the bar cannot buckle thermally. Let the temperature increase be ($\Delta T = t^0$) when (P) reaches the critical (Euler) value. Then, by Hooke's law, in this case

$$\frac{AE}{L}\alpha_T L t^0 = P_E = \frac{\pi^2 EI}{L^2} \qquad \text{(Euler load)}$$

and

$$t^0 = \frac{\pi^2 I}{\alpha_T A L^2} = \frac{\pi^2 r^2}{\alpha_T L^2} = \frac{\pi^2}{\alpha_T S^2} \tag{1.4.1}$$

where

$S = (L/r)$:	is the slenderness of the bar
(L):	the initial length
(r):	the radius of gyration of the cross-section A
(I):	the moment of inertia of the cross-section A
(E):	the elastic modulus of the bar
(α_T):	the positive thermal expansion coefficient

Any further extension or flexural deformation of this bar is due to an additional rise (t) in the temperature. A further discussion of the thermal buckling process or of the unloading of the thermally critically stressed bar depends on the kinematics of the geometrical changes in a post-buckling mechanism of which the bar is a part.

In the case of an axially *fully* constrained bar connected by pins to two opposite rigid walls, Fig. 1.4.3, whose ends do not move in thermal post-buckling, the compatability condition requires that the visible flexural shortening of the member chord of length ($L' = L$) from its critical state is zero, i.e. ($e = 0$). The thermoelastic flexural contraction (e_T) under a temperature increase (t) in this case can be calculated from Eq. (1.3.2) on putting ($e = 0$), i.e.

$$e_T = \alpha_T L t \tag{1.4.1b}$$

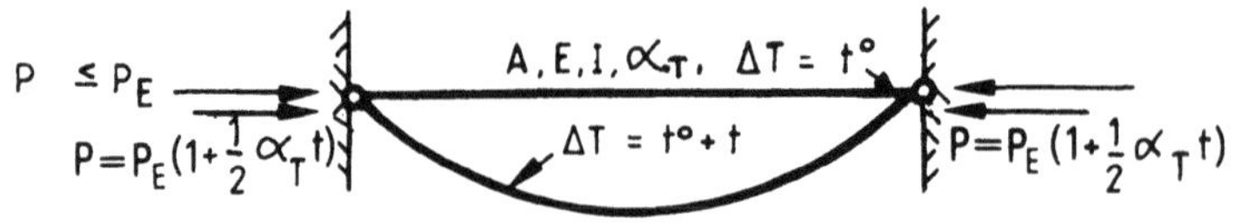

Figure 1.4.3
Axially fully constrained bar under a temperature increase ($\Delta T = t^0 + t$)

The increase (δP) in the critical axial load under the temperature increase (t) above the critical temperature (t^0) can then be calculated from Eq. (1.3.18), i.e.

$$\frac{\delta P}{P^0} = K\alpha_T t = \frac{1}{2}\alpha_T t \tag{1.4.1c}$$

where ($P^0 = P_E = \pi^2 EI/L^2 = EA\alpha_T t^0$) is the thermal critical load given by Eq. (1.4.1) and the critical temperature ($t^0 = \pi^2/\alpha_T S^2$).

The geometrical compatibility conditions in buckled pin-jointed lattices in a plane are comprehensively discussed in Ref. 15, Chapter 3 where the general geometrical compatibility conditions are derived (see also Section 1.2, page 16). These are the same in this case, as in the post-buckling process under external loads, which are applied centrally at the joints. We reproduce these equations here and refer the reader to the mentioned source for proof.[1] If (ψ_i) denotes the body rotation of the bar (i) (positive counterclockwise), (e_i) its *visible* flexural contraction, i.e. the relative displacements between the member ends at the opposite nodes measured from the thermally critical state or position (in the case of a visible *linear* contraction, (e_i) is replaced by ($\bar{e}_i$), as defined in the last section). Then, for the geometrical compatibility of each closed loop of members, before and after buckling, in a braced plane framework

$$\sum_{i=1}^{m} (e_i \cos \alpha_i + L_i\psi_i \sin \alpha_i) = 0 \tag{1.4.2a}$$

$$\sum_{i=1}^{m} (e_i \sin \alpha_i - L_i\psi_i \cos \alpha_i) = 0 \tag{1.4.2b}$$

where (e_i) now denote the *visible* or *actual* flexural or axial contractions, and where the summation is taken around a closed loop. Here (m) denotes the number of members in each loop, (α_i) the angular coordinate of each bar in the global coordinate system and (L_i) the initial length of each bar. The number of such independent loops is always sufficient to calculate the body rotations (ψ_i) in terms of the (e_i) from these equations. Such additional relations, may be obtained from the supporting conditions of the structure, as shown in Ref. 15, p. 201.

The solution of the problem may now be carried out. For this purpose it suffices to consider first the lower half of the structure in Fig. 1.4.2, reproduced in Fig. 1.4.4.

---

[1] See also Eqs. (1.2.12) in Section 1.2 on p. 16.

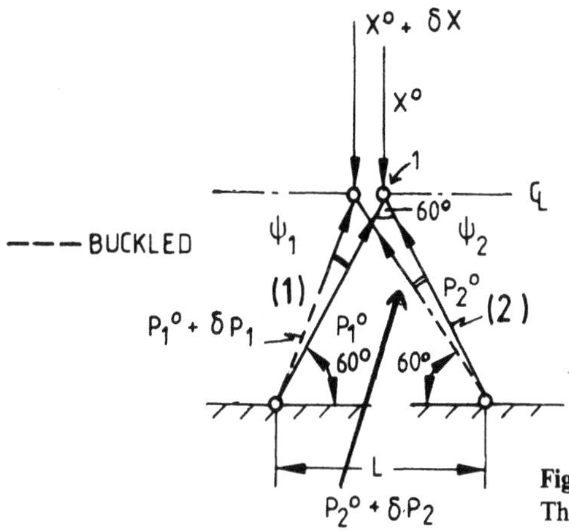

Figure 1.4.4
The lower part of the structure in Figure 1.4.2

Case (i): *Members (1) and (4) Buckling simultaneously*

Consider that in post-buckling only members (1) and (4) shorten flexurally and that members (2) and (3) remain straight. The joint (1) then moves horizontally along the centerline (line of symmetry), as indicated in the figure. All four members are then on the point of buckling simultaneously, while in the undisplaced position. In that critical state

$$P_1^0 = P_2^0 = P_3^0 = P_4^0 = P_E \qquad \text{(Euler)} \tag{1.4.3}$$

It follows from the symmetry of the upper and lower parts that there is only a vertical force of contact (X^0) between these two parts, where

$$X^0 = \sqrt{3}P_1^0 = \sqrt{3}P_E = \sqrt{3}EA\alpha_T t^0 \tag{1.4.4}$$

which is transmitted to the foundations of the two parts of the structure.

We assume that in postbuckling equilibrium caused by a temperature increase (t), this vertical force changes in magnitude by an amount (δX) and that equilibrium of the thermally buckled structure is possible under a vertical force of contact ($X^0 + \delta X$). Theoretically, an incremental horizontal force of contact (δP) say, could be imagined. However, the horizontal incremental component acting on the upper part of the structure would have to be in that case ($-\delta Y$), and the buckled form of the entire structure could not be symmetrical as assumed. So, in the absence of an external horizontal restraint, (δY) can be set equal to zero. On the other hand, we know from the elementary stability analysis[9] (see Chapter 3, page 206, Example 1), that the equilibrium of the buckled shape, such as that assumed in Fig. (1.4.4), is possible under one vertical variable load and that, the symmetrical mode in which members (1) and (2) buckle simultaneously under an increasing vertical load embodies higher energy

[9] "The Stability of Elastic Systems" by S.J. Britvec, Pergamon Press, Oxford, New York, 1973 (Chapter 3). See also, Test 1, Section 1.2, pp. 20, 21.

levels and that it is statically unstable and therefore, *does not occur in a test* on the model structure.

Geometry After Buckling

We proceed with the analysis by applying the compatibility equations (1.4.2) to the closed loops formed by the two bars and the rigid foundation before and after buckling. This operation results in two equations, which, when solved for the body rotations (ψ_1) and (ψ_2) of the bars (1) and (2) yield

$$\psi_1 L\sqrt{3} = e_1 - 2\bar{e}_2 \tag{1.4.5a}$$

$$\psi_2 L\sqrt{3} = 2e_1 - \bar{e}_2 \tag{1.4.5b}$$

where ($\bar{e}_2$) is the linear or visible contraction of the straight bar (2). In this case an *additional compatibility condition* is imposed by the *external hyperstatic constraint* in the vertical direction, namely, that joint (1) suffers no vertical displacement, i.e.

$$L \sin 60 = (L - e_1) \sin(60 + \psi_1) \tag{1.4.6}$$

(e_1) represents here the *visible flexural contraction* of the buckling bar (1) (or (4)), measured from the thermal critical state of this bar, which, in this case, is identical to the initial length (L). Expanding the last result for small values of (e_1) and (ψ_1),

$$L \sin 60 = L \sin 60 - e_1 \sin 60 + \psi_1 L \cos 60$$

or

$$\psi_1 L\sqrt{3} = 3e_1 \tag{1.4.7}$$

Substituting now for (ψ_1) from the first compatibility condition we get the condition to calculate ($\bar{e}_2$), i.e.

$$3e_1 = e_1 - 2\bar{e}_2 \tag{1.4.8}$$

and, hence

$$(-\bar{e}_2) = e_1 \tag{1.4.8a}$$

i.e. the axial or linear extension ($-\bar{e}_2$) of bar (2) is equal the flexural (visible) contraction (e_1) of bar (1) caused by thermal buckling under a total temperature increase ($\Delta T = t^0 + t$).

Equilibrium of Forces at the Top Joint After Buckling

From Fig. 1.4.4, the equilibrium of the contact force and the internal axial forces is given by the following two equations

$$(P_1^0 + \delta P_1) \cos(60 + \psi_1) - (P_2^0 + \delta P_2) \cos(60 - \psi_2) = 0 \tag{1.4.9a}$$

$$(P_1^0 + \delta P_1) \sin(60 + \psi_1) + (P_2^0 + \delta P_2) \sin(60 - \psi_2) = X^0 + \delta X \tag{1.4.9b}$$

It is assumed that the post-buckling situation arises throughout the overheating of the frame by the temperature (t) over the critical temperature (t^0), given by Eq. (1.4.1). On expanding and linearizing these equations, one gets after the usual procedure

$$\frac{1}{2}\delta P_1 - \frac{1}{2}\delta P_2 - \frac{\sqrt{3}}{2}P_1^0\psi_1 - \frac{\sqrt{3}}{2}P_2^0\psi_2 = 0 \qquad (1.4.10a)$$

$$\frac{\sqrt{3}}{2}\delta P_1 + \frac{\sqrt{3}}{2}\delta P_2 + \frac{1}{2}P_1^0\psi_1 - \frac{1}{2}P_2^0\psi_2 = \delta X \qquad (1.4.10b)$$

where Eq. (1.4.4) in the critical state is also accounted for. Replacing (ψ_1) and (ψ_2) by the Eqs. (1.4.5) and putting ($\lambda_1 = e_1/L$, $\bar{\lambda}_2 = \bar{e}_2/L$) we get

$$\tfrac{1}{2}\delta P_1 - \tfrac{1}{2}\delta P_2 - \tfrac{1}{2}P_1^0(\lambda_1 - 2\bar{\lambda}_2) - \tfrac{1}{2}P_2^0(2\lambda_1 - \bar{\lambda}_2) = 0 \qquad (1.4.11a)$$

$$\tfrac{3}{2}\delta P_1 + \tfrac{3}{2}\delta P_2 + \tfrac{1}{2}P_1^0(\lambda_1 - 2\bar{\lambda}_2) - \tfrac{1}{2}P_2^0(2\lambda_1 - \bar{\lambda}_2) = \delta X\sqrt{3} \qquad (1.4.11b)$$

Dividing by ($P_1^0 = P_2^0 = P_E$) and substituting for (δP_1) and (δP_2) according to the general law, for compressive buckling and non-buckling members, Eqs. (1.3.18) and (1.3.23) respectively, explained in last section, i.e.

$$\frac{\delta P_1}{P_1^0} = K_1(\alpha_T t + \lambda_1), \qquad K_1 = \frac{1}{2} \qquad (1.4.12a)$$

$$\frac{\delta P_2}{P_2^0} = \bar{K}_2(\alpha_T t + \bar{\lambda}_2), \qquad \bar{K}_2 = \frac{S^2}{\pi^2} \qquad (1.4.12b)$$

we get

$$K_1(\alpha_T t + \lambda_1) - \bar{K}_2(\alpha_T t + \bar{\lambda}_2) - (\lambda_1 - 2\bar{\lambda}_2) - (2\lambda_1 - \bar{\lambda}_2) = 0$$

$$3K_1(\alpha_T t + \lambda_1) + 3\bar{K}_2(\alpha_T t + \bar{\lambda}_2) + (\lambda_1 - 2\bar{\lambda}_2) - (2\lambda_1 - \bar{\lambda}_2) = 2\frac{\delta X}{P_E}\sqrt{3}$$

Using now the *additional* compatibility condition (1.4.8a) in non-dimensional form, i.e. that

$$(-\bar{\lambda}_2) = \lambda_1 \qquad (1.4.8a)\text{bis}$$

these equations reduce to

$$K_1(\alpha_T t + \lambda_1) - \bar{K}_2(\alpha_T t - \lambda_1) - 3\lambda_1 - 3\lambda_1 = 0 \qquad (1.4.13a)$$

$$3K_1(\alpha_T t + \lambda_1) + 3\bar{K}_2(\alpha_T t - \lambda_1) + 3\lambda_1 - 3\lambda_1 = 2\sqrt{3}\frac{\delta X}{P_E} \qquad (1.4.13b)$$

In these two equations the unknowns are (λ_1) and the incremental contact force (δX) and the independent variable is the temperature increase (t). Solving for (λ_1) we get from (1.4.13a), Fig. (1.4.5c),

$$\lambda_1 = \frac{\bar{K}_2 - K_1}{K_1 + \bar{K}_2 - 6}\alpha_T t \qquad (1.4.14)$$

As, usually $\bar{K}_2 \gg K_1 = 1/2$, in a first approximation,

$$\lambda_1 \simeq \alpha_T t \qquad (1.4.15)$$

6. *Data Available*

1. Overall statistics are available from inception to date. Records of expenditures and number of items sent to libraries have been maintained along with other statistics over the years from 1949 to date. It is therefore possible to determine the average cost of an item sent to participating libraries by dividing total expenditures of USBE during any given year by the number of items sent during that year. The annual average costs per item processed for shipment range from .98 cents in 1970 to $ 2.90 in 1979. The average fee charged libraries is at about these same levels although some variance is present, as exact expenditures and volume of business cannot be determined at the beginning of each budget year. The following chart summarizes annual costs over the latest ten year period.

	Total Annual Expenses	Publications Processed for Shipment	Average Expenditure per item
1970	530,745	542,215	.98
1971	574,317	473,050	1.21
1972	607,448	483,963	1.26
1973	667,881	490,207	1.36
1974	698,579	444,059	1.57
1975	732,503	378,838	1.93
1976	846,133	363,542	2.33
1977	850,317	337,454	2.52
1978	913,590	332,494	2.75
1979 (est.)	960,000	330,000	2.90

It should be noted that from time to time USBE has received a few foundation and other grants to support special studies and activities; in fact a grant from the Rockefeller Foundation of $ 90,000 enabled USBE to begin operations. In 1971/72 a grant of $ 48,408 was received from the Council on Library Resources, Inc. to help assist operations through a difficult financial period. These two grants are the only ones ever received by USBE to support general operations although there have been other grants to pay handling fees for individual institutions. Support from these and similar sources has been included when determining the average expenditure per item. It should also be noted that the new fee schedule effective in November of 1979 is only the tenth general increase in fees in the thirty-three year history of USBE.

In other words, the visible nondimensional thermal flexural contraction of the buckling bar is equal to the thermal strain of this bar under a temperature increase (t). The incremental contact force (δX) is found from Eq. (1.4.13b), after substituting for (λ_1) from the last result

$$\frac{\delta X}{\sqrt{3}P_E} = \frac{2K_1\bar{K}_2 - 3(K_1 + \bar{K}_2)}{K_1 + \bar{K}_2 - 6}\alpha_T t \tag{1.4.16}$$

Since $\bar{K}_2 \gg K_1$

$$\frac{\delta X}{\sqrt{3}P_E} = \frac{\delta X}{X^0} = \frac{2K_1 - 3\left(\frac{K_1}{\bar{K}_2} + 1\right)}{\frac{K_1}{\bar{K}_2} + 1 - \frac{6}{\bar{K}_2}}\alpha_T t \tag{1.4.17}$$

or approximately, since ($\bar{K}_2 \gg K_1$) and ($K_1 = 1/2$), Fig. (1.4.5b),

$$\frac{\delta X}{\sqrt{3}P_E} = \frac{\delta X}{X^0} = -2\alpha_T t \tag{1.4.18}$$

Since ($\lambda_1 \simeq \alpha_T t$), the last result may be written alternatively

$$\frac{\delta X}{\sqrt{3}P_E} = \frac{\delta X}{X^0} \simeq -2\lambda_1 \tag{1.4.19}$$

Thus the contact force ($X = X^0 + \delta X$) is a decreasing force with the increasing temperature change (t), Fig. (1.4.5a). It is of interest to compare the rate of change in the contact force with the rate of change of the axial force in the buckling members.

$$\delta X = -2\sqrt{3}P_E\lambda_1 \tag{1.4.20a}$$

and the change in the axial force of member (1),

$$\delta P_1 = \tfrac{1}{2}P_E(\alpha_T t + \lambda_1) = P_E\lambda_1 \tag{1.4.20b}$$

Thus, the rate of change in the contact force is *negative* and ($2\sqrt{3}$) times as large as the *apparent* or visible rate of flexural shortening of member (1) on account of thermal buckling! A buckling member no longer supports an axial load at the usual axial elastic rate (EA/L), but at the much smaller rate of its post-buckling flexural stiffness

$$\frac{\delta P_1}{e_1} = K_1\frac{P_E}{L} = \frac{\pi^2}{2}\frac{EA}{LS^2} \tag{1.4.21}$$

Then

$$\frac{\delta P_1}{e_1}\bigg/\frac{EA}{L} = \frac{\pi^2}{2}\frac{1}{S^2} \tag{1.4.21a}$$

Since, the slenderness ratio (S^2) is a very large number, the post-buckling flexural stiffness ($\delta P_1/e_1$) is very small compared to the axial stiffness (EA/L) of a straight elastic bar. A plot of the contact force (X) against the flexural contraction (λ_1) and the temperature increase (t) is shown in Fig. (1.4.5a).

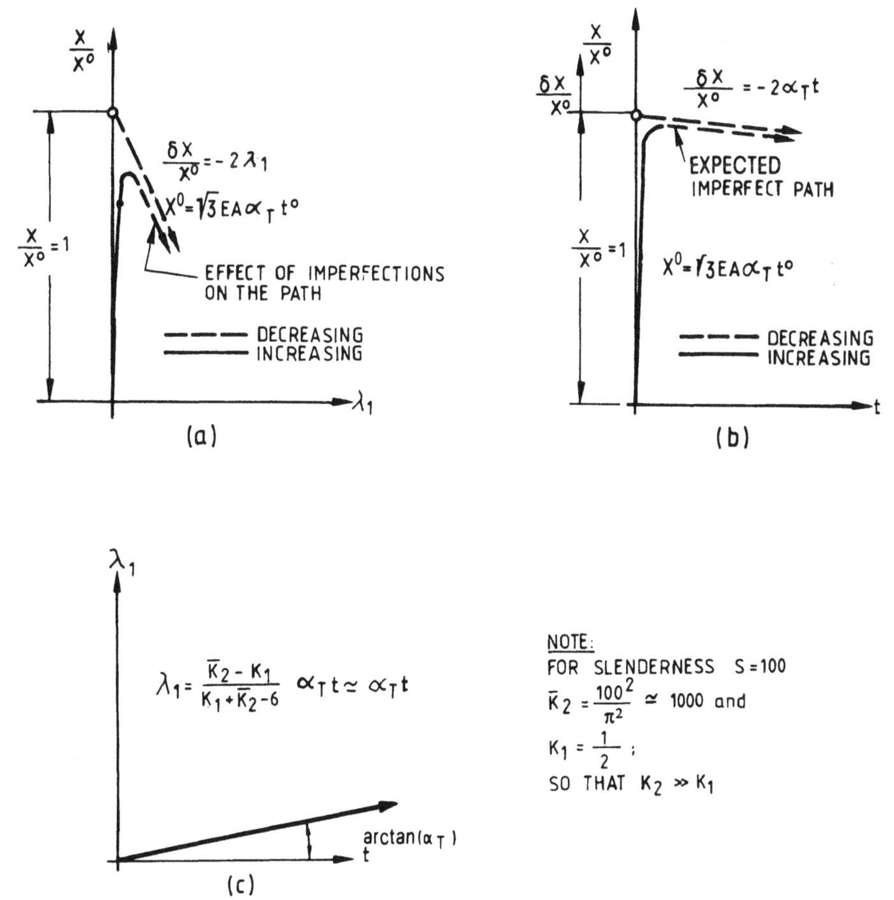

Figure 1.4.5
(a) Vertical contact force at the apex versus the *visible* flexural shortening (λ_1) of the buckling members. (b) Vertical contact force at the apex versus temperature. (c) Visible flexural shortening of the buckling members plotted against temperature increase (t)

It is of interest at this point to illustrate numerically the thermal buckling effect on the thermal distortion of member (1). The total thermoelastic contraction parameter (λ_{T1}) of member (1) is, according to (1.3.2) and (1.4.15), equal to

$$\lambda_{T1} = \lambda_t + \lambda_1 = \alpha_T t + \alpha_T t = 2\alpha_T t \qquad (1.3.2)\text{bis}$$

The end-tangent rotation (α_1) of the thermally distorted member is, according to Eq. (16.3d) in Ref. 15,

$$\alpha_1^2 = 4\lambda_{1T} + \cdots = 8\alpha_T t + \cdots \qquad (1.4.21\text{b})$$

or, approximately,

$$\alpha_1 = 2\sqrt{2\alpha_T t} \text{ radians}$$

Suppose that for a steel member ($\alpha_T = 126 \cdot 10^{-7} \cdot 1/°C$) and that the temperature increase (t) above the initial ambient temperature is equal to ($t = 36°C$). This gives for

$$\alpha_1 = 2\frac{180}{\pi}\sqrt{2\alpha_T t} = \frac{360}{\pi}\sqrt{2 \cdot 126 \cdot 10^{-7} \cdot 36} = 3.4515 \text{ degrees} \tag{1.4.21c}$$

This distortion is sufficiently large, so that the thermally buckled shape of the bars would be visible to the naked eye.

Changes in the Axial Loads

We can now compute the changes in the axial forces (P_1) and (P_2) from the critical values. Thus, the actual thermoelastic flexural contraction of bar (1) ($e_{T1} = \alpha_T Lt + e_1$) is related to the change (δP_1) by the equation

$$\frac{\delta P_1}{P_E} = \frac{K_1}{L}(\alpha_T Lt + e_1)$$

Substituting for (e_1) from (1.4.14)

$$\frac{\delta P_1}{P_E} = K_1\left(1 + \frac{\bar{K}_2 - K_1}{K_1 + \bar{K}_2 - 6}\right)\alpha_T t = 2K_1\frac{\bar{K}_2 - 3}{\bar{K}_2 + K_1 - 6}\alpha_T t \tag{1.4.22}$$

Since ($\bar{K}_2 \gg K_1 = 1/2$), on dividing by ($\bar{K}_2$) and letting it become large, the last result reduces approximately to

$$\frac{\delta P_1}{P_E} = 2K_1\alpha_T t = \alpha_T t \tag{1.4.23}$$

In view of Eq. (1.4.15), this is equivalent to

$$\frac{\delta P_1}{P_E} = \lambda_1 \tag{1.4.24}$$

Since the actual thermoelastic flexural shortening (e_{T1}) of bar (1), in this case, is equal to

$$e_{T1} = \alpha_T Lt + e_1 = 2e_1 \tag{1.4.25}$$

or

$$\lambda_{T1} = 2\lambda_1 \tag{1.4.25a}$$

it is equal to *twice* the visible flexural contraction (e_1). The amount (e_1) thereof represents the *lengthening* of the bowed-out member (1). The thermoelastic rate of change ($\delta P_1/e_{T1}$) in the axial load (P_1) is then

$$\frac{\delta P_1}{e_{T1}} = \frac{1}{2}\frac{\delta P_1}{e_1} = \frac{1}{2}K_1\frac{P_E}{L} = \frac{\pi^2}{4}\frac{EA}{LS^2} \tag{1.4.26}$$

which is only (1/2) of the apparent or visible rate of change, given by Eq. (1.4.21).

Similarly, the change (δP_2) in the axial load in bar (2), which does not buckle thermally and remains straight is equal to

$$\frac{\delta P_2}{P_2^0} = \frac{\delta P_2}{P_E} = \bar{K}_2(\alpha_T t + \bar{\lambda}_2) \tag{1.4.27}$$

Substituting now for ($\bar{\lambda}_2 = -\lambda_1$) from (1.4.8a) and for (λ_1) the *more accurate* expression in Eq. (1.4.14), we get

$$\frac{\delta P_2}{P_E} = \bar{K}_2(\alpha_T t - \lambda_1) = \bar{K}_2\left(\alpha_T t - \frac{\bar{K}_2 - K_1}{K_1 + \bar{K}_2 - 6}\alpha_T t\right)$$

or

$$\frac{\delta P_2}{P_E} = \bar{K}_2 \frac{2(K_1 - 3)}{\bar{K}_2 + K_1 - 6}\alpha_T t \tag{1.4.28}$$

Dividing now the denominator by ($\bar{K}_2$)

$$\frac{\delta P_2}{P_E} = \frac{2(K_1 - 3)}{1 + (K_1 - 6)/\bar{K}_2}\alpha_T t$$

Letting, now ($\bar{K}_2$) become large compared to ($K_1 - 6 = 1/2 - 6 = -11/2$), where ($K_1 = 1/2$), the last result reduces approximately to

$$\frac{\delta P_2}{P_E} \simeq 2(K_1 - 3)\alpha_T t = -5\alpha_T t \tag{1.4.29}$$

Thus, the axial compressive load in bar (2) *decreases*, as the bar unloads in the buckling mechanism under an *increase* (t) in the ambient temperature. This permits the bar to remain straight under the circumstances and the buckling mechanism is *kinematically and statically* admissible.

A plot of the axial load in bar (2) against the flexural contraction of bar (1) is shown in Fig. 1.4.6. The axial force in the straight compressive member remains *subcritical* after buckling and, therefore, this mode is physically possible. The last result may be written alternatively by virtue of the Eq. (1.4.15) as

$$\frac{\delta P_2}{P_2^0} = -5\lambda_1 \tag{1.4.30}$$

where ($P_2^0 = P_E$). The buckling mechanism for the entire substructure in Fig. 1.4.2 is now clear, as the upper and the lower parts distort symmetrically about the horizontal axis of symmetry (y-axis). Therefore

$$\lambda_1 = e_1/L = \lambda_4 = e_4/L = \alpha_T t \tag{1.4.31}$$

and

$$\frac{\delta P_1}{P_E} = \frac{\delta P_4}{P_E} = \alpha_T t = \lambda_1 = \lambda_4 \tag{1.4.32}$$

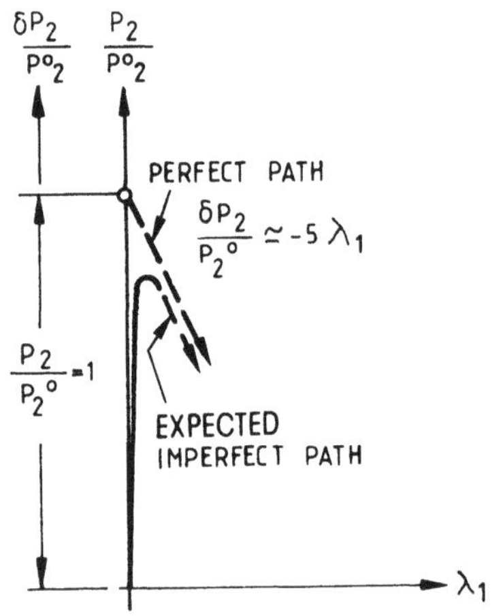

Figure 1.4.6
Decreasing axial load in the straight compressive member

Similarly

$$\frac{\delta P_2}{P_E} = \frac{\delta P_3}{P_E} = -5\lambda_1 = -5\lambda_4 \tag{1.4.33}$$

so that both members (2) and (3) unload axially under the temperature increase (t).

Case (ii): *Members (3) and (4) Buckling*

Consider now the possibility that, instead of the members (1) and (4) in Fig. (1.4.2), the bars (3) and (4) buckle simultaneously when the thermal critical state is reached, Fig. 1.4.7. The thermal critical state is reached as before, when the temperature change ($\Delta T = t^0$), where (t^0) is given by Eq. (1.4.1). Now, the joint (1) moves on the vertical line of symmetry (negative x-axis in Fig. 1.4.2). In this case the compatibility conditions are different and these may be obtained by applying the Eqs. (1.4.2a) and (1.4.2b) to the closed loop formed before and after buckling by the members (1) and (4) and the distance between the joints on the opposite walls, Fig. 1.4.7. Members (1) and (2) now remain straight, while members (3) and (4) contract flexurally.

Geometry After Buckling

The compatibility conditions relating the body rotations (ψ_1) and (ψ_4) to the flexural contraction ($e_4 = \lambda_4 L$) and the linear contraction ($\bar{e}_1 = \bar{\lambda}_1 L$) are given, in this case, by

$$\sqrt{3}\psi_1 = 2\lambda_4 + \bar{\lambda}_1 \tag{1.4.34a}$$

$$\sqrt{3}\psi_4 = -\lambda_4 - 2\bar{\lambda}_1 \tag{1.4.34b}$$

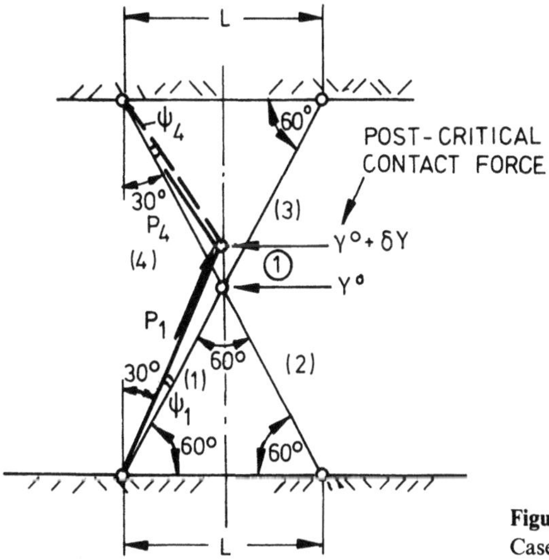

Figure 1.4.7
Case (ii): Members (3) and (4) buckling

Here, (e_4) is the apparent or visible flexural contraction of bar (4) and $(-\bar{e}_1)$ the visible linear elongation of bar (1).

An additional geometrical compatibility equation in this case can be derived from the condition that joint (1) now moves along the vertical line of symmetry, as members (1) and (2) extend by equal amounts. Then, before and after buckling, the horizontal projection of member (4) is constant, i.e.

$$L \sin 30 = (L - e_4) \sin(30 + \psi_4) \tag{1.4.35}$$

or, after linearizing,

$$L \sin 30 = L \sin 30 - e_4 \sin 30 + \psi_4 L \cos 30 \tag{1.4.35a}$$

Hence

$$\sqrt{3}\psi_4 = \lambda_4 \tag{1.4.36}$$

where $(\lambda_4 = e_4/L)$. Substituting into (1.4.34b) one gets

$$\lambda_4 = -\lambda_4 - 2\bar{\lambda}_1$$

or

$$(-\bar{\lambda}_1) = \lambda_4 \tag{1.4.37}$$

The Critical State

In the critical state, the contact force (Y^0) equals

$$Y^0 = P_1^0 = P_4^0 = P_E = EA\alpha_T t^0 \tag{1.4.38}$$

where (P_E) is assumed to be identical for all four bars.

Equilibrium of Forces at the Contact Joint After Buckling

Resolving the forces in the vertical and horizontal directions at the contact joint (1) we get

$$(P_4^0 + \delta P_4)\cos(30 + \psi_4) - (P_1^0 + \delta P_1)\cos(30 - \psi_1) = 0 \tag{1.4.39a}$$

$$(P_4^0 + \delta P_4)\sin(30 + \psi_4) + (P_1^0 + \delta P_1)\sin(30 - \psi_1) = Y^0 + \delta Y \tag{1.4.39b}$$

On expanding the trigonometric functions and linearizing

$$3\delta P_4 - 3\delta P_1 - \sqrt{3}\psi_4 P_4^0 - \sqrt{3}\psi_1 P_1^0 = 0 \tag{1.4.40a}$$

$$\delta P_4 + \delta P_1 + \sqrt{3}\psi_4 P_4^0 - \sqrt{3}\psi_1 P_1^0 = 2\delta Y \tag{1.4.40b}$$

Substituting now for (ψ_1) and (ψ_4) from (1.4.34a) and (1.4.34b)

$$3\delta P_4 - 3\delta P_1 - P_4^0(-\lambda_4 - 2\bar{\lambda}_1) - P_1^0(2\lambda_4 + \bar{\lambda}_1) = 0 \tag{1.4.41a}$$

$$\delta P_4 + \delta P_1 + P_4^0(-\lambda_4 - 2\bar{\lambda}_1) - P_1^0(2\lambda_4 + \bar{\lambda}_1) = 2\delta Y \tag{1.4.41b}$$

Dividing by $P_1^0 = P_2^0 = P_E$ and writing according (1.3.18) and (1.3.23),

$$\frac{\delta P_4}{P_4^0} = \frac{\delta P_4}{P_E} = K_4(\alpha_T t + \lambda_4); \qquad K_4 = \frac{1}{2} \tag{1.4.42}$$

$$\frac{\delta P_1}{P^0} = \frac{\delta P_1}{P_E} = \bar{K}_1(\alpha_T t + \bar{\lambda}_1); \qquad \bar{K}_1 = \frac{S_1^2}{\pi^2} \tag{1.4.43}$$

where, in this case ($S_1 = S_2 = S$) is the slenderness ratio for the members. One finds

$$3K_4(\alpha_T t + \lambda_4) - 3\bar{K}_1(\alpha_T t + \bar{\lambda}_1) - \lambda_4 + \bar{\lambda}_1 = 0 \tag{1.4.44a}$$

$$K_4(\alpha_T t + \lambda_4) + \bar{K}_1(\alpha_T t + \bar{\lambda}_1) - 3\lambda_4 - 3\bar{\lambda}_1 = 2\frac{\delta Y}{P_E} \tag{1.4.44b}$$

Putting now $(-\bar{\lambda}_1) = \lambda_4$, we get

$$3K_4(\alpha_T t + \lambda_4) - 3\bar{K}_1(\alpha_T t - \lambda_4) - 2\lambda_4 = 0 \tag{1.4.45a}$$

$$K_4(\alpha_T t + \lambda_4) + \bar{K}_1(\alpha_T t - \lambda_4) = 2\frac{\delta Y}{P_E} \tag{1.4.45b}$$

or

$$[3(\bar{K}_1 + K_4) - 2]\lambda_4 + 3(K_4 - \bar{K}_1)\alpha_T t = 0 \tag{1.4.46a}$$

$$(K_4 - \bar{K}_1)\lambda_4 + (K_4 + \bar{K}_1)\alpha_T t = 2\frac{\delta Y}{P_E} \tag{1.4.46b}$$

From the Eq. (1.4.46a) it follows then

$$\lambda_4 = 3\frac{\bar{K}_1 - K_4}{3(\bar{K}_1 + K_4) - 2}\alpha_T t \tag{1.4.47}$$

Since, usually, $\bar{K}_1 \gg K_4 = 1/2$; the last results reduce to

$$\lim \lambda_4 = 3 \frac{1 - K_4/\bar{K}_1}{3(1 + K_4/\bar{K}_1) - 2/\bar{K}_1} \alpha_T t = \alpha_T t; \quad \bar{K}_1 \text{ large} \tag{1.4.48}$$

The increment (δY) in the contact force is,

$$2\frac{\delta Y}{P_E} = (K_4 - \bar{K}_1)\lambda_4 + (K_4 + \bar{K}_1)\alpha_T t$$

Substituting for (λ_4) from (1.4.47), this simplifies to

$$\frac{\delta Y}{P_E} = \frac{6\bar{K}_1 K_4 - (\bar{K}_1 + K_4)}{3(\bar{K}_1 + K_4) - 2} \alpha_T t \tag{1.4.49}$$

or in terms of (λ_4) from (1.4.47),

$$\frac{\delta Y}{P_E} = \frac{6\bar{K}_1 K_4 - (\bar{K}_1 + K_4)}{3(\bar{K}_1 - K_4)} \lambda_4 \tag{1.4.50}$$

when $\bar{K}_1 \gg K_4 = 1/2$, and allowing for (1.4.38), the last result becomes

$$\frac{\delta Y}{P_E} = \frac{\delta Y}{Y^0} = \frac{6/2 - 1}{3} \lambda_4 = \frac{2}{3} \lambda_4 \simeq \frac{2}{3} \alpha_T t \tag{1.4.51}$$

which agrees with Eq. (1.4.48) as approximately $(\lambda_4 = \alpha_T t)$. In this case, the contact force (Y) increases after buckling. The result of Eq. (1.4.51) is displayed graphically in Fig. 1.4.8, where also the expected influence of initial imperfections on this path is indicated.

The changes in the axial loads (P_1) and (P_4) from the critical values can be found from (1.4.42) and (1.4.47),

$$\frac{\delta P_4}{P_4^0} = \frac{\delta P_4}{P_E} = K_4(\alpha_T t + \lambda_4) = 2K_4 \frac{3\bar{K}_1 - 1}{3(\bar{K}_1 + K_4) - 2} \alpha_T t \tag{1.4.53}$$

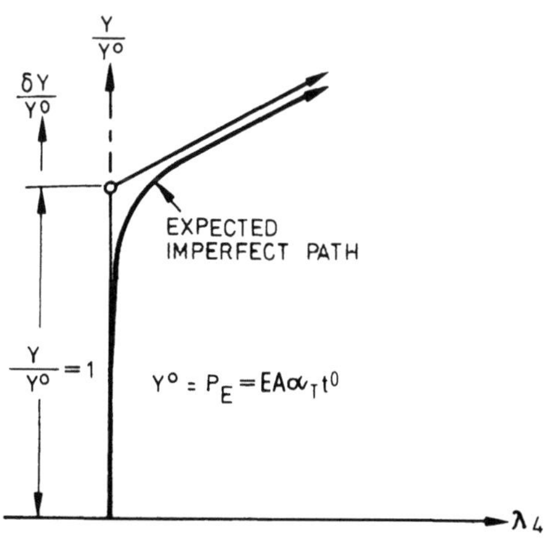

Figure 1.4.8
Contact force (Y/Y^0) versus the apparent shortening (λ_4)

or, in terms of (λ_4)

$$\frac{\delta P_4}{P_4^0}=\frac{\delta P_4}{P_E}=\frac{2}{3}K_4\frac{3\bar{K}_1-1}{\bar{K}_1-K_4}\lambda_4 \tag{1.4.54}$$

Since, $\bar{K}_1 \gg K_4 = 1/2$, approximately

$$\frac{\delta P_4}{P_4^0}=\frac{\delta P_4}{P_E}=\frac{1}{3}\frac{3}{1}\lambda_4=\lambda_4=\alpha_T t \tag{1.4.55}$$

This agrees again with Eq. (1.4.48). Again, by virtue of Eq. (1.4.48) the actual thermoelastic flexural shortening,

$$e_{T4}=\alpha_T t+e_4=2e_4 \tag{1.4.56}$$

is equal to twice the apparent or visible flexural contraction of bar (4), where $(e_{T4}=\lambda_{T4}L)$ and $(e_4=\lambda_4 L)$.

Similarly, using Eq. (1.4.43),

$$\frac{\delta P_1}{P_1^0}=\frac{\delta P_1}{P_E}=\bar{K}_1(\alpha_T t+\bar{\lambda}_1); \qquad \text{where } \bar{K}_1=\frac{S_1^2}{\pi^2} \tag{1.4.57}$$

and $(S_1=S_2=S)$ is the slenderness ratio of the bars. Substituting for $(\bar{\lambda}_1=-\lambda_4)$ and for (λ_4) from (1.4.47), we get

$$\frac{\delta P_1}{P_1^0}=\frac{\delta P_1}{P_E}=\bar{K}_1(\alpha_T t-\lambda_4)=2\bar{K}_1\frac{3K_4-1}{3(\bar{K}_1+K_4)-2}\alpha_T t \tag{1.4.58}$$

or, in terms of (λ_4),

$$\frac{\delta P_1}{P_1^0}=\frac{\delta P_1}{P_E}=\frac{2}{3}\bar{K}_1\frac{3K_4-1}{\bar{K}_1-K_4}\lambda_4 \tag{1.4.59}$$

Approximately, if $\bar{K}_1 \gg K_4=1/2$,

$$\frac{\delta P_1}{P_1^0}=\frac{\delta P_1}{P_E}=\frac{1}{3}\alpha_T t=\frac{1}{3}\lambda_4>0 \tag{1.4.60}$$

which is again in agreement with Eq. (1.4.48). The axial load in the straight bar (1) increases slightly beyond the critical value, Fig. 1.4.9, which means that a slightly *supercritically stressed member* (1) is *unstable* in the straight configuration and, given the choice between the two modes examined in cases (i) and (ii), the system is expected to "select" the mode evaluated in case (i).

The vertical contact force $(X^0+\delta X)$ between the upper and lower parts, formed by members (3) and (4) and (1) and (2) respectively, may be now computed. From symmetry.

$$X^0+\delta X=2(P_4^0+\delta P_4)\sin(60-\psi_4) \tag{1.4.61}$$

or, after expanding and linearizing

$$X^0+\delta X=\sqrt{3}P_4^0+\sqrt{3}\,\delta P_4-P_4^0\psi_4$$

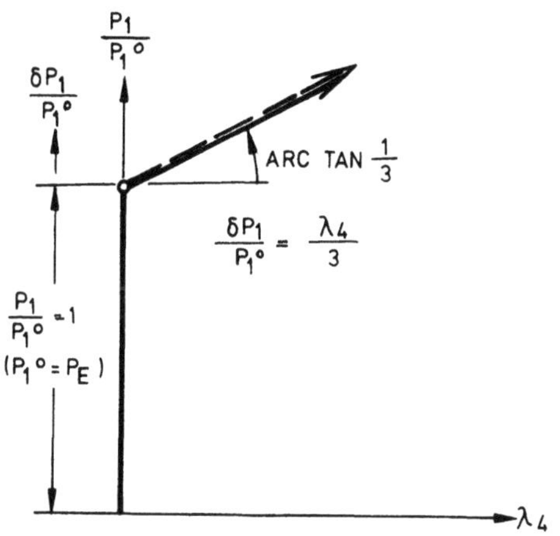

Figure 1.4.9
Slightly supercritical axial load in the straight bars (1) and (2) in the thermal buckling mode in case (ii)

Hence

$$X^0 = \sqrt{3} P_4^0 = \sqrt{3} P_E \tag{1.4.62}$$

and

$$\delta X = \sqrt{3}\, \delta P_4 - P_4^0 \psi_4 \tag{1.4.63}$$

Substituting for (X^0),

$$\frac{\delta X}{X^0} = \frac{\delta P_4}{P_4^0} - \frac{\sqrt{3}\psi_4}{3} = \frac{\delta P_4}{P_4^0} - \frac{1}{3}(-\lambda_4 - 2\bar{\lambda}_1) = \frac{\delta P_4}{P_4^0} - \frac{\lambda_4}{3} \tag{1.4.64}$$

where Eq. (1.4.34b) and (1.4.37) were used. From (1.4.42)

$$\frac{\delta X}{X^0} = K_4(\alpha_T t + \lambda_4) - \frac{\lambda_4}{3}$$

and from (1.4.48), for $K_1 \gg K_4 = 1/2$

$$\frac{\delta X}{X^0} = K_4(\lambda_4 + \lambda_4) - \frac{\lambda_4}{3} = \left(2K_4 - \frac{1}{3}\right)\lambda_4 = \frac{2}{3}\lambda_4 = \frac{2}{3}\alpha_T t \tag{1.4.65}$$

The contact force $(X = X^0 + \delta X)$, between the upper and lower parts, now *increases* with the temperature increase (t). The mode is *unstable*, as the lower straight part, comprising straight bars (1) and (2), is supercritically stressed under this increase.

Case (iii): *Asymmetric Modes*

The question arises whether an asymmetric mode can form from the thermal critical state under a temperature rise $(t^0 + t)$, so that only one member, say bar (4), buckles thermally, the other bars remaining straight, Fig. 1.4.10. To verify this we first derive the geometrical compatibility equations for this case. Apply-

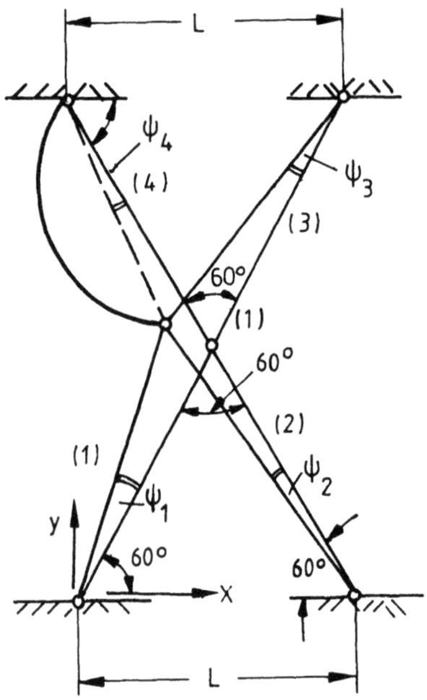

Figure 1.4.10
An asymmetric mode

ing now the Eqs. (1.4.2a and b) to the four loops formed by the members (1), (2); (1), (4); (2), (3); and (3), (4) and the distances between the four joints on the rigid foundation, we get

Loop (1), (2):

$$\sqrt{3}\psi_1 L = \bar{e}_1 - 2\bar{e}_2 \tag{1.4.66a}$$

$$\sqrt{3}\psi_2 L = 2\bar{e}_1 - \bar{e}_2 \tag{1.4.66b}$$

where counterclockwise body rotations (ψ) are reckoned positive and ($\bar{e}_1$) and ($\bar{e}_2$) are *linear contractions*, if positive.

Loop (1), (4):

$$e_4 \cos 300 + L\psi_4 \sin 300 + \bar{e}_1 \cos 240 + L\psi_1 \sin 240 = 0$$

$$e_4 \sin 300 - L\psi_4 \cos 300 + \bar{e}_1 \sin 240 - L\psi_1 \cos 240 = 0$$

On expanding, linearizing and solving for (ψ_1) and (ψ_4),

$$\sqrt{3}\psi_4 L = -2\bar{e}_1 - e_4 \tag{1.4.67a}$$

$$\sqrt{3}\psi_1 L = \bar{e}_1 + 2e_4 \tag{1.4.67b}$$

where ($\bar{e}_1$) is the linear contraction of straight member (1) and (e_4) the flexural contraction of the member (4), which is supposed to buckle thermally.

Loop (3), (4):

Similarly,

$$\bar{e}_3 \cos 240 + L\psi_3 \sin 240 + e_4 \cos 120 + L\psi_4 \sin 120 = 0$$

$$\bar{e}_3 \sin 240 - L\psi_3 \cos 240 + e_4 \sin 120 - L\psi_4 \cos 120 = 0$$

gives, ultimately, in a similar procedure,

$$\sqrt{3}\psi_4 L = 2\bar{e}_3 - e_4 \quad (1.4.68a)$$

$$\sqrt{3}\psi_3 L = \bar{e}_3 - 2e_4 \quad (1.4.68b)$$

Loop (2), (3):

$$\bar{e}_2 \cos 120 + L\psi_2 \sin 120 + \bar{e}_3 \cos 60 + L\psi_3 \sin 60 = 0$$

$$\bar{e}_2 \sin 120 - L\psi_2 \cos 120 + \bar{e}_3 \sin 60 - L\psi_3 \cos 60 = 0$$

which yields,

$$\sqrt{3}\psi_2 L = -\bar{e}_2 - 2\bar{e}_3 \quad (1.4.69a)$$

$$\sqrt{3}\psi_3 L = 2\bar{e}_2 + \bar{e}_3 \quad (1.4.69b)$$

Only two contractions (say $\bar{e}_1$ and $\bar{e}_2$) may be considered independent while other quantities are dependent. Equating (1.4.66a) and (1.4.67b), we get

$$\sqrt{3}\psi_1 L = \bar{e}_1 - 2\bar{e}_2 = \bar{e}_1 + 2e_4$$

Hence

$$e_4 = -\bar{e}_2 \quad (1.4.70)$$

Equating (1.4.66b) and (1.4.69a),

$$\sqrt{3}\psi_2 L = 2\bar{e}_1 - \bar{e}_2 = -\bar{e}_2 - 2\bar{e}_3$$

Hence

$$\bar{e}_3 = -\bar{e}_1 \quad (1.4.71)$$

From (1.4.66a), (1.4.66b), (1.4.69b), (1.4.67a),

$$\sqrt{3}\psi_1 L = \bar{e}_1 - 2\bar{e}_2 \quad (1.4.72)$$

$$\sqrt{3}\psi_2 L = 2\bar{e}_1 - \bar{e}_2 \quad (1.4.73)$$

$$\sqrt{3}\psi_3 L = 2\bar{e}_2 + \bar{e}_3 = 2\bar{e}_2 - \bar{e}_1 = -\sqrt{3}\psi_1 L \quad (1.4.74)$$

$$\psi_3 = -\psi_1 \quad (1.4.74a)$$

Also

$$\sqrt{3}\psi_4 L = -2\bar{e}_1 - e_4 = -2\bar{e}_1 + \bar{e}_2 = -\sqrt{3}\psi_2 L \quad (1.4.75)$$

Hence

$$\psi_4 = -\psi_2 \quad (1.4.75a)$$

The two remaining relations from the loop (3), (4) serve as a check. Evaluating, we find from (1.4.68a) and (1.4.68b)

$$\sqrt{3}\psi_4 L = 2\bar{e}_3 - e_4 = -2\bar{e}_1 + \bar{e}_2 \qquad (1.4.76), (1.4.75)\text{bis}$$

$$\sqrt{3}\psi_3 L = \bar{e}_3 - 2e_4 = -\bar{e}_1 + 2\bar{e}_2 \qquad (1.4.77), (1.4.74)\text{bis}$$

The last two results are then identical to those in (1.4.74) and (1.4.75), as two of the eight compatibility equations are not independent.

If the initially critical bars (1) and (2) are assumed to extend under a temperature increase (t) from (t^0) while remaining straight, then ($\bar{e}_1$) and ($\bar{e}_2$) must be negative contractions. Therefore (e_4) can become a positive flexural contraction as ($-e_2$) is now positive. However, since also ($\bar{e}_1$) is an extension under this assumption and, therefore, negative, ($-\bar{e}_1$) must be a positive number and so ($\bar{e}_3$) must be also positive, Eq. (1.4.71), and therefore a *contraction*. This would mean that under the circumstances at least one straight member, in this case the critically stressed compressive bar (3), would contract axially and that the axial load in this straight bar would become *supercritical*. But, a straight member under a supercritical load is not statically stable and the member would buckle thermally. For this reason the asymmetric mode does not develop if the system is not perfect, as the system would snap through into the mode considered in case (ii). But that mode is also not stable and, therefore, *the mode considered, in case (i) is the only one among the three mode choices considered which represents a kinematically and statically admissible primary thermal buckling mode* of the system, *in which no straight bars are supercritically stressed.*

Thus, using this theory, thermal buckling modes or patterns of a hyperstatic lattice become *predictable*. This basic theory may be extended to more complex cases. However, these are beyond the scope and the size of the present book and, therefore, more complex problems of this kind are not discussed at this time.

2 The Post-Buckling Equilibrium of Isostatic Hinge-Connected Space Structures Composed of Slender Members

Elastic space structures, composed of pin-jointed slender members with symmetrical flexural properties, loaded centrally at the joints by conservative loads represent a class of symmetrical elastic systems discussed before.[1] The behavior of such structures in the critical and post-critical statical equilibrium states is similar to that of plane pin-jointed frameworks, discussed in Chapter 1 and at length in Ref. 15. Also, the statements made in regard to the onset of unstable motion for plane isostatic frameworks, apply equally to space frameworks or lattices. In hyperstatic lattices the onset of unstable motion may take place only in certain restrained mechanisms. Hyperstatic systems are discussed later.

Differences occur in the formulation of the geometrical conditions of compatibility of flexural contractions. Also, the equilibrium conditions must be referred to three coordinate axes in space, which makes the analysis of space structures more complex. We proceed to develop the theory in a systematic way extending the analysis to matrix formulation for three-dimensional cases.

2.1 General Force-Displacement Equilibrium Paths for Perfect Members

Buckling of single pin-ended members in space is summarized by the general law. This law remains valid for all types of pin-jointed elastic systems. It may be restated for three-dimensional systems by the relation

$$\frac{\delta P}{P^0} = K\lambda \tag{2.1.1}$$

where ($\lambda = e/L$) represents the total flexural contraction parameter of a member

[1] See "The Stability of Elastic Systems" by S. J. Britvec, Pergamon Press, 1973, Ch. 1–8 and Ch. 3 and 6.

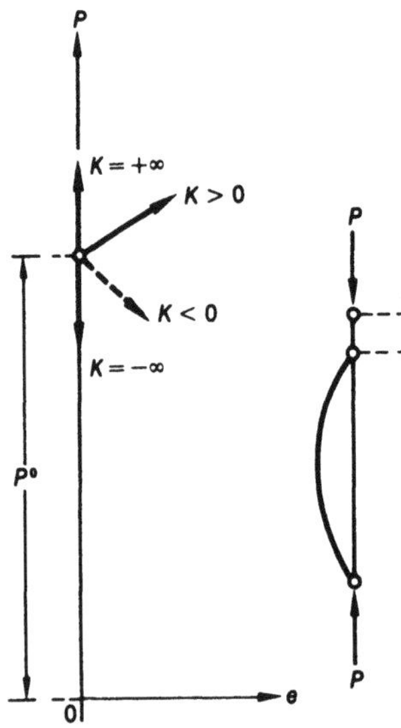

Figure 2.1.1
Slopes to the load-displacement equilibrium paths of perfect pin-jointed flexible members in elastic post-buckling

in space. Here,

$e =$ the flexural contraction of a bar of initial length L
$K = 1/2$ represents the buckling constant for prismatic bars in compression ($P^0 = P_E$, Euler)
$K = +\infty$ or $-\infty$ for all tensile bars (P^0 is tensile)
$K = -\infty$ for compressive bars when ($\lambda = 0$) so that the axial force decreases from P^0.

Different possible paths are summarized in Fig. 2.1.1.

2.2 Geometrical Compatibility Conditions in Space

To arrive at the geometrical compatibility conditions in three dimensions we first define the position, displacement and rotation of a single member, Fig. 2.2.1. Let ($\mathbf{r}_i$) be the position vector of joint (i), to which a member (i, j) is attached. Let (i') (j') be the displaced position of this member. In this position the chord of the member has contracted by an amount (e_{ij}). This chord rotates, with respect to the initial position (i) (j) by the total angle (ψ_{ij}). The object is to find first the direction cosines of the chords (i') (j') in the new positions, and rotations in space (ψ_{ij}) in terms of the flexural contraction parameters ($\lambda_{ij} = e_{ij}/L_{ij}$), where ($L_{ij}$) is the initial length (i, j).

Consider first the relationship between the number (J_F) of free joints (joints not attached to the foundation) and members (N_F) (attached to the free joints) in a statically determinate (isostatic) space framework. We start with three fixed points on the foundation. If these are not connected by members,

$$3J_F = N_F \tag{2.2.1}$$

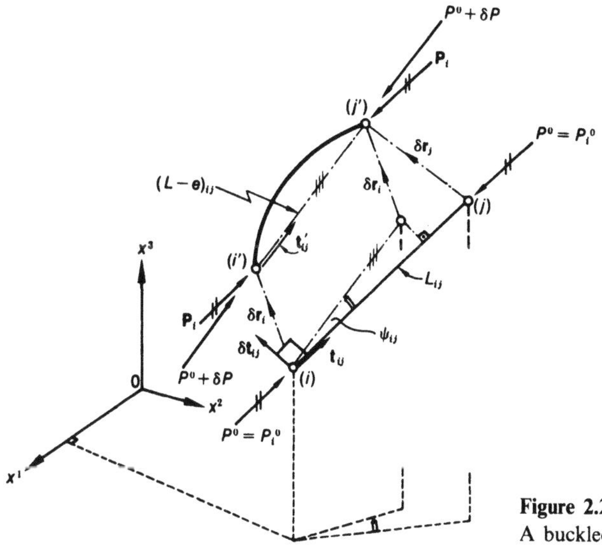

Figure 2.2.1
A buckled pin-jointed member in space

If they are mutually connected, three pin-jointed bars must be added to the frame, reducing the number of statically determinate external restraints to six. In that case

$$3J = N + 6 \tag{2.2.2}$$

where $(N = N_F + 3)$ is the total number of members, and $(J = J_F + 3)$ the total number of joints. Either type of system is acceptable in a generalized analysis, since the number of flexural contraction-compatibility conditions is the same in either case.

We associate three displacement components (u, v, w) with each free joint. One flexural contraction parameter (λ) and a body rotation in space (ψ) is associated with each member. There are, therefore, $(3J_F)$ free displacement-components and $(2N_F)$ body rotations (ψ) and contraction parameters (λ) in an isostatic space system, referred to the three fixed points on the foundation.

The concept of independent closed circuits of members, adopted in plane systems, leading to the required compatibility conditions is not the best one to apply to space systems, since, in more complex cases, these circuits are difficult to visualize. An easier approach to evaluate the compatibility conditions in space is to introduce first the joint displacements (u, v, w) into the analysis.

We refer a joint (i) to a global Cartesian coordinate system by the coordinates (x_i, y_i, z_i). The displacement of a joint (i) along the x_i-axis is then (u_i) and similarly (v_i) and (w_i) along the y- and z-axes, respectively. The initial length (L_{ij}) of member (i, j) may be written

$$L_{ij}^2 = (x_j - x_i)^2 + (y_j - y_i)^2 + (z_j - z_i)^2 \tag{2.2.3}$$

After buckling, the joints are displaced into the position (i') and (j'). Then

$$(L_{ij} - e_{ij})^2 = (x_j + u_j - x_i - u_i)^2 + (y_j + v_j - y_i - v_i)^2 + (z_j + w_j - z_i - w_i)^2 \tag{2.2.4}$$

Since the displacements are small, each term may be linearized. Retaining after multiplication only linear terms in (u, v, and w), the last result gives

$$-e_{ij} = \frac{x_j - x_i}{L_{ij}}(u_j - u_i) + \frac{y_j - y_i}{L_{ij}}(v_j - v_i) + \frac{z_j - z_i}{L_{ij}}(w_j - w_i) \tag{2.2.5}$$

Dividing this further by (L_{ij}), we get

$$-\lambda_{ij} = \frac{x_j - x_i}{L_{ij}}\frac{u_j - u_i}{L_{ij}} + \frac{y_j + y_i}{L_{ij}}\frac{v_j - v_i}{L_{ij}} + \frac{z_j - z_i}{L_{ij}}\frac{w_j - w_i}{L_{ij}}$$

Denoting by (l, m, n) the direction cosines of the members in the initial position,

$$-\lambda_{ij} = l_{ij}\frac{u_j - u_i}{L_{ij}} + m_{ij}\frac{v_j - v_i}{L_{ij}} + n_{ij}\frac{w_j - w_i}{L_{ij}} \tag{2.2.6}$$

There are all together $(N_F = 3J_F)$ equations of this type in an isostatic space system. Since only three independent displacements (u, v, w) of each free joint are possible, there are, all together $(N_F = 3J_F)$ such independent displacements in the system. Therefore, the $(3J_F)$ linear Eqs. (2.2.6) are sufficient to find these $(3J_F)$ displacement components in terms of the flexural contraction parameters (λ_{ij}).

We next determine the direction-cosines (l'_{ij}, m'_{ij}, and n'_{ij}) in the displaced positions of the members (i, j). These are given by

$$\begin{aligned} l'_{ij} &= \frac{x_j + u_j - x_i - u_i}{L_{ij} - e_{ij}} \\ m'_{ij} &= \frac{y_j + v_j - y_i - v_i}{L_{ij} - e_{ij}} \\ n'_{ij} &= \frac{z_j + w_j - z_i - w_i}{L_{ij} - e_{ij}} \end{aligned} \tag{2.2.7}$$

The displacements (u, v, w) and the contractions (e_{ij}) are small in relation to $(x_j - x_i)$ $(y_j - y_i)$ etc. Thus, in the resolution of axial member-forces in the displaced positions, only the linear terms involving the small quantities are retained. After expanding in Taylor series, the expressions become:

$$\begin{aligned} l'_{ij} &= l_{ij} + l_{ij}\lambda_{ij} + \frac{u_j - u_i}{L_{ij}} \\ m'_{ij} &= m_{ij} + m_{ij}\lambda_{ij} + \frac{v_j - v_i}{L_{ij}} \\ n'_{ij} &= n_{ij} + n_{ij}\lambda_{ij} + \frac{w_j - w_i}{L_{ij}} \end{aligned} \tag{2.2.8}$$

where $(\lambda_{ij} = e_{ij}/L_{ij})$.

It remains to determine the body rotations in space (ψ_{ij}). Let ($\mathbf{t}_{ij}$) be the unit vector along member (i, j) in its initial position, Fig. 2.2.1, and ($\mathbf{t}'_{ij}$) the unit vector along the chord of the displaced member (i', j'). Then

$$\mathbf{t}'_{ij} = \mathbf{t}_{ij} + \delta\mathbf{t}_{ij} \tag{2.2.9}$$

where ($\delta\mathbf{t}_{ij}$) represents the difference of the vectors ($\mathbf{t}'_{ij}$) and ($\mathbf{t}_{ij}$). Multiplying the last equation by $(L_{ij} - e_{ij})$ we obtain a vectorial expression for the chord (i', j'),

$$(L_{ij} - e_{ij})\mathbf{t}'_{ij} = (L_{ij} - e_{ij})\mathbf{t}_{ij} + (L_{ij} - e_{ij})\,\delta\mathbf{t}_{ij} \tag{2.2.10a}$$

The same chord-vector is given by,

$$\mathbf{r}'_j - \mathbf{r}'_i = \mathbf{r}_j - \mathbf{r}_i + \delta\mathbf{r}_j - \delta\mathbf{r}_i \tag{2.2.10b}$$

On equating the last two expressions, we get

$$\delta\mathbf{r}_j - \delta\mathbf{r}_i = -e_{ij}\mathbf{t}_{ij} + (L_{ij} - e_{ij})\,\delta\mathbf{t}_{ij} \tag{2.2.11}$$

Introducing a new unit vector ($\boldsymbol{\tau}_{ij}$) in the direction of ($\delta\mathbf{t}_{ij}$), gives in the first approximation

$$\delta\mathbf{t}_{ij} = \psi_{ij}\boldsymbol{\tau}_{ij} \tag{2.2.12}$$

Substituting this result into Eq. (2.2.11), and neglecting the second-order terms

$$\delta\mathbf{r}_j - \delta\mathbf{r}_i = -e_{ij}\mathbf{t}_{ij} + L_{ij}\psi_{ij}\boldsymbol{\tau}_{ij} \tag{2.2.13}$$

If the direction cosines of the unit vector ($\boldsymbol{\tau}_{ij}$) are denoted by (l^1, m^1, n^1), then, writing the last result in terms of the components in non-dimensional form

$$\begin{aligned}
\frac{u_j - u_i}{L_{ij}} &= -\lambda_{ij}l_{ij} + \psi_{ij}l^1_{ij} \\
\frac{v_j - v_i}{L_{ij}} &= -\lambda_{ij}m_{ij} + \psi_{ij}m^1_{ij} \\
\frac{w_j - w_i}{L_{ij}} &= -\lambda_{ij}n_{ij} + \psi_{ij}n^1_{ij}
\end{aligned} \tag{2.2.14}$$

There are all together ($3N_F$) such equations. The left hand sides of these equations represent known numbers, since the displacements u_i, v_i, w_i; $i = 1, 2, \ldots, N$ can be found from the Eqs. (2.2.6). Therefore, Eqs. (2.2.14) determine the body rotations in space (ψ_{ij}) in terms of the flexural contraction-parameters (λ_{ij}). By squaring and adding we get

$$\psi_{ij} = \left[\left(\frac{u_j - u_i}{L_{ij}} + \lambda_{ij}l_{ij}\right)^2 + \left(\frac{v_j - v_i}{L_{ij}} + \lambda_{ij}m_{ij}\right)^2 + \left(\frac{w_j - w_i}{L_{ij}} + \lambda_{ij}n_{ij}\right)^2\right]^{1/2} \tag{2.2.15}$$

There are all together (N_F) such relations in an axially-isostatic framework.

If the framework is simply supported on a three-point base, consisting of three additional members, then there are $[N = (N_F + 3)]$ Eqs. (2.2.15). In that case the number of Eqs. (2.2.6) is also increased to $(N_F + 3 = N)$. Since the simple supporting of the base triangle requires only six specified constraints the total number of free joint-displacements becomes $(3J - 6)$, and this, according to Eq. (2.2.2) is precisely equal to (N). Then, these (N) displacements can be found out of the (N) Eqs. (2.2.6) and the body rotations from the (N) Eqs. (2.2.15).

2.3 Initial Kinematic Relations

The Eq. (2.2.6) relating the end displacements of the member (i, j) to its flexural contraction (e_{ij}) may be rewritten in a more convenient form. Let the displacement vector $(\mathbf{u}_i)$ be represented by its Cartesian components (u_i, v_i, w_i) in the column vector

$$\begin{pmatrix} u_i \\ v_i \\ w_i \end{pmatrix} = \begin{pmatrix} u \\ v \\ w \end{pmatrix}_i = \mathbf{u}_i\dagger \tag{2.3.1a}$$

and similarly

$$\begin{pmatrix} u_j \\ v_j \\ w_j \end{pmatrix} = \begin{pmatrix} u \\ v \\ w \end{pmatrix}_j = \mathbf{u}_j \tag{2.3.1b}$$

Let also the direction cosines of member (i, j), referred to the local Cartesian coordinate system parallel to the global system at (i), be assembled in a column vector $(\mathbf{C}_{ij})$, such that

$$\mathbf{C}_{ij} = \begin{pmatrix} l_{ij} \\ m_{ij} \\ n_{ij} \end{pmatrix} = \begin{pmatrix} l \\ m \\ n \end{pmatrix}_{ij} \tag{2.3.2}$$

An alternative notation may be used by denoting the member (i, j) by the index (α).[2] Then, if the member is referred to a Cartesian coordinate system at (i), we may write

$$\mathbf{C}_{ij} = \mathbf{C}_{\alpha i} = \begin{pmatrix} l_{\alpha i} \\ m_{\alpha i} \\ n_{\alpha i} \end{pmatrix} = \begin{pmatrix} l_\alpha \\ m_\alpha \\ n_\alpha \end{pmatrix}_i \tag{2.3.2a}$$

If a local Cartesian system is placed at end (j), such that it is parallel to the global Cartesian frame, then the x-axes of the two systems and the member axis (i, j) define one and the same plane. The angle between the x-axis and the member axis, say (γ_{xi}), at (i) is then complementary to the angle (γ_{xj}) from the x-axis at (j) to the axis (j, i) at (j), and $(\gamma_{xj} = \pi - \gamma_{xi})$. Thus

$$l_{ij} = \cos \gamma_{xi} = \cos(\gamma_{xj} - \pi) = -\cos \gamma_{xj} = -l_{ji}$$

---

[2] Here, Greek indices are adopted for member designation and Latin indices for the designation of the joints.

† Symbols in bold type denote matrices and vectors.

Similarly:

$$m_{ij} = -m_{ji}; \quad n_{ij} = -n_{ji} \text{ or } l_{\alpha i} = -l_{\alpha j}; \quad m_{\alpha i} = -m_{\alpha j}; \quad n_{\alpha i} = -n_{\alpha j} \tag{2.3.3}$$

Eq. (2.2.6) may now be written

$$\lambda_\alpha = \frac{1}{L_\alpha}(l_{\alpha i}\ m_{\alpha i}\ n_{\alpha i})\begin{pmatrix} u_i \\ v_i \\ w_i \end{pmatrix} - \frac{1}{L_\alpha}(l_{\alpha i}\ m_{\alpha i}\ n_{\alpha i})\begin{pmatrix} u_j \\ v_j \\ w_j \end{pmatrix}$$

or

$$\lambda_\alpha = \frac{1}{L_\alpha}(\mathbf{C}_{\alpha i}^T \mathbf{u}_i - \mathbf{C}_{\alpha i}^T \mathbf{u}_j) = \frac{1}{L_\alpha}(\mathbf{C}_{\alpha i}^T \mid -\mathbf{C}_{\alpha i}^T)\begin{pmatrix} \mathbf{u}_i \\ \hline \mathbf{u}_j \end{pmatrix}$$

or, simply

$$e_\alpha = (\mathbf{C}_{\alpha i}^T \mid -\mathbf{C}_{\alpha i}^T)\begin{pmatrix} \mathbf{u}_i \\ \hline \mathbf{u}_j \end{pmatrix} \tag{2.3.4a}$$

Since ($\mathbf{C}_{\alpha i} = -\mathbf{C}_{\alpha j}$), this result may also be written

$$e_\alpha = (\mathbf{C}_{\alpha i}^T \mid \mathbf{C}_{\alpha j}^T)\begin{pmatrix} \mathbf{u}_i \\ \hline \mathbf{u}_j \end{pmatrix} \tag{2.3.4b}$$

The relations (2.3.4a) and (2.3.4b) assume that the member designated by the index (α) is situated between the joints designated by the indices (i) and (j).

Alternatively, if only the joint indices (i) and (j) are used, the Eqs. (2.3.4a) and (2.3.4b) may be written

$$e_{ij} = (\mathbf{C}_{ij}^T \mid -C_{ij}^T)\begin{pmatrix} \mathbf{u}_i \\ \hline \mathbf{u}_j \end{pmatrix} \tag{2.3.4c}$$

or

$$e_{ij} = (\mathbf{C}_{ij}^T \mid \mathbf{C}_{ji}^T)\begin{pmatrix} \mathbf{u}_i \\ \hline \mathbf{u}_j \end{pmatrix} \tag{2.3.4d}$$

since ($\mathbf{C}_{ij} = -\mathbf{C}_{ji}$), as explained by the Eqs. (2.3.3).

This supposition implies that the joints are numbered from (1) to (J) and the members from (1) to (N). If the nodes are partitioned into free and attached nodes and the members into those supporting only the free nodes (not attached to a foundation) and necessary to make the lattice just-stiff or isostatic, which we may denote by (N_F), then it is readily verified that

$$3J_F = N_F$$

It is also easily verified that the total number of members must be ($N = N_F + 3$) if the lattice is internally isostatic and the total number of joints (J) is then

$$J = J_F + 3$$

Thus, regardless of whether an isostatic lattice is attached to a foundation or free in space the total number (J) of the nodes and the number (N) of its members are always given by

$$3(J-3) = N - 3$$

or (2.3.5)

$$3J = N + 6$$

Since (N) of the members may be assumed to shorten flexurally (buckle) in a flexible lattice, also the number of the degrees of freedom is (N). Thus, from (2.3.5)

$$N = 3J - 6 \tag{2.3.5a}$$

In a detached lattice there are 3 degrees of freedom at each node, corresponding to the 3 components $(u, v, w)_i$ of the node displacement vectors ($\mathbf{u}_i$; $i = 1, 2, \ldots, J$). But, 6 of these ($3J$) degrees of freedom are suppressed by the isostatic supports which also define a fixed global coordinate system from which all the relative displacement components in the vectors ($\mathbf{u}_i$) are measured. Therefore, the number ($3J - 6$) of relative freedoms of a simply supported lattice is equal to the number (N) of members in the lattice, assuming all its nodes are connected by these members. Then, according to Eq. (2.3.5a), the number of the members (N) is identical to the number of independent relative displacement components in the vectors $[\mathbf{u}_i = (u_i, v_i, w_i)^T; i = 1, \ldots, J]$.

Therefore, if the flexural contractions (e_α) are assembled into a one-dimensional array or column vector ($\mathbf{e}$) according to the numbering sequence of the members and the nodal displacements are arranged in another one-dimensional array in column vector ($\mathbf{u}$) according to the numbering sequence of the nodes and each displacement resolved according to the x, y, z-axes of the global coordinate system, then the two column vectors can be related by an $N \times (N+6)$ matrix ($\mathbf{C}$) such that

$$\mathbf{e} = \underbrace{\begin{bmatrix} e_1 \\ e_2 \\ \cdot \\ \cdot \\ \cdot \\ \cdot \\ \cdot \\ e_N \end{bmatrix}}_{=\mathbf{e}} = \underbrace{\begin{bmatrix} C_{11} & C_{12} & C_{13} & \cdot\ \cdot & C_{1(N+4)} & C_{1(N+5)} & C_{1(N+6)} \\ C_{21} & C_{22} & C_{23} & \cdot\ \cdot & C_{2(N+4)} & C_{2(N+5)} & C_{2(N+6)} \\ \cdot & \cdot & \cdot & \cdot\ \cdot & \cdot & \cdot & \cdot \\ \cdot & \cdot & \cdot & \cdot\ \cdot & \cdot & \cdot & \cdot \\ \cdot & \cdot & \cdot & \cdot\ \cdot & \cdot & \cdot & \cdot \\ \cdot & \cdot & \cdot & \cdot\ \cdot & \cdot & \cdot & \cdot \\ \cdot & \cdot & \cdot & \cdot\ \cdot & \cdot & \cdot & \cdot \\ C_{N1} & C_{N2} & C_{N3} & \cdot\ \cdot & \cdot & \cdot & C_{N(N+6)} \end{bmatrix}}_{=\mathbf{C}[N \times (N+6)]} \underbrace{\begin{bmatrix} \left.\begin{matrix} u_1 \\ v_1 \\ w_1 \end{matrix}\right\} = \mathbf{u}_1 \\ \left.\begin{matrix} u_2 \\ v_2 \\ w_2 \end{matrix}\right\} = \mathbf{u}_2 \\ \cdot \\ \cdot \\ \cdot \\ \left.\begin{matrix} u_J \\ v_J \\ w_J \end{matrix}\right\} = \mathbf{u}_J \end{bmatrix}}_{=\mathbf{u}}$$

However, since 6 displacement components in the vector (**u**) at the simple supports are zero, the above matrix reduces to a $(N \times N)$ matrix (**C**), such that

$$\mathbf{e} = \mathbf{C}\mathbf{u} \tag{2.3.6}$$

where (**C**) now represents the reduced matrix, which is called the kinematic matrix. The elements of (**C**) are filled using the equations (2.3.4a) or (2.3.4b) in such a way, that the (α^{th}) member is located between the nodes (i) and (j), represented by 6 elements (direction cosines) in the αth row of (**C**) corresponding to the 6 elements in the column vector (**u**). A dot may represent a non-zero or a zero element. So, once the initial geometry of the lattice is defined and the numbering sequences for the joints and the members of the lattice established, the compilation of the (**C**) matrix is a relatively straightforward task. In large structures this process should be done on the computer using a program to generate the matrix (**C**) from a data file. If the lattice shows some symmetry, or if its space elements are repeated in some fashion, which is often the case in space lattices, the numbering of the members and the nodes should be undertaken in a systematic sequence so that the repetitive pattern of the elements and numbers in the space lattice is reflected in the kinematic (**C**) matrix. For example, in the case of a reticulated beam composed of repetitive space elements the numbering of the nodes should proceed from one end of the beam in a clockwise or counter clockwise sense around the beam, embodying all the peripheral and intermediate nodes and members, until all the nodes and members to the opposite end are numbered in a consistent manner. An example of this sequencing is shown in Fig. 2.3.1. The kinematic matrix for

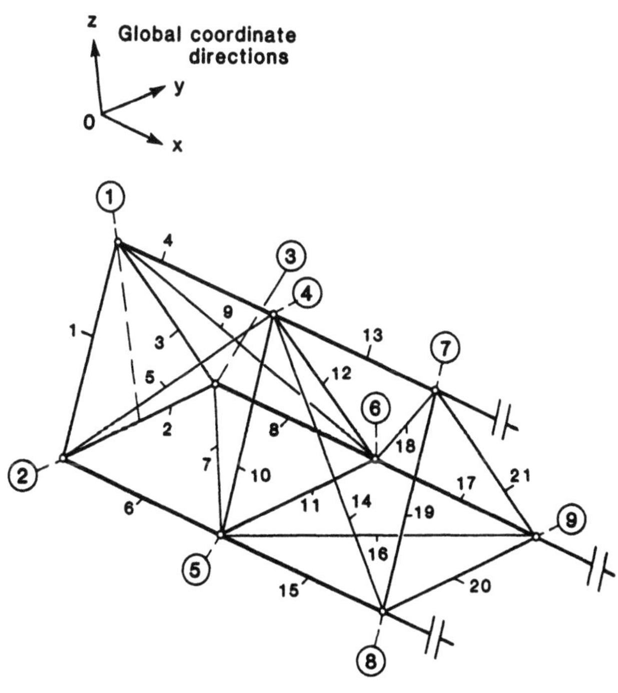

Figure 2.3.1
Joint- and member-numbering sequences of a reticulated iso-static beam

$$
\begin{bmatrix} e_1\\ e_2\\ e_3\\ e_4\\ e_5\\ e_6\\ e_7\\ e_8\\ e_9\\ e_{10}\\ e_{11}\\ e_{12}\\ \hline e_{13}\\ e_{14}\\ e_{15}\\ e_{16}\\ e_{17}\\ e_{18} \end{bmatrix}
=
\left[\begin{array}{ccccccccc:ccccccccc}
u_1 & v_1 & w_1 & u_2 & v_2 & w_2 & u_3 & v_3 & w_3 & u_4 & v_4 & w_4 & u_5 & v_5 & w_5 & u_6 & v_6 & w_6 \\
0 & -1/2 & -\sqrt{3}/2 & 0 & 1/2 & \sqrt{3}/2 & 0 & 0 & 0 & 0 & 0 & 0 & 0 & 0 & 0 & 0 & 0 & 0 \\
0 & 0 & 0 & 0 & 1 & 0 & 0 & -1 & 0 & 0 & 0 & 0 & 0 & 0 & 0 & 0 & 0 & 0 \\
0 & 1/2 & -\sqrt{3}/2 & 0 & 0 & 0 & 0 & -1/2 & \sqrt{3}/2 & 0 & 0 & 0 & 0 & 0 & 0 & 0 & 0 & 0 \\
1 & 0 & 0 & 0 & 0 & 0 & 0 & 0 & 0 & -1 & 0 & 0 & 0 & 0 & 0 & 0 & 0 & 0 \\
0 & 0 & 0 & \sqrt{2}/2 & \sqrt{2}/4 & \sqrt{6}/4 & 0 & 0 & 0 & -\sqrt{2}/2 & -\sqrt{2}/4 & -\sqrt{6}/4 & 0 & 0 & 0 & 0 & 0 & 0 \\
0 & 0 & 0 & 1 & 0 & 0 & 0 & 0 & 0 & 0 & 0 & 0 & -1 & 0 & 0 & 0 & 0 & 0 \\
0 & 0 & 0 & 0 & 0 & 0 & \sqrt{2}/2 & -\sqrt{2}/2 & 0 & 0 & 0 & 0 & -\sqrt{2}/2 & \sqrt{2}/2 & 0 & 0 & 0 & 0 \\
0 & 0 & 0 & 0 & 0 & 0 & 1 & 0 & 0 & 0 & 0 & 0 & 0 & 0 & 0 & -1 & 0 & 0 \\
\sqrt{2}/2 & \sqrt{2}/4 & -\sqrt{6}/4 & 0 & 0 & 0 & 0 & 0 & 0 & 0 & 0 & 0 & 0 & 0 & 0 & \sqrt{2}/2 & \sqrt{2}/4 & \sqrt{6}/4 \\
0 & 0 & 0 & 0 & 0 & 0 & 0 & 0 & 0 & 0 & -1/2 & -\sqrt{3}/2 & 0 & 1/2 & \sqrt{3}/2 & 0 & 0 & 0 \\
0 & 0 & 0 & 0 & 0 & 0 & 0 & 0 & 0 & 0 & 0 & 0 & 1 & 0 & 0 & -1 & 0 & 0 \\
0 & 0 & 0 & 0 & 0 & 0 & 0 & 0 & 0 & 0 & 1/2 & -\sqrt{3}/2 & 0 & 0 & 0 & 0 & -1/2 & \sqrt{3}/2 \\
\hdashline
0 & 0 & 0 & 0 & 0 & 0 & 0 & 0 & 0 & 1 & 0 & 0 & 0 & 0 & 0 & 0 & 0 & 0 \\
0 & 0 & 0 & 0 & 0 & 0 & 0 & 0 & 0 & \sqrt{2}/2 & -\sqrt{2}/4 & -\sqrt{6}/4 & 0 & 0 & 0 & 0 & 0 & 0 \\
0 & 0 & 0 & 0 & 0 & 0 & 0 & 0 & 0 & 0 & 0 & 0 & 1 & 0 & 0 & 0 & 0 & 0 \\
0 & 0 & 0 & 0 & 0 & 0 & 0 & 0 & 0 & 0 & 0 & 0 & \sqrt{2}/2 & \sqrt{2}/2 & 0 & 0 & 0 & 0 \\
0 & 0 & 0 & 0 & 0 & 0 & 0 & 0 & 0 & 0 & 0 & 0 & 0 & 0 & 0 & 1 & 0 & 0 \\
0 & 0 & 0 & 0 & 0 & 0 & 0 & 0 & 0 & 0 & 0 & 0 & 0 & 0 & 0 & \sqrt{2}/2 & \sqrt{2}/4 & \sqrt{6}/4
\end{array}\right]
\begin{bmatrix}
\left.\begin{matrix} u_1\\ v_1\\ w_1 \end{matrix}\right\} = \mathbf{u}_1 \\
\left.\begin{matrix} u_2\\ v_2\\ w_2 \end{matrix}\right\} = \mathbf{u}_2 \\
\left.\begin{matrix} u_3\\ v_3\\ w_3 \end{matrix}\right\} = \mathbf{u}_3 \\
\left.\begin{matrix} u_4\\ v_4\\ w_4 \end{matrix}\right\} = \mathbf{u}_4 \\
\left.\begin{matrix} u_5\\ v_5\\ w_5 \end{matrix}\right\} = \mathbf{u}_5 \\
\left.\begin{matrix} u_6\\ v_6\\ w_6 \end{matrix}\right\} = \mathbf{u}_6
\end{bmatrix}
$$

Eq. (2.3.7)

the first two bays of the reticulated beam in Fig. 2.3.1, in the case where the sides of the triangles and the longerons between the triangles are equal, is given by Eq. (2.3.7). It can be obtained using either Eq. (2.3.4a) or (2.3.4b) by translating the origin of the global coordinate system to the respective joint of the beam and writing the direction cosines of the members in the appropriate slots of each row the angles always being swept from the positive direction of each coordinate axis to the respective member direction counting the node where the coordinate system is placed as the node (i) (i.e. the near node) and the opposite node as the node (j) (i.e. the far node). The joints 1 to 6 are connected only by the members 1 to 12. Therefore, if only one end of the beam is supported 6 of the 18 displacements in the vector $(\mathbf{u})$ are zero. The horizontal partition indicates the flexural contractions of the members 1 to 12 in the first bay. If the beam is supported as a cantilever at the joints 1, 2 and 3, for example, then the displacements $(\mathbf{u}_1)$, $(\mathbf{u}_2)$ and $(\mathbf{u}_3)$ are zero, as well as the shortenings (e_1), (e_2) and (e_3). In this case the reduced $(\mathbf{C})$ matrix for the first bay would involve only rows 4 to 12 and columns 10 to 18 partitioned by the vertical broken line and it would be of the size (9×9).

2.4 Kinematic Relations in Post-Buckling

The direction cosines of a displaced and buckled member (i, j) were derived in section 1.2 in Eqs. (2.2.8). We define the (3×1) vector $(\mathbf{C}'_{ij})$ as

$$\mathbf{C}'_{ij} = \begin{pmatrix} l'_{ij} \\ m'_{ij} \\ n'_{ij} \end{pmatrix} \tag{2.4.1}$$

Then, in matrix form the Eqs. (2.2.8) become

$$\mathbf{C}'_{ij} = \begin{pmatrix} l'_{ij} \\ m'_{ij} \\ n'_{ij} \end{pmatrix} = \begin{pmatrix} l_{ij} \\ m_{ij} \\ n_{ij} \end{pmatrix} + \lambda_{ij} \begin{pmatrix} l_{ij} \\ m_{ij} \\ n_{ij} \end{pmatrix} + \frac{1}{L_{ij}} \begin{pmatrix} u_j \\ v_j \\ w_j \end{pmatrix} - \frac{1}{L_{ij}} \begin{pmatrix} u_i \\ v_i \\ w_i \end{pmatrix} \tag{2.4.2}$$

or

$$\mathbf{C}'_{ij} = \mathbf{C}_{ij} + \mathbf{C}_{ij}\lambda_{ij} + \frac{1}{L_{ij}} (\mathbf{I} \mid -\mathbf{I}) \begin{pmatrix} \mathbf{u}_j \\ \hline \mathbf{u}_i \end{pmatrix} \tag{2.4.2a}$$

Introducing the notation (L_α), (λ_α) etc. where the $(\alpha)^{\text{th}}$ member is situated between the nodes (i) and (j), (i) being the *near* joint, where the local coordinate system parallel to the global system has its origin, and (j) being the opposite node, the last result may be written alternatively

$$\mathbf{C}'_{\alpha i} = \mathbf{C}_{\alpha i} + \mathbf{C}_{\alpha i}\lambda_\alpha + \frac{1}{L_\alpha} (\mathbf{I}_{\alpha i} \mid -\mathbf{I}_{\alpha i}) \begin{pmatrix} \mathbf{u}_j \\ \hline \mathbf{u}_i \end{pmatrix} \tag{2.4.2b}$$

where $(\mathbf{C}_{\alpha i} = -\mathbf{C}_{\alpha j})$ and $(\mathbf{I}_{\alpha i})$ is a (3×3) unit matrix. Substituting for $(\lambda_\alpha =$

e_α/L_α) from (2.3.4b) in (2.4.2a) and (2.4.2b), we get

$$\mathbf{C}'_{ij} = \mathbf{C}_{ij} + \frac{1}{L_{ij}}[\mathbf{C}_{ij}(\mathbf{C}^T_{ij} \mid -\mathbf{C}^T_{ij}) - (\mathbf{I} \mid -\mathbf{I})]\left(\frac{\mathbf{u}_i}{\mathbf{u}_j}\right) \tag{2.4.3}$$

or, using the index (α),

$$\mathbf{C}'_{\alpha i} = C_{\alpha i} + \frac{1}{L_\alpha}[\mathbf{C}_{\alpha i}(\mathbf{C}^T_{\alpha i} \mid -\mathbf{C}^T_{\alpha i}) - (\mathbf{I}_{\alpha i} \mid -\mathbf{I}_{\alpha i})]\left(\frac{\mathbf{u}_i}{\mathbf{u}_j}\right) \tag{2.4.4}$$

2.5 Initial Equilibrium States

A space structure made of flexible members suffers negligible axial deformations *before* the onset of bending or buckling in the members. This is why also the critical state is considered to take place on the undeformed lattice while the lattice is in its initial or original straight configuration. If one considers the equilibrium of a typical node (i) with the member (i, j) in the positive quadrant of the translated global coordinate system with the origin at (i) and the positive compressive axial load ($\mathbf{P}_{ij} = \mathbf{P}_\alpha$) pointing towards this center (the components of (P_α) pointing in the negative directions of the coordinate axes), then one can write the Cartesian components of the adjoining forces ($\mathbf{P}_\alpha$) assumed positive in compression as:

$$\mathbf{F}_{\alpha i} = \begin{pmatrix} X_{\alpha i} \\ Y_{\alpha i} \\ Z_{\alpha i} \end{pmatrix} = \begin{pmatrix} P_\alpha l_{\alpha i} \\ P_\alpha m_{\alpha i} \\ P_\alpha n_{\alpha i} \end{pmatrix} = P_\alpha \mathbf{C}_{\alpha i} \tag{2.5.1}$$

where ($X_{\alpha i}$), ($Y_{\alpha i}$), ($Z_{\alpha i}$) are the (x, y, z) components of ($\mathbf{P}_\alpha$) referred to the coordinate system at the near joint (i), to which the member (α) is attached. The equilibrium equations can then be written in the:

$$x\text{-direction} \quad \sum_{\alpha=1}^{m_i} P_\alpha l_{\alpha i} = W_{ix}$$

$$y\text{-direction} \quad \sum_{\alpha=1}^{m_i} P_\alpha m_{\alpha i} = W_{iy}$$

$$z\text{-direction} \quad \sum_{\alpha=1}^{m_i} P_\alpha n_{\alpha i} = W_{iz}$$

(m_i) being the number of the adjoining members at joint (i), or in matrix formulation at a node (i):

$$\sum_{\alpha=1}^{m_i} \mathbf{F}_{\alpha i} = \sum_{\alpha=1}^{m_i} \begin{pmatrix} X_{\alpha i} \\ Y_{\alpha i} \\ Z_{\alpha i} \end{pmatrix} = \sum_{\alpha=1}^{m_i} P_\alpha \mathbf{C}_{\alpha i}; \qquad i = 1, 2, \ldots, J$$

Denoting

$$\begin{pmatrix} W_{ix} \\ W_{iy} \\ W_{iz} \end{pmatrix} = \mathbf{W}_i$$

the equilibrium equations *at a node* become:

$$\sum_{\alpha=1}^{m_i} \mathbf{F}_{\alpha i} = \sum_{\alpha=1}^{m_i} P_\alpha \mathbf{C}_{\alpha i} = \mathbf{W}_i \tag{2.5.2}$$

where $(l_{\alpha i}, m_{\alpha i}, n_{\alpha i})$ represent the direction cosines of member (α) situated between the joints (i) and (j). (These can be written alternatively (l_{ij}, m_{ij}, n_{ij})). Note that the components of (P_α) point in the negative directions of the axes (x), (y) and (z). (W_{ix}), (W_{iy}), (W_{iz}) are the components of an external load (W_i) applied at this node. These components are assumed to be acting in the positive directions of the coordinate axes (x, y, z).

If the Eq. (2.5.1) is evaluated at different nodes, equilibrium equations for the entire space structure are generated such that on the right side of the equations, a column vector of external loads (assumed to be applied at the free joints) is generated i.e.

$$\mathbf{W} = \begin{bmatrix} \left.\begin{matrix} W_{1x} \\ W_{1y} \\ W_{1z} \end{matrix}\right\} = \mathbf{W}_1 \\ --- \\ \mathbf{W}_2 \\ --- \\ \cdot \\ --- \\ \mathbf{W}_J \end{bmatrix} = \begin{bmatrix} \left.\begin{matrix} \beta_{1x} \\ \beta_{1y} \\ \beta_{1z} \end{matrix}\right\} = \boldsymbol{\beta}_1 \\ --- \\ \boldsymbol{\beta}_2 \\ --- \\ \cdot \\ --- \\ \boldsymbol{\beta}_J \end{bmatrix} w = \boldsymbol{\beta} w \tag{2.5.3}$$

where $(\boldsymbol{\beta})$ is the vector of load-proportionality constants and (w) the *load-parameter* in the case of *proportionate* external loading, and on the left side of the equation a matrix and a column vector of internal loads $(P_\alpha;\ \alpha = 1, 2, \ldots, N)$ are obtained. This matrix is denoted by $(\mathbf{A})$ and it represents the *statical matrix* for the lattice. We write, therefore, that the initial equilibrium states are defined by the matrix equation

$$\mathbf{AP} = \mathbf{W} = \boldsymbol{\beta} w \tag{2.5.4}$$

where $(\mathbf{P})$ is the column vector of the internal axial loads in the members starting with member (1) and ending with member (N).‡ Thus

$$\mathbf{P} = (P_1 \ P_2 \ \cdots \ P_N)^T \tag{2.5.4a}$$

This relationship is valid also in the critical state if this state takes place in the initial configuration of the lattice in which the nodes are *not* displaced, which is reasonably true in the case of flexible slender members. Then a displacement field $(\mathbf{u})$ of the nodal points is generated if the members *contract flexurally* by the amounts $(e_\alpha;\ \alpha = 1, \ldots, N)$ (some of which may be zero). The corresponding vector $(\mathbf{u})$ for this displacement field is then related to the vector

‡ In statically determinate lattices these equations are always sufficient to determine the axial loads in terms of the load-parameter (w).

$[\mathbf{e} = (e_1\ e_2\ \cdots\ e_N)^T]$ by the kinematic relation

$$\mathbf{e} = \mathbf{C}\mathbf{u} \tag{2.5.5}$$

derived previously in Eq. (2.3.6).

If (**e**) is caused by an independent group of forces (unrelated to the group ($\mathbf{AP} = \mathbf{W}$) in Eq. (2.5.4)), then the virtual work of the internal forces contained in the vector (**P**) on the contractions (**e**) must equal to the virtual work of the external forces (**W**) on the corresponding displacements of the nodes contained in the vector (**u**). Thus, in matrix form, this may be expressed

$$\mathbf{P}^T\mathbf{e} = \mathbf{W}^T\mathbf{u}$$

Substituting for (**e**) and (**W**) from (2.5.5) and (2.5.4) respectively, we get

$$\mathbf{P}^T\mathbf{C}\mathbf{u} = \mathbf{P}^T\mathbf{A}^T\mathbf{u}$$

Hence

$$\mathbf{C} = \mathbf{A}^T$$

or (2.5.6)

$$\mathbf{A} = \mathbf{C}^T$$

and the equations (2.5.4) in critical equilibrium may be written

$$\mathbf{C}^T\mathbf{P} = \mathbf{W} = \boldsymbol{\beta}w \tag{2.5.7}$$

where the kinematic matrix (**C**) is considered to be known from the previous analysis. We obtain the statical matrix (**A**) by simply transposing (**C**). The internal forces (**P**) are then obtained by computing the inverse matrix ($\mathbf{A}^{-1}$). But

$$\mathbf{A}^{-1} = (\mathbf{C}^T)^{-1} = (\mathbf{C}^{-1})^T = \mathbf{C}_1^T \tag{2.5.8}$$

where ($\mathbf{C}_1$) is the inverse of the kinematic matrix (**C**). Thus only the kinematic matrix for a particular isostatic lattice need be inverted. Then

$$\mathbf{P} = \mathbf{A}^{-1}\boldsymbol{\beta}w = \mathbf{C}_1^T\boldsymbol{\beta}w \tag{2.5.9}$$

If equilibrium is correctly defined, (**A**) is always nonsingular, i.e. $|\mathbf{A}| \neq 0$ and also $|\mathbf{C}^T| = |\mathbf{C}| \neq 0$ and, therefore, also (**C**) is a nonsingular matrix.

By increasing (w) from zero, eventually (c) out of (N) internal loads will become critically stressed when ($w = w^0$). (c may be a very small number compared to (N), i.e. (c) may be 1 or 5 or 10 while (N) may be a large integer such as 324 or 506) or higher.

The critical loads for all the compressive (and tensile) members must be defined and entered with the appropriate plus sign into the vector.

$$\mathbf{P}^0 = (P_1^0\ P_2^0\ \cdots\ P_N^0)^T \tag{2.5.10}$$

according to the numbering sequence of the members. A comparison of the vectors ($\mathbf{P}^0$) and the vector $[\mathbf{C}_1^T\ \boldsymbol{\beta}\ w]$ for each increment (Δw) in (w), provides a check on the attainment of the critical state, characterized by ($\mathbf{P}^0$) and (w^0).

Thus, the determination of the initial critical equilibrium state requires the inversion of the kinematic matrix (**C**) which is nonsingular, as shown. This inversion can be carried out using any standard technique on a digital computer, such as a Gaussian method of inversion, for instance. The critical state itself is then determined by comparing the column vector (**P**) and the column vector of critical loads ($\mathbf{P}^0$). When at least one value of a compressive load in (**P**) equals that of the corresponding critical load in ($\mathbf{P}^0$), the critical state of the isostatic system has been reached.

2.6 Unsupported Structures—Initial Kinematic and Equilibrium Conditions

In large space lattices, such as a space station, the structure may be isostatic, yet not supported on a foundation but free in space while subjected to an equilibrated system of quasi-static loads applied centrally at its nodes. In that case the number of nodes (J) and the number of members (N) are in relation to each other according to the well known formula

$$3J = N + 6 \tag{2.6.1}$$

Now the kinematic matrix (**C**) has (N) rows as before but ($N + 6$) columns, 6 of these being associated with 6 nodal displacements in the vector (**u**) related to the rigid-body freedoms of the structure. To evaluate the displacements (**u**) of the nodes in post-buckling it is necessary to fix a coordinate system in space relative to the structure by specifying 6 of the displacements. To do this, it is convenient to zero 6 displacements in the vector (**u**) and thereby to fix a Cartesian coordinate system such that the displacement field (**u**) is then evaluated with respect to this fixed Cartesian system. We illustrate the point in Fig. 2.6.1. The origin of the coordinate system is placed at joint (1). Then, ($\mathbf{u}_1 = \mathbf{0}$) i.e. ($u_1 = v_1 = w_1 = 0$). Joint (2) is free to move only in the y-direction. Thus ($u_2 = w_2 = 0$) or [$\mathbf{u}_2^T = (0 \;\; v_2 \;\; 0)$]. Joint (3) is restrained from moving in the z-direction. Thus [$\mathbf{u}_3^T = (u_3 \;\; v_3 \;\; 0)$]. The 6 displacements ($u_1$), ($v_1$), ($w_1$), ($u_2$), ($v_2$) and ($w_3$) in the vector (**u**) of ($N + 6$) elements are thus zero. The corresponding columns in the (**C**) matrix in the kinematic relation

$$\mathbf{e} = \mathbf{C}\mathbf{u} \tag{2.6.2}$$

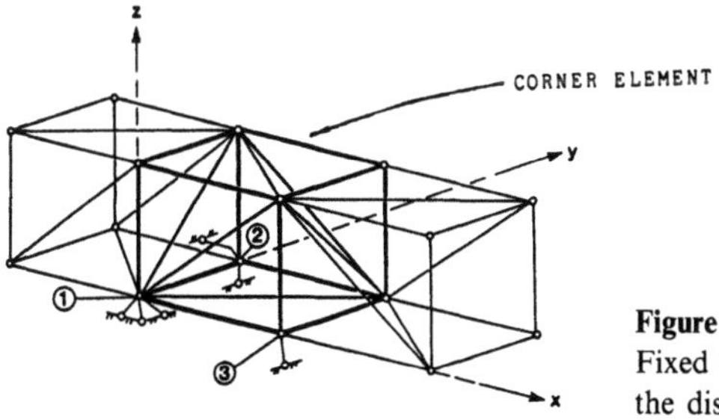

Figure 2.6.1
Fixed Cartesian coordinate system in space to which the displacement field vector (**u**) is referred

are simply deleted, since a multiplication by zeros eliminates these columns from this relation. This kinematic relation, however, will serve only to evaluate the remaining nodal displacements in a post-buckling mode of deformation with reference to this particular coordinate system. The reduced matrix (**C**) is now a square nonsingular matrix which can be inverted to yield the vector (**u**) of the remaining nodal displacements i.e.

$$\mathbf{u} = \mathbf{C}^{-1}\mathbf{e} = \mathbf{C}_{\mathrm{I}}\mathbf{e} \tag{2.6.3}$$

If the statical matrix (**A**) is derived from this reduced (**C**) matrix i.e.

$$\mathbf{A} = \mathbf{C}^{T} \tag{2.6.4}$$

then, the components of the external equilibrated loads applied at the nodes (1), (2), (3) in the respective directions in which the displacements were set equal to zero, can be considered as the 6 "reactive" load components of an externally statically determinate structure. In this case, these "reactive" forces are

$$W_{1x},\ W_{1y},\ W_{1z},\ W_{2x},\ W_{2z} \text{ and } W_{3z}$$

But, for the statical solution of the internal forces in the bars in the initial configuration, in which the critical state occurs, any other 6 statically compatible components may be chosen and the statical (**A**) matrix derived accordingly.

2.7 Equilibrium in Post-Buckling

Let the Cartesian components of an axial bar force ($\mathbf{P}_\alpha$), positive in compression and situated between the nodes (i) and (j) and referred to the global coordinate system (x, y, z), be denoted by

$$\begin{pmatrix} X_{ij} \\ Y_{ij} \\ Z_{ij} \end{pmatrix} = \begin{pmatrix} X \\ Y \\ Z \end{pmatrix}_{ij} = \mathbf{F}_{ij} = \begin{pmatrix} P_{ij}l_{ij} \\ P_{ij}m_{ij} \\ P_{ij}n_{ij} \end{pmatrix} = P_{ij}\begin{pmatrix} l \\ m \\ n \end{pmatrix}_{ij} = P_{ij}\mathbf{C}_{ij} = \mathbf{C}_{ij}P_{ij} \tag{2.7.1}$$

Alternatively, this statement can be written using the index (α) of the bar and the index of the "near" joint (i), as stated in Eq. (2.5.1) in the last section.

On the deformed structure this identity becomes

$$\mathbf{F}_{ij} = \mathbf{F}^0_{ij} + \delta\mathbf{F}_{ij} = \mathbf{C}'_{ij}(P^0_{ij} + \delta P_{ij}) \tag{2.7.2}$$

where ($\mathbf{C}'_{ij}$) is the vector of the direction cosines of the displaced chord of the member (i, j). Substituting for ($\mathbf{C}'_{ij}$) from Eq. (2.4.3) in Section 2.4 (or from (2.4.4), if using the alternative index notation) and retaining only linear small terms, we get

$$\mathbf{F}^0_{ij} + \delta\mathbf{F}_{ij} = \mathbf{C}_{ij}P^0_{ij} + \frac{P^0_{ij}}{L_{ij}}[\mathbf{C}_{ij}(\mathbf{C}^T_{ij} \mid -\mathbf{C}^T_{ij}) + (-\mathbf{I} \mid \mathbf{I})]\begin{pmatrix} \mathbf{u}_i \\ \hline \mathbf{u}_j \end{pmatrix} + \mathbf{C}_{ij}\,\delta P_{ij} \tag{2.7.3}$$

But, for equilibrium in the critical state

$$\mathbf{F}_{ij}^{0} = \mathbf{C}_{ij} P_{ij}^{0} \tag{2.7.4}$$

Hence, in a post-buckling equilibrium state

$$\delta \mathbf{F}_{ij} = \mathbf{C}_{ij}\, \delta P_{ij} + \frac{P_{ij}^{0}}{L_{ij}} [(\mathbf{C}_{ij}\mathbf{C}_{ij}^{T} \mid -\mathbf{C}_{ij}\mathbf{C}_{ij}^{T}) - (\mathbf{I} \mid -\mathbf{I})] \left(\frac{\mathbf{u}_i}{\mathbf{u}_j} \right) \tag{2.7.5}$$

This result may be written, using the alternative indices, as

$$\delta \mathbf{F}_{\alpha i} = \mathbf{C}_{\alpha i}\, \delta P_{\alpha} + \frac{P_{\alpha}^{0}}{L_{\alpha}} [\mathbf{C}_{\alpha i}(\mathbf{C}_{\alpha i}^{T} \mid -\mathbf{C}_{\alpha i}^{T}) - (\mathbf{I}_{\alpha i} \mid -\mathbf{I}_{\alpha i})] \left(\frac{\mathbf{u}_i}{\mathbf{u}_j} \right) \tag{2.7.5a}$$

where

$$\mathbf{C}_{\alpha i} \equiv \mathbf{C}_{ij} \quad \text{and} \quad \mathbf{C}_{\alpha j} \equiv \mathbf{C}_{ji} \tag{2.7.5b}$$

(Here (i) may be called the "near" joint and (j) the "far" joint of member (α)).

The Eqs. (2.7.5) can be developed further if now the general law relating the incremental axial loads (δP_{ij}) or (δP_{α}) to the flexural contractions in the buckled or unbuckled members of the lattice is introduced in the form

$$\delta P_{ij} = K_{ij} \frac{P_{ij}^{0}}{L_{ij}} e_{ij} \tag{2.7.6}$$

where the buckling constant (K_{ij}) has the usual values i.e. $(K_{ij} = 1/2)$ for all prismatic members which *actually* shorten flexurally in the post-buckling process, or, $(K_{ij} = \pm\infty)$ for all other members, as explained earlier. Then combining this law with the kinematic relation (2.3.4c),

$$e_{ij} = (\mathbf{C}_{ij}^{T} \mid -\mathbf{C}_{ij}^{T}) \left(\frac{\mathbf{u}_i}{\mathbf{u}_j} \right) \tag{2.7.7}$$

one can write the first term in Eq. (2.7.5)

$$\mathbf{C}_{ij}\, \delta P_{ij} = K_{ij} \frac{P_{ij}^{0}}{L_{ij}} \mathbf{C}_{ij}(\mathbf{C}_{ij}^{T} \mid -\mathbf{C}_{ij}^{T}) \left(\frac{\mathbf{u}_i}{\mathbf{u}_j} \right) \tag{2.7.8}$$

and so Eq. (2.7.5) can be modified as follows

$$\delta \mathbf{F}_{ij} = \frac{P_{ij}^{0}}{L_{ij}} [(1 + K_{ij})(\mathbf{C}_{ij}\mathbf{C}_{ij}^{T} \mid -\mathbf{C}_{ij}\mathbf{C}_{ij}^{T}) - (\mathbf{I} \mid -\mathbf{I})] \left(\frac{\mathbf{u}_i}{\mathbf{u}_j} \right) \tag{2.7.9}$$

where $(\mathbf{C}_{ij}\mathbf{C}_{ij}^{T})$ is a (3×3) matrix and $(\mathbf{I})$ a (3×3) unit matrix. This equation may be written further

$$\delta \mathbf{F}_{ij} = [\mathbf{H}_{ij} \mid -\mathbf{H}_{ij}] \left(\frac{\mathbf{u}_i}{\mathbf{u}_j} \right) \tag{2.7.9a}$$

where

$$\mathbf{H}_{ij} = \frac{P_{ij}^{0}}{L_{ij}} [(1 + K_{ij})\mathbf{C}_{ij}\mathbf{C}_{ij}^{T} - \mathbf{I}] = \mathbf{H}_{ji} \tag{2.7.10}$$

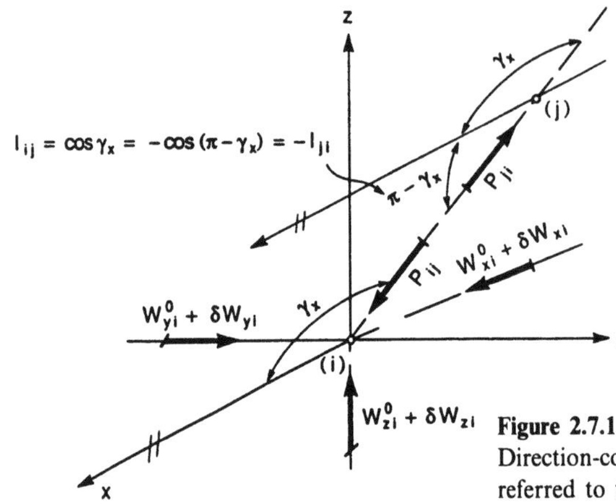

Figure 2.7.1
Direction-cosines of a typical member (i, j) in space referred to the opposite joints (i) and (j)

Since $(P^0_{ij} = P^0_{ji})$ and $(L_{ij} = L_{ji};\ K_{ij} = K_{ji})$, and since the direction consines at the opposite ends of the member (i, j) have opposite signs, Fig. (2.7.1), i.e.

$$\mathbf{C}_{ij} = -\mathbf{C}_{ji} \quad \text{and} \quad \mathbf{C}^T_{ij} = -\mathbf{C}^T_{ji}$$

the result of Eq. (2.7.10) follows from the identity

$$\mathbf{C}_{ij}\mathbf{C}^T_{ij} = \mathbf{C}_{ji}\mathbf{C}^T_{ji}$$

so that

$$\mathbf{H}_{ji} = \frac{P^0_{ji}}{L_{ji}}[(1 + K_{ji})\mathbf{C}_{ji}\mathbf{C}^T_{ji} - \mathbf{I}] = \mathbf{H}_{ij} \tag{2.7.10b}$$

The EQUILIBRIUM conditions at node (i) in a post-buckling state are then written

$$\sum_{j=1}^{m_i} \delta\mathbf{F}_{ij} = \begin{pmatrix} \delta W_{xi} \\ \delta W_{yi} \\ \delta W_{zi} \end{pmatrix} = \delta\mathbf{W}_i \tag{2.7.11}$$

where the internal compressive incremental axial force components are assumed positive in the negative directions of the coordinate axes and the external incremental loads (δW_{xi}), (δW_{yi}) and (δW_{zi}) are shown positive in the positive directions of these axes.

If the external loads are proportionate and vector of load-proportionality constants in post-buckling is $(\boldsymbol{\beta}')$, then the last result may be written also

$$\sum_{j=1}^{m_i} \delta\mathbf{F}_{ij} = \begin{pmatrix} \delta W_{xi} \\ \delta W_{yi} \\ \delta W_{zi} \end{pmatrix} = \begin{pmatrix} \beta'_{xi} \\ \beta'_{yi} \\ \beta'_{zi} \end{pmatrix} \delta w = \boldsymbol{\beta}'_i\, \delta w \tag{2.7.12}$$

(m_i) is the number of the adjoining members at joint (i) and (δw) a change in the load parameter (w).

The totality of the equilibrium equations (2.7.12) combined with the expressions for $(\delta \mathbf{F}_{ij})$ in Eq. (2.7.9) leads to a relationship between the displacement vector $(\mathbf{u})$ and the load vector $(\boldsymbol{\beta}')$. The matrix connecting these two vectors in equilibrium state of the buckled lattice is the statical matrix which we denote by the symbol $(\mathbf{A}')$. This matrix consists of submatrices $(\mathbf{H}_{ij})$ for all the members defined in Eqs. (2.7.9a) and (2.7.10b). Each element of the $(\mathbf{H}_{ij})$ matrix can be separated into two parts

$$\mathbf{H}_{ij} = \frac{P_{ij}^0}{L_{ij}}(\mathbf{C}_{ij}\mathbf{C}_{ij}^T - \mathbf{I}) + K_{ij}\frac{P_{ij}^0}{L_{ij}}\mathbf{C}_{ij}\mathbf{C}_{ij}^T \tag{2.7.13}$$

or

$$\mathbf{H}_{ij} = \mathbf{a}_{ij} + K_{ij}\mathbf{b}_{ij} \tag{2.7.14}$$

with

$$\mathbf{a}_{ij} = \frac{P_{ij}^0}{L_{ij}}(\mathbf{C}_{ij}\mathbf{C}_{ij}^T - \mathbf{I}) \tag{2.7.14a}$$

$$\mathbf{b}_{ij} = \frac{P_{ij}^0}{L_{ij}}\mathbf{C}_{ij}\mathbf{C}_{ij}^T \tag{2.7.14b}$$

where $(\mathbf{a}_{ij})$ and $(\mathbf{b}_{ij})$ are (3×3) matrices.

Substituting (2.7.9) into (2.7.12) in which $(\delta \mathbf{F}_{ij})$ is written in the form (2.7.14) one gets

$$\delta \mathbf{F}_{ij} = \sum_{j=1}^{m_i} [\mathbf{a}_{ij} + K_{ij}\mathbf{b}_{ij} \mid -(\mathbf{a}_{ij} + K_{ij}\mathbf{b}_{ij})]\left(\frac{\mathbf{u}_i}{\mathbf{u}_j}\right) = \boldsymbol{\beta}_i' \,\delta w \tag{2.7.15a}$$

or, after multiplication and rearrangement of terms

$$\delta \mathbf{F}_{ij} = \sum_{j=1}^{m_i} [\mathbf{a}_{ij} + K_{ij}\mathbf{b}_{ij}](\mathbf{u}_i - \mathbf{u}_j) = \boldsymbol{\beta}_i' \,\delta w \tag{2.7.15b}$$

If these equations are written at each node (i), $(i = 1, \ldots, N)$, then the totality of these equilibrium equations can be reduced to the form

$$\mathbf{S}\mathbf{u} = \boldsymbol{\beta}' \,\delta w \tag{2.7.16}$$

where $(\mathbf{S})$ is a post-buckling stiffness matrix related to the nodal displacement vector $(\mathbf{u})$. $(\mathbf{S})$ is composed of the elements contained in the Eqs. (2.7.15) and $(\mathbf{u})$ represents the augmented node-displacement vector, in which the nodal displacements are ordered according to the numbering sequence of the nodes and members, which they connect, referred to the (x, y, z) directions of the same global coordinate system. The vector $(\boldsymbol{\beta}')$ of load-proportionality constants in the post-buckling range [note that this vector is denoted by $(\boldsymbol{\beta}')$ and $(\boldsymbol{\beta}')$ is generally not equal to $(\boldsymbol{\beta})$] must be arranged according to the same numbering sequence.

The augmented displacement vector $(\mathbf{u})$ may be expressed in terms of the flexural contractions from the kinematic relation (2.3.6). This requires inverting the matrix $(\mathbf{C})$, which is now nonsingular, as the initial equilibrium of the system is uniquely defined by the nonsingular statical matrix (**A**), and (**C**) and (**A**) are related by ($\mathbf{C} = \mathbf{A}^T$) (see Fig. 2.5.6).

If next the inverse kinematic relation

$$\mathbf{u} = \mathbf{C}^{-1}\mathbf{e} = \mathbf{C}_1\mathbf{e}$$

is introduced in (2.7.16), we get

$$\mathbf{S}\mathbf{C}_1\mathbf{e} = \boldsymbol{\beta}'\,\delta w$$

which may be written

$$\mathbf{E}\mathbf{e} = \boldsymbol{\beta}'\,\delta w \tag{2.7.17}$$

with

$$\mathbf{E} = \mathbf{S}\mathbf{C}_1 \tag{2.7.17a}$$

(**E**) is called the post-buckling stiffness matrix related to the vector (**e**) of flexural contractions, where, as this was done earlier in the approach used in Ch. III, of Ref. 15, *all* members are assumed to shorten flexurally. It can be shown that (**E**) reduces to a matrix in which each element is of the form

$$a'_{ij} + K_j b'_{ij}$$

similarly as in the case of plane isostatic frames discussed in Section 1.2 and in Ref. 15, so that the buckling constants (K_j) are the same in each column of (**E**). This follows actually from the structure of Eq. (2.7.5) after expressing the displacements in terms of the flexural shortenings from the inverse relation

$$\mathbf{u} = \mathbf{C}_1\mathbf{e}$$

and introducing the general law (see Eq. 2.1.1) in the first term involving (δP_{ij}). Once (δP_{ij}) is replaced by the corresponding shortening ($\mathbf{e}_{ij}$) and the constant ($K_{ij}(P_{ij}/L_{ij})$), it is easily seen that the post-buckling equilibrium equations in an isostatic pin-jointed three-dimensional lattice assume the form

$$\mathbf{E}\mathbf{e} = [a'_{ij} + K_j b'_{ij}]\mathbf{e} = \boldsymbol{\beta}'\,\delta w \tag{2.7.18}$$

in which (a'_{ij}) and (b'_{ij}) may be non-zero or zero coefficients in the matrix contained within the square brackets. Also, the vector (**e**) contains the flexural shortenings of all members in the appropriate sequence and the question of zero shortenings is resolved by giving the constant (K_j) values of ($\pm\infty$), similarly as explained for plane systems in Section 1.2 and in Ref. 15, Chapter III. Thus, it is seen that the post-buckling equilibrium equations for three-dimensional systems can be reduced to the form obtained earlier for plane systems and these equations may be solved similarly as in the case of two-dimensional isostatic trusses, in Ref. 15, Chapter 3.

However, this method leads to a very large set of simultaneous equations which must be solved for each buckling mode by giving the constants (K_j; $j = 1, \ldots, N$), appropriate values. We show next that an *alternative* formulation of these equilibrium equations is preferable in *complex* three-dimensional lattices, involving a large number of flexible members. It will be apparent also later when the post-buckling equilibrium of statically indeterminate lattices is discussed, that the alternative method offers several other advantages. We discuss this alternative formulation next.

2.8 Alternative Derivation of the Post-Buckling Equilibrium Equations

There is an alternative way to derive the post-buckling equilibrium equations of an iso-static (or a hyperstatic) system following **Britvec** and **Davister** (Refs. 20 and 23), embodying a pin-connected space lattice made of elastic flexible members, which is preferable for reasons of ease of computation and programming when these equations are solved on a digital computer. Namely, in this way the equations can be reduced to much smaller sets of equations where each set applies to a particular post-buckling mode, and all these sets are part of a larger set, which is essentially governed by the set of the critical members for the entire system. Then, in each post-buckling mode a subset of the equilibrium equations contained in matrix (**E**), Eq. (2.7.18), applies, involving only the flexural shortenings of those buckling critical members which actually develop in the statically and kinematically admissible buckling modes or mechanisms.

The first essential step in this approach is the inversion of the kinematic matrix (**C**). This is also the only inversion of a matrix required in this reduction to smaller sets of equations, other than the inversion of the reduced post-buckling stiffness matrices (**E**) or ($\mathbf{E}_c$ or $\mathbf{E}_b$) in the solution of the final reduced equations which are usually small in number, as we shall see later. After defining the global coordinate system and numbering the joints from (1) to (J) and the members from (1) to ($n = N$), we form the (**C**) matrix using the equations

$$\mathbf{e}_{ij} = (\mathbf{C}_{ij}^T \mid -\mathbf{C}_{ij}^T)\left(\frac{\mathbf{u}_i}{\mathbf{u}_j}\right) \qquad \text{(2.3.4c)bis}$$

Here it is assumed that the initial geometry of the structure is defined with respect to a global coordinate system (x, y, z). This means that all the column matrices ($\mathbf{C}_\alpha$; $\alpha = 1, \ldots, n = N$), can be evaluated and we arrive at the relation

$$\mathbf{e} = \mathbf{C}\mathbf{u} \qquad \text{(2.3.6)bis}$$

in which the flexural shortenings (e_α) are numbered according to the numbering sequence of the members.

An alternative way to derive (**C**) is to derive the statical matrix (**A**) for the equilibrium equations in the initial state, connecting the internal axial loads (**P**) and the vector of applied external loads (**W**). (Here (**P**) must be numbered according to the numbering sequence of the members and ($\mathbf{W}$) according to the numbering sequence of the nodes allowing consistently for the three global coordinate directions (x, y, z) at each node.) Then,

$$\mathbf{C} = \mathbf{A}^T \tag{2.5.6)bis}$$

as explained before.

The inversion of the kinematic matrix makes it possible to find the global displacement vector ($\mathbf{u}$), where

$$\mathbf{u} = \mathbf{C}^{-1}\mathbf{e} \tag{2.8.1}$$

or

$$\mathbf{u} = \mathbf{C}_1\mathbf{e} \qquad \text{where } \mathbf{C}_1 = \mathbf{C}^{-1} \tag{2.8.1a}$$

We next partition the ($\mathbf{e}$) vector according to the shortenings pertaining to the critically stressed members from those of the noncritically stressed or tensile members, which remain straight in post-buckling, the flexural shortenings of which are zero. The set of the critical members is a known set, which we find from the results of the equilibrium analysis in the *initial* configuration in which the critical state is determined, when in one or more of the compressive iso-static members the axial loads reach the critical values in the loading process.

Let the number of compressive critical members be denoted by (c) and the number of the remaining compressive and tensile members by (s). Then $c + s = N$ and

$$\mathbf{e} = \begin{pmatrix} \mathbf{e}_c \\ \hline \mathbf{e}_s \end{pmatrix} \tag{2.8.2}$$

A consequence of this partition is that also the inverse matrix ($\mathbf{C}^{-1} = \mathbf{C}_1$) must be paritioned according to the columns of ($\mathbf{e}_c$) and ($\mathbf{e}_s$), so that the columns involving the shortenings of the critical members come into the leading position, (in ascending order) and the kinematic relation becomes

$$\mathbf{u} = \mathbf{C}^{-1}\mathbf{e} = (\mathbf{C}_{1c} \mid \mathbf{C}_{1s})\begin{pmatrix} \mathbf{e}_c \\ \hline \mathbf{e}_s \end{pmatrix}$$

or

$$\mathbf{u} = \mathbf{C}_{1c}\mathbf{e}_c + \mathbf{C}_{1s}\mathbf{e}_s = \mathbf{C}_{1c}\mathbf{e}_c \tag{2.8.3}$$

since ($\mathbf{e}_s = \mathbf{0}$). ($\mathbf{C}_{1c}$) is now a ($N \times c$) matrix where ($\mathbf{e}_c$) is a column vector of a dimension ($c \times 1$) where (c) may be much smaller than ($n = N$). If this partitioning is desired with respect to only (b) out of (c) members, which may actually shorten in a buckling mechanism, then

$$\mathbf{u} = \mathbf{C}^{-1}\mathbf{e} = (\mathbf{C}_{1b} \mid \mathbf{C}_{1s'})\begin{pmatrix} \mathbf{e}_b \\ \hline \mathbf{e}_{s'} \end{pmatrix} = \mathbf{C}_{1b}\mathbf{e}_b \tag{2.8.4}$$

Since all the shortenings ($\mathbf{e}_{s'} \equiv 0$) and the vector ($\mathbf{e}_{s'}$) contains the vector ($\mathbf{e}_s$) plus those elements of the vector ($\mathbf{e}_c$) which are zero (i.e. the shortenings of those critical compressive members which do not buckle but remain straight and unload, as their axial loads decrease), then, also ($b + s' = N$).

2.9 Alternative Derivation of the Post-Buckling Equilibrium Equations. Matrix Formulation of the General Law

The general law introduced in Section 2.1, Eq. (2.1.1), may be written for any member (α) within the lattice. This is so regardless whether this member is compressive or tensile, as explained earlier. (See Section 2.1.1 and Ref. 15.) In that case we get

$$\delta P_\alpha = K_\alpha P_\alpha L_\alpha^{-1} e_\alpha; \qquad \alpha = 1, 2, \ldots, N \tag{2.9.1}$$

To generalize this law and to make it applicable to the whole lattice we define a diagonal matrix of order ($N \times N$), where, in the isostatic lattice the number (n) of freedoms and the number (N) of members are the same so that ($n = N$). Thus

$$\mathbf{T}(\text{diag.}) = \mathbf{K}\mathbf{P}^0\mathbf{L}^{-1} \tag{2.9.2}$$

in which ($\mathbf{K}$), ($\mathbf{P}^0$) and ($\mathbf{L}^{-1}$) are also diagonal matrices of the same order, defined as follows

$$\mathbf{K}(\text{diag.}) = \begin{bmatrix} K_1 & 0 & 0 & \mathbf{0} & 0 \\ 0 & K_2 & 0 & \mathbf{0} & 0 \\ 0 & 0 & K_3 & \mathbf{0} & 0 \\ \mathbf{0} & \mathbf{0} & \mathbf{0} & \cdot & \mathbf{0} \\ 0 & 0 & 0 & \mathbf{0} & K_n \end{bmatrix} \tag{2.9.3}$$

where (K_α; $\alpha = 1, \ldots, n = N$) is the buckling constant for the αth member. Then,

$$\mathbf{P}^0(\text{diag.}) = \begin{bmatrix} P_1^0 & 0 & 0 & \mathbf{0} & 0 \\ 0 & P_2^0 & 0 & \mathbf{0} & 0 \\ 0 & 0 & P_3^0 & \mathbf{0} & 0 \\ \mathbf{0} & \mathbf{0} & \mathbf{0} & \cdot & \mathbf{0} \\ 0 & 0 & 0 & \mathbf{0} & P_n^0 \end{bmatrix} \tag{2.9.4}$$

Similarly,

$$\mathbf{L}^{-1}(\text{diag.}) = \begin{bmatrix} L_1^{-1} & 0 & 0 & \mathbf{0} & 0 \\ 0 & L_2^{-1} & 0 & \mathbf{0} & 0 \\ 0 & 0 & L_3^{-1} & \mathbf{0} & 0 \\ \mathbf{0} & \mathbf{0} & \mathbf{0} & \cdot & \mathbf{0} \\ 0 & 0 & 0 & \mathbf{0} & L_n^{-1} \end{bmatrix} \tag{2.9.5}$$

Then, from (2.9.1),

$$\mathbf{T} = \mathbf{KP}^0\mathbf{L}^{-1} = \begin{bmatrix} K_1 P_1^0 L_1^{-1} & 0 & \mathbf{0} & 0 \\ 0 & K_2 P_2^0 L_2^{-1} & \mathbf{0} & 0 \\ \mathbf{0} & \mathbf{0} & \cdot & \mathbf{0} \\ 0 & 0 & \mathbf{0} & K_n P_n^0 L_n^{-1} \end{bmatrix} \tag{2.9.6}$$

It is convenient to define the product in (**T**), which does not depend on the matrix of the buckling constants (K), by the matrix (**T**′), i.e. let

$$\mathbf{T}' = \mathbf{P}^0\mathbf{L}^{-1} \tag{2.9.6a}$$

Then the matrix (**T**) can be written alternatively

$$\mathbf{T} = \mathbf{KT}' \tag{2.9.6b}$$

where, clearly, both (**T**) and (**T**′) are diagonal matrices. If we define ($\delta\mathbf{P}$) as a column vector, i.e.

$$\delta\mathbf{P} = (\delta P_1 \;\; \delta P_2 \;\; \cdots \;\; \delta P_n)^T \tag{2.9.7}$$

then, the totality of the equations expressing the general law for each member in the lattice according to the numbering sequence for the members, may be written in matrix form, as

$$\delta\mathbf{P} = \mathbf{KP}^0\mathbf{L}^{-1}\mathbf{e} \tag{2.9.8}$$

where (**e**) is the vector of the flexural contractions of all the members in the lattice regardless whether these are compressive or tensile, critical or not. This concept and approach was first introduced in Refs. 19, 20 and later in Ref. 23. Using Eq. (2.9.2), the last result may be written alternatively

$$\delta\mathbf{P} = \mathbf{Te} = \mathbf{KT}'\mathbf{e} \tag{2.9.9}$$

It is of interest to note here that the matrix (**K**) is in general singular, if any of the constants (K_α; $\alpha = 1, 2, \ldots, n = N$) tends to either ($+\infty$) or ($-\infty$) which happens if any one member remains straight in post-bucking. This is practically always the case since there are tensile or subcritical compressive members within the lattice. However, the inverse matrix ($\mathbf{K}^{-1}$) is generally nonsingular, as the opposite is true. Thus

$$\mathbf{K}^{-1}(\text{diag.}) = \begin{bmatrix} K_1^{-1} & 0 & 0 & \mathbf{0} & 0 \\ 0 & K_2^{-1} & 0 & \mathbf{0} & 0 \\ 0 & 0 & K_3^{-1} & \mathbf{0} & 0 \\ \mathbf{0} & \mathbf{0} & \mathbf{0} & \cdot & \mathbf{0} \\ 0 & 0 & 0 & \mathbf{0} & K_n^{-1} \end{bmatrix} \tag{2.9.10}$$

is certainly not singular if ($K_\alpha \neq 0, \alpha = 1, \ldots, n = N$). We may thus invert Eq. (2.9.9) to obtain the *inverse general law* in the form

$$\mathbf{e} = \mathbf{T}^{-1}\,\delta\mathbf{P} = \mathbf{K}^{-1}(\mathbf{T}')^{-1}\,\delta\mathbf{P} \tag{2.9.11}$$

where [$\mathbf{T}^{-1}$(diag.)] is generally nonsingular, since the matrix ($\mathbf{K}^{-1}$) is nonsingular, as are also generally the matrices ($\mathbf{P}^0$) and ($\mathbf{L}^{-1}$), where ($\mathbf{T}' = \mathbf{P}^0\mathbf{L}^{-1}$).

Suppose that we regroup the vector (**e**) according to the members which are critical and which may be (c) in number and those which are noncritical of which there may be (s) in the lattice, such that ($c + s = n = N$). If we put the (c) flexural contractions of the critical members in the leading or top position of the vector (**e**) in the ascending numbering order, then, from (2.9.11), we get

$$\mathbf{e} = \begin{pmatrix} \mathbf{e}_c \\ \hline \mathbf{e}_s \end{pmatrix} = \left(\begin{array}{c|c} \mathbf{T}_c^{-1} & \mathbf{0} \\ \hline \mathbf{0} & \mathbf{T}_s^{-1} \end{array}\right) \begin{pmatrix} \delta\mathbf{P}_c \\ \hline \delta\mathbf{P}_s \end{pmatrix} \tag{2.9.12}$$

where the matrix ($\mathbf{T}^{-1}$) and the vector ($\delta\mathbf{P}$) have been partitioned accordingly. Carrying out the multiplication, one finds

$$\mathbf{e}_c = \mathbf{T}_c^{-1}\,\delta\mathbf{P}_c \tag{2.9.12a}$$

and

$$\mathbf{e}_s = \mathbf{T}_s^{-1}\,\delta\mathbf{P}_s \tag{2.9.12b}$$

Since the (s) members in Eq. (2.9.12) may be either subcritical compressive or tensile members and these actually remain straight, the buckling constants (K_α; $\alpha = 1, \ldots, s$), all tend to plus or minus infinity as required by the general law in Eq. (2.1.1). It follows then that ($\mathbf{T}_s^{-1}$) is a zero matrix as ($\mathbf{K}^{-1}$) is a zero matrix. The changes ($\delta\mathbf{P}_s$) in the straight members are certainly *not* zero, but *finite*, and the pre-multiplication of ($\delta\mathbf{P}_s$) by the zero matrix ($\mathbf{T}_s^{-1}$) in this case gives ($\mathbf{e}_s = \mathbf{0}$), as required. Thus, Eq. (2.9.12b) is identically zero and Eq. (2.9.12) reduces to (2.9.12a) in which ($\mathbf{T}_c^{-1}$) is assumed to be nonsingular provided all the critical members in the system shorten flexurally.

If this is not the case, and the buckling modes are generated in which only (b) out of (c) critical members buckle and shorten flexurally, the vector (**e**) may be partitioned similarly using the subvector ($\mathbf{e}_b$) of the actual shortenings of the buckling members and the vector ($\mathbf{e}_{s'}$), where (s') embodies all straight bars; i.e.

$$\mathbf{e} = \begin{pmatrix} \mathbf{e}_b \\ \hline \mathbf{e}_{s'} \end{pmatrix} \tag{2.9.13}$$

where now ($\mathbf{e}_b > 0$) and ($b + s' = n = N$). The subsequent development is the same as in the previous case except that now Eq. (2.9.12a) assumes the form

$$\mathbf{e}_b = \mathbf{T}_b^{-1}\,\delta\mathbf{P}_b = \mathbf{K}_b^{-1}(\mathbf{T}_b')^{-1}\,\delta\mathbf{P}_b \tag{2.9.14}$$

where ($\mathbf{T}_b^{-1}$) is certainly nonsingular. This set of equations actually represents what may be called the reduced inverse *law* for a pin-jointed lattice embodying the critical members which actually shorten flexurally. This equation may be inverted to give the *direct reduced general law* in the form

$$\delta\mathbf{P}_b = \mathbf{T}_b\mathbf{e}_b = \mathbf{K}_b\mathbf{T}_b'\mathbf{e}_b \tag{2.9.15}$$

where

$$\mathbf{T}_b = \mathbf{K}_b\mathbf{P}_b^0\mathbf{L}_b^{-1} = \mathbf{K}_b\mathbf{T}_b' \tag{2.9.15a}$$

We shall make use of these two equations later, when we discuss the solution of the reduced post-buckling equilibrium equations.

Combining now Eq. (2.8.3) in Section 2.8 and the last result expressed first in the form of Eq. (2.9.12a), we get the nodal displacements (u)

$$\mathbf{u} = \mathbf{C}_{1c}\mathbf{e}_c = \mathbf{C}_{1c}\mathbf{T}_c^{-1}\,\delta\mathbf{P}_c \tag{2.9.16}$$

and if not all (c) members actually shorten or buckle, Eq. (2.8.3) may be written analogously as Eq. (2.8.4), i.e.

$$\mathbf{u} = \mathbf{C}_{1b}\mathbf{e}_b \qquad \text{(2.8.4)bis}$$

in which now the partitioning of the matrix ($\mathbf{C}^{-1} = \mathbf{C}_1$) is according to the partitioned vector ($\mathbf{e}$) given in Eq. (2.9.13). Then, finally,

$$\mathbf{u} = \mathbf{C}_{1b}\mathbf{T}_b^{-1}\,\delta\mathbf{P}_b; \qquad b = 1, 2, \ldots, c \tag{2.9.17}$$

where use was made of Eq. (2.9.14). Substituting for ($\mathbf{T}_b^{-1}$)

$$\mathbf{u} = \mathbf{C}_{1b}\mathbf{K}_b^{-1}(\mathbf{P}_b^0)^{-1}\mathbf{L}_b\,\delta\mathbf{P}_b; \qquad b = 1, 2, \ldots, c \tag{2.9.17a}$$

or

$$\mathbf{u} = \mathbf{C}_{1b}\mathbf{K}_b^{-1}(\mathbf{T}_b')^{-1}\,\delta\mathbf{P}_b \tag{2.9.17b}$$

where use was made also of Eq. (2.9.6a). Here, ($\delta\mathbf{P}_b$) may be regarded as independent variables.

2.10 Alternative Derivation of the Post-buckling Equilibrium Equations. Matrix Formulation of the Equilibrium Equations on the Distorted Geometry

In the Section 2.7 we derived the basic expression of Eq. (2.7.5) for the increment in the internal force components at any joint (i) in the three directions of the global coordinate system (x, y, z). We showed that these 3 incremental internal force components, partly conditioned by the changes (δP_α) in the axial loads of the adjoining members and partly by the body rotations of these members, are given by the (3×1) vector ($\delta\mathbf{F}_{ij}$), where

$$\delta\mathbf{F}_{ij} = -\delta\mathbf{F}_{ji} = \mathbf{C}_{ij}\,\delta P_{ij} + \frac{P_{ij}^0}{L_{ij}}(\mathbf{C}_{ij}\mathbf{C}_{ij}^T \mid -\mathbf{C}_{ij}\mathbf{C}_{ij}^T)\begin{pmatrix}\mathbf{u}_i\\ \hline \mathbf{u}_j\end{pmatrix} - \frac{P_{ij}^0}{L_{ij}}(\mathbf{I} \mid -\mathbf{I})\begin{pmatrix}\mathbf{u}_i\\ \hline \mathbf{u}_j\end{pmatrix} \qquad \text{(2.7.5)bis, (2.10.1)}$$

where ($\mathbf{I}$) is a (3×3) unit matrix ($\mathbf{C}_{ij}$) is given by Eq. (2.3.2).

The next step is to express ($\delta\mathbf{F}_{ij}$) in terms of the entire displacement vector ($\mathbf{u}$). To this end the (3×6) matrices ($\mathbf{C}_{ij}\mathbf{C}_{ij}^T \mid -\mathbf{C}_{ij}\mathbf{C}_{ij}^T$) of the direction cosine vectors ($\mathbf{C}_{ij}$) and the two (3×3) unit submatrices ($\mathbf{I}_{ij} \mid -\mathbf{I}_{ij}$) may be augmented by ($n - 6$) null column vectors of order (3×1), where usually ($n = N$), to account for the displacements in the vector ($\mathbf{u}$) which are alien to the member (i, j). This results in the $(3 \times n)$ once augmented matrices $(\mathbf{C}_{ij}^A)$ and $(\mathbf{I}^A)$ (the letter (A) indicates that the (3×3) matrices $(\mathbf{C}_{ij}\mathbf{C}_{ij}^T)$ or $(\mathbf{I})$ have been augmented by $(n-6)$ (3×1) null column vectors), where now

$$\mathbf{C}_{ij}^A = (\mathbf{0} \mid \mathbf{C}_{ij}\mathbf{C}_{ij}^T \mid \mathbf{0} \mid -\mathbf{C}_{ij}\mathbf{C}_{ij}^T \mid \mathbf{0}) \tag{2.10.2}$$

$$\mathbf{I}^A = (\mathbf{0} \mid \mathbf{I} \mid \mathbf{0} \mid -\mathbf{I} \mid \mathbf{0}) \tag{2.10.3}$$

in which the zero matrices account for the $(n-6)$ null column vectors.

Similarly, we may write the vector $(\mathbf{u})$ of the nodal displacements in the form

$$\mathbf{u} = (\cdot \mid \mathbf{u}_i^T \mid \cdot \mid \mathbf{u}_j^T \mid \cdot)^T \tag{2.10.4}$$

where the dots stand for the displacements of the nodes other than (i) and (j). These displacements are entered in the numbering sequence of the nodes and in the (x, y, z) directions of the global coordinate system. There are altogether $(n-6)$ alien displacement components (elements) in the vector $(\mathbf{u})$ represented by these dots.

A similar operation is performed on the term $(\mathbf{C}_{ij}\,\delta\mathbf{P}_{ij})$ in Eq. (2.10.1). First, the vector $(\delta\mathbf{P})$ is formed containing the incremental axial forces for all the members in the lattice. It is now the vector $(\mathbf{C}_{ij})$ (not the (3×3) matrix $(\mathbf{C}_{ij}\mathbf{C}_{ij}^T)$ that is augmented by $(n-1)$ null column vectors of the dimension (3×1) to account for the members which are alien to member (i, j) in the vector $(\delta\mathbf{P})$, to give an augumented matrix $(\mathbf{C}_{ij}^a)$ (note that now lower case (a) is used to designate this augmentation), where the single superscript (a) indicates a once augmented matrix, i.e.

$$\mathbf{C}_{ij}^a = (\mathbf{0} \mid \mathbf{C}_{ij} \mid \mathbf{0} \mid \mathbf{0} \mid \mathbf{0}) \tag{2.10.5}$$

and

$$\delta\mathbf{P} = (\cdot \mid \delta P_{ij} \mid \cdot \mid \cdot \mid \cdot)^T \tag{2.10.6}$$

where a dot stands for axial load increments in members other than the member (i, j). The (3×1) vector $(\delta\mathbf{F}_{ij})$ may now be expressed symbolically in terms of the augmented displacement vector $(\mathbf{u})$ as,

$$\delta\mathbf{F}_{ij} = -\delta\mathbf{F}_{ji} = \mathbf{C}_{ij}^a\,\delta\mathbf{P} + \frac{P_{ij}^0}{L_{ij}}\mathbf{C}_{ij}^A\mathbf{u} - \frac{P_{ij}^0}{L_{ij}}\mathbf{I}^A\mathbf{u} \tag{2.10.7}$$

Substituting formally for $(\mathbf{C}_{ij}^a)$, $(\mathbf{C}_{ij}^A)$, $(\delta\mathbf{P})$ and $(\mathbf{u})$ from equations (2.10.2) to (2.10.6) and carrying out the multiplication of the partitioned submatrices results in the original equation (2.7.5) or (2.10.1).

We see that for the member (i, j) two such incremental vectors namely $(\delta\mathbf{F}_{ij})$ and $(\delta\mathbf{F}_{ji})$ exist. Let us define a number (α) which is the sequence number for this member (according to any numbering sequence of the members in the entire lattice) and postulate that the member (α) is situated between the joints (i) and (j). Then, the last result becomes

$$\delta\mathbf{F}_{ij} \equiv \delta\mathbf{F}_{\alpha i} = \mathbf{C}_{\alpha i}^a\,\delta\mathbf{P} + \frac{P_\alpha^0}{L_\alpha}\mathbf{C}_{\alpha i}^A\mathbf{u} - \frac{P_\alpha^0}{L_\alpha}\mathbf{I}_{\alpha i}^A\mathbf{u} \tag{2.10.8}$$

and

$$\delta\mathbf{F}_{ji} = \delta\mathbf{F}_{\alpha j} = -\mathbf{C}^a_{\alpha i}\,\delta\mathbf{P} - \frac{P^0_\alpha}{L_\alpha}\,\mathbf{C}^A_{\alpha i}\mathbf{u} + \frac{P^0_\alpha}{L_\alpha}\,\mathbf{I}^A_{\alpha i}\mathbf{u} \tag{2.10.9}$$

or

$$\delta\mathbf{F}_{ji} = \delta\mathbf{F}_{\alpha j} = \mathbf{C}^a_{\alpha j}\,\delta\mathbf{P} + \frac{P^0_\alpha}{L_\alpha}\,\mathbf{C}^A_{\alpha j}\mathbf{u} - \frac{P_\alpha}{L_\alpha}\,\mathbf{I}^A_{\alpha j}\mathbf{u} \tag{2.10.10}$$

where

$$\begin{aligned} \mathbf{C}^a_{\alpha j} &= -\mathbf{C}^a_{\alpha i} \\ \mathbf{C}^A_{\alpha j} &= -\mathbf{C}^A_{\alpha i} \end{aligned} \tag{2.10.11}$$

and

$$\mathbf{I}^A_{\alpha j} = -\mathbf{I}^A_{\alpha i} = -\mathbf{I}^A \quad \text{so that } \mathbf{I}^A_{\alpha i} = \mathbf{I}^A.$$

The last two equations (2.10.8) and (2.10.9) can be combined into one matrix equation, always referred to the same global coordinate system (x, y, z) and the adopted initial numbering sequences for the joints and the members. Then, for a member (α), the vector $(\delta\mathbf{F}_\alpha)$ associated with this member, can be written, by adding $(n-6)$ null elements on the L.H.S. and $(n-6)$ null rows on the R.H.S. of the combined Eqs. (2.10.8) and (2.10.9), as follows,

$$\delta\mathbf{F}_\alpha = \begin{Bmatrix} \mathbf{0} \\ \delta\mathbf{F}_{\alpha i} \\ \mathbf{0} \\ \delta\mathbf{F}_{\alpha j} \\ \mathbf{0} \end{Bmatrix} = \underbrace{\begin{bmatrix} \mathbf{0} & \mathbf{0} & \mathbf{0} & \mathbf{0} & \mathbf{0} \\ \mathbf{0} & \mathbf{C}_{\alpha i} & \mathbf{0} & \mathbf{0} & \mathbf{0} \\ \mathbf{0} & \mathbf{0} & \mathbf{0} & \mathbf{0} & \mathbf{0} \\ \mathbf{0} & -\mathbf{C}_{\alpha i} & \mathbf{0} & \mathbf{0} & \mathbf{0} \\ \mathbf{0} & \mathbf{0} & \mathbf{0} & \mathbf{0} & \mathbf{0} \end{bmatrix}}_{=\mathbf{C}^{aa}_\alpha(n\times n)} \begin{Bmatrix} \cdot \\ \delta P_\alpha \\ \cdot \\ \cdot \\ \cdot \end{Bmatrix}$$

$$+\frac{P^0_\alpha}{L_\alpha} \underbrace{\begin{bmatrix} \mathbf{0} & \mathbf{0} & \mathbf{0} & \mathbf{0} & \mathbf{0} \\ \mathbf{0} & \mathbf{C}_{\alpha i}\mathbf{C}^T_{\alpha i} & \mathbf{0} & -\mathbf{C}_{\alpha i}\mathbf{C}^T_{\alpha i} & \mathbf{0} \\ \mathbf{0} & \mathbf{0} & \mathbf{0} & \mathbf{0} & \mathbf{0} \\ \mathbf{0} & -\mathbf{C}_{\alpha i}\mathbf{C}^T_{\alpha i} & \mathbf{0} & \mathbf{C}_{\alpha i}\mathbf{C}^T_{\alpha i} & \mathbf{0} \\ \mathbf{0} & \mathbf{0} & \mathbf{0} & \mathbf{0} & \mathbf{0} \end{bmatrix}}_{=\mathbf{C}^{AA}_\alpha(n\times n)} \begin{Bmatrix} \cdot \\ \mathbf{u}_i \\ \cdot \\ \mathbf{u}_j \\ \cdot \end{Bmatrix}$$

$$-\frac{P^0_\alpha}{L_\alpha} \underbrace{\begin{bmatrix} \mathbf{0} & \mathbf{0} & \mathbf{0} & \mathbf{0} & \mathbf{0} \\ \mathbf{0} & \mathbf{I} & \mathbf{0} & -\mathbf{I} & \mathbf{0} \\ \mathbf{0} & \mathbf{0} & \mathbf{0} & \mathbf{0} & \mathbf{0} \\ \mathbf{0} & -\mathbf{I} & \mathbf{0} & \mathbf{I} & \mathbf{0} \\ \mathbf{0} & \mathbf{0} & \mathbf{0} & \mathbf{0} & \mathbf{0} \end{bmatrix}}_{=\mathbf{I}^{AA}_\alpha(n\times n)} \begin{Bmatrix} \cdot \\ \mathbf{u}_i \\ \cdot \\ \mathbf{u}_j \\ \cdot \end{Bmatrix} \tag{2.10.12}$$

where each horizontal partition in the first $(n \times n)$ matrix on the right side of this identity either represents a zero submatrix or the (3×1) direction cosine vector

$(\mathbf{C}_{\alpha i})$ or $(-\mathbf{C}_{\alpha i})$. A dot in the vector $(\delta\mathbf{P})$ implies any internal axial load increment or a subvector of such increments, other than that in the member (α) itself. The remaining partitions represent zero submatrices. Similarly, in the other two $(n \times n)$ matrices, the submatrices in the partitions in which the (3×3) submatrices $(\mathbf{C}_{\alpha i}\mathbf{C}_{\alpha i}^T)$, $(-\mathbf{C}_{\alpha i}\mathbf{C}_{\alpha i}^T)$ or $(\mathbf{I})$ appear, have 3 columns or 3 rows respectively, as indicated.

We designate the twice augmented matrices on the right side of Eq. (2.10.12) by the symbols $(\mathbf{C}_\alpha^{aa})$, $(\mathbf{C}_\alpha^{AA})$ and $(\mathbf{I}_\alpha^{AA})$, consistent with the notation introduced for the once augmented matrices defined in Eqs. (2.10.2), (2.10.3) and (2.10.5). Note that in the $(n \times n)$ augmented matrix $(\mathbf{I}_\alpha^{AA})$ the (3×3) unit and minus unit submatrices fit into the same slots or partitions as do the (3×3) $(\mathbf{C}_{\alpha i}\mathbf{C}_{\alpha i}^T)$ and $(-\mathbf{C}_{\alpha i}\mathbf{C}_{\alpha i}^T)$ submatrices in the $(n \times n)$ matrix $(\mathbf{C}_\alpha^{AA})$. We note that $(\mathbf{C}_\alpha^{AA})$ and $(\mathbf{I}_\alpha^{AA})$ are *symmetrical* matrices. This rule is important to remember when these matrices are coded for machine computation for a particular problem. We may thus write Eq. (2.10.12) symbolically, using these matrices, in the form,

$$\delta\mathbf{F}_\alpha = \mathbf{C}_\alpha^{aa}\,\delta\mathbf{P} + \frac{P_\alpha^0}{L_\alpha}\mathbf{C}_\alpha^{AA}\mathbf{u} - \frac{P_\alpha^0}{L_\alpha}\mathbf{I}_\alpha^{AA}\mathbf{u} \tag{2.10.13}$$

or

$$\delta\mathbf{F}_\alpha = \mathbf{C}_\alpha^{aa}\,\delta\mathbf{P} + \frac{P_\alpha^0}{L_\alpha}(\mathbf{C}_\alpha^{AA} - \mathbf{I}_\alpha^{AA})\mathbf{u} \tag{2.10.13a}$$

The multiplication of these matrices, as defined in Eq. (2.10.12), then results in the vectors $(\delta\mathbf{F}_{ij})$ (or $\delta\mathbf{F}_{\alpha i}$) and $(\delta\mathbf{F}_{ji})$ (or $\delta\mathbf{F}_{\alpha j}$), defined by Eq. (2.10.1).

Clearly, the alternative notation for the submatrices $(C_{\alpha i})$ and $(\mathbf{C}_{\alpha i}\mathbf{C}_{\alpha i}^T)$ would be $(\mathbf{C}_{ij})$ and $(\mathbf{C}_{ij}\mathbf{C}_{ij}^T)$, etc., as explained by the definitions in Sections 2.3 and 2.7, e.g. Eq. (2.7.5b). The matrices $(\mathbf{C}_\alpha^{aa})$, $(\mathbf{C}_\alpha^{AA})$ and $(\mathbf{I}_\alpha^{AA})$ associated with any member (α) are always $(n \times n)$ matrices, where (n) denotes the number of degrees of freedom of the isostatic lattice, as well as the number of its members. The vector $(\mathbf{u})$ of the nodal displacements has also the dimension $(n \times 1)$, which is the number of the displacements of the free (unsupported) joints (or unrestrained coordinate joint-directions). If the number of free joints is (J_F), then $(n = 3J_F)$ in a statically determinate lattice. It is readily verified that the matrices $(\mathbf{C}_\alpha^{aa})$ and $(\mathbf{C}_\alpha^{AA})$ are related, in fact

$$\mathbf{C}_\alpha^{aa}\mathbf{C}_\alpha^{aaT} = \mathbf{C}_\alpha^{AA} \tag{2.10.14}$$

To prove this, we write these matrices as in Eq. (2.10.12), omitting the zero submatrices, which do not affect this proof. Thus

$$\underbrace{\begin{bmatrix} \mathbf{0} & \mathbf{0} & \mathbf{0} \\ \mathbf{0} & \mathbf{C}_{\alpha i} & \mathbf{0} \\ \mathbf{0} & \mathbf{0} & \mathbf{0} \\ \mathbf{0} & -\mathbf{C}_{\alpha i} & \mathbf{0} \\ \mathbf{0} & \mathbf{0} & \mathbf{0} \end{bmatrix}}_{=\mathbf{C}_\alpha^{aa}(n\times n)} \underbrace{\begin{pmatrix} \mathbf{0}^T & \mathbf{0}^T & \mathbf{0}^T & \mathbf{0}^T & \mathbf{0}^T \\ \mathbf{0}^T & \mathbf{C}_{\alpha i}^T & \mathbf{0}^T & -\mathbf{C}_{\alpha i} & \mathbf{0}^T \\ \mathbf{0}^T & \mathbf{0}^T & \mathbf{0}^T & \mathbf{0}^T & \mathbf{0}^T \end{pmatrix}}_{=\mathbf{C}_\alpha^{aaT}(n\times n)} = \underbrace{\begin{bmatrix} \mathbf{0} & \mathbf{0} & \mathbf{0} & \mathbf{0} & \mathbf{0} \\ \mathbf{0} & \mathbf{C}_{\alpha i}\mathbf{C}_{\alpha i}^T & \mathbf{0} & -\mathbf{C}_{\alpha i}\mathbf{C}_{\alpha i}^T & \mathbf{0} \\ \mathbf{0} & \mathbf{0} & \mathbf{0} & \mathbf{0} & \mathbf{0} \\ \mathbf{0} & -\mathbf{C}_{\alpha i}^T\mathbf{C}_{\alpha i} & \mathbf{0} & \mathbf{C}_{\alpha i}\mathbf{C}_{\alpha i} & \mathbf{0} \\ \mathbf{0} & \mathbf{0} & \mathbf{0} & \mathbf{0} & \mathbf{0} \end{bmatrix}}_{=\mathbf{C}_\alpha^{AA}(n\times n)}$$

and the result on the right side is identical to the result given in Eq. (2.10.14).

$$
\mathbf{C}_\alpha^{aa}\mathbf{T}' =
\begin{array}{c}
\begin{array}{ccc} \overbrace{\text{Zero Cols.}}^{(\alpha-1)} & \overbrace{\text{Col.}}^{(\alpha^{\text{th}})} & \overbrace{\text{Zero Cols.}}^{(n-\alpha)} \end{array} \\
\left[\begin{array}{c|c|c}
\mathbf{0} & \mathbf{0} & \mathbf{0} \\ \hline
3\text{ rows}\left\{\mathbf{0}\right. & \mathbf{C}_{\alpha i} & \mathbf{0} \\ \hline
\mathbf{0} & \mathbf{0} & \mathbf{0} \\ \hline
3\text{ rows}\left\{\mathbf{0}\right. & -\mathbf{C}_{\alpha i} & \mathbf{0} \\ \hline
\mathbf{0} & \mathbf{0} & \mathbf{0}
\end{array}\right]
\end{array}
\begin{array}{c}
\overbrace{\text{Col.}}^{\alpha^{\text{th}}} \\
\left[\begin{array}{ccc|c|ccc}
P_1^0 L_1^{-1} & \mathbf{0} & 0 & 0 & 0 & \mathbf{0} & 0 \\
\mathbf{0} & \cdot & \mathbf{0} & \mathbf{0} & \mathbf{0} & \mathbf{0} & \mathbf{0} \\
0 & \mathbf{0} & P_{\alpha-1} L_{\alpha-1}^{-1} & 0 & 0 & \mathbf{0} & 0 \\ \hline
\alpha^{\text{th}}\text{ row}\left\{0\right. & \mathbf{0} & 0 & P_\alpha L_\alpha^{-1} & 0 & \mathbf{0} & 0 \\ \hline
0 & \mathbf{0} & 0 & 0 & P_{\alpha+1} L_{\alpha+1}^{-1} & \mathbf{0} & 0 \\
\mathbf{0} & \mathbf{0} & \mathbf{0} & \mathbf{0} & \mathbf{0} & \cdot & \mathbf{0} \\
0 & \mathbf{0} & 0 & 0 & 0 & \mathbf{0} & P_n L_n^{-1}
\end{array}\right]
\end{array}
=
\begin{array}{c}
\begin{array}{ccc} \overbrace{\text{Zero Cols.}}^{(\alpha-1)} & \overbrace{\text{Col.}}^{\alpha^{\text{th}}} & \overbrace{\text{Zero Cols.}}^{(n-\alpha)} \end{array} \\
\left[\begin{array}{c|c|c}
\mathbf{0} & \mathbf{0} & \mathbf{0} \\ \hline
3\text{ rows}\left\{\mathbf{0}\right. & \mathbf{C}_{\alpha i} P_\alpha L_\alpha^{-1} & \mathbf{0} \\ \hline
\mathbf{0} & \mathbf{0} & \mathbf{0} \\ \hline
3\text{ rows}\left\{\mathbf{0}\right. & -\mathbf{C}_{\alpha i} P_\alpha L_\alpha^{-1} & \mathbf{0} \\ \hline
\mathbf{0} & \mathbf{0} & \mathbf{0}
\end{array}\right]
\end{array}
= \frac{P_\alpha^0}{L_\alpha} C_\alpha^{aa}
$$

Eq. (2.10.17)

Consider next the product

$$\frac{P_\alpha^0}{L_\alpha}\mathbf{C}_\alpha^{AA} = \frac{P_\alpha^0}{L_\alpha}\mathbf{C}_\alpha^{aa}\mathbf{C}_\alpha^{aaT} \tag{2.10.15}$$

Now, the product

$$\frac{P_\alpha^0}{L_\alpha}\mathbf{C}_\alpha^{aa} = P_\alpha^0 L_\alpha^{-1}\mathbf{C}_\alpha^{aa}$$

may be written in matrix form, using Eq. (2.9.6a), as follows,

$$\frac{P_\alpha^0}{L_\alpha}\mathbf{C}_\alpha^{aa} = \mathbf{C}_\alpha^{aa}\mathbf{T}' \tag{2.10.16}$$

as only the (α^{th}) column of ($\mathbf{C}_\alpha^{aa}$) has non-zero elements.

Namely, post-multiplying a matrix (in this case the matrix $\mathbf{C}_\alpha^{aa}$) by a diagonal matrix, results in a matrix in which the columns of the new matrix are the columns of the original matrix multiplied by the respective elements of the diagonal matrix. Thus, the first column by the first diagonal element, the second column by the second element etc. Then, substituting for ($\mathbf{C}_\alpha^{aa}$) from Eq. (2.10.12) and for ($\mathbf{T}'$) from (2.9.6a) one gets the Eq. (2.10.17) (see page 98) where a dot implies non-zero or zero elements in a row or column (or a two-dimensional submatrix of such elements. Here, the (3×1) vector ($\mathbf{C}_{\alpha i}$) and ($-\mathbf{C}_{\alpha i}$) of direction cosines for member (α) appear in the αth column of the matrix ($\mathbf{C}_a^{aa}$). The result of this multiplication then proves the result of Eq. (2.10.16).

Therefore, Eq. (2.10.15) may now be written, as follows,

$$\frac{P_\alpha^0}{L_\alpha}\mathbf{C}_\alpha^{AA} = \frac{P_\alpha^0}{L_\alpha}\mathbf{C}_\alpha^{aa}\mathbf{C}_\alpha^{aaT} = \mathbf{C}_\alpha^{aa}\mathbf{T}'\mathbf{C}_\alpha^{aaT} \tag{2.10.18}$$

where ($\mathbf{T}'$) is the diagonal matrix given by Eq. (2.9.6a). With this modification the expression for the vector ($\delta\mathbf{F}_\alpha$) in Eq. (2.10.13) may be finally written,

$$\delta\mathbf{F}_\alpha = \mathbf{C}_\alpha^{aa}\,\delta\mathbf{P} + \mathbf{C}_\alpha^{aa}\mathbf{T}'\mathbf{C}_\alpha^{aaT}\mathbf{u} - \frac{P_\alpha^0}{L_\alpha}\mathbf{I}_\alpha^{AA}\mathbf{u} \tag{2.10.19}$$

which now depends only on the matrices ($\mathbf{C}_\alpha^{aa}$), ($\mathbf{T}'$) and ($\mathbf{I}_\alpha^{AA}$) and the critical loads (P_α^0; $\alpha = 1, 2, \ldots, n = N$) in the members and their initial lengths. Except for the loads, all the quantities depend on the initial geometry of the lattice. The axial load-vector ($\delta\mathbf{P}$) and the displacement vector ($\mathbf{u}$) are to be considered as yet undetermined variables.

2.11 The Post-Buckling Equilibrium States

The statical equilibrium equations for the distorted and buckled lattice are now written on the displaced configuration of the lattice. The ($3J_F = N = n$) external load increments in the load vector ($\mathbf{W}$), along the (x, y, z)-axes of the global coordinate system applied at the free joints (which are assumed to be J_F in number), are represented by the $(n \times 1)$ load vector

$$\delta \mathbf{W} = \boldsymbol{\beta}' \, \delta w$$

where ($\boldsymbol{\beta}'$) is the vector of load proportionality constants, which may be generally different from the vector ($\boldsymbol{\beta}$) of load-proportionality constants in the pre-buckling range of loading, defined in Eq. (2.5.3) in Sections 2.5, i.e. before the critical state is reached and before there is an onset of buckling in the lattice. We recall that in the pre-buckling range of proportionate loading the load vector ($\mathbf{W}$) is defined by Eq. (2.5.3) i.e.

$$\mathbf{W} = \boldsymbol{\beta} w \tag{2.5.3)bis}$$

So, in the critical state, before the onset of flexural distortions due to buckling the perfect lattice is in its initial configuration in which equilibrium is governed by the equation (2.5.4), Section 2.5, i.e.

$$\mathbf{AP} = \mathbf{W} = \boldsymbol{\beta} w \tag{2.5.4)bis}$$

where ($\mathbf{A}$) is the statical matrix, ($\mathbf{P}$) the vector of internal loads in the members, which must be entered consistently according to the initially defined numbering sequence for the members and the joints of a particular lattice. Similarly, the elements of the load vectors ($\mathbf{W}$), ($\boldsymbol{\beta}$) or ($\boldsymbol{\beta}'$) must be entered according to the defined sequence of joint numbering always in the order of the (x, y, z) of the global coordinate system (not, for instance, in a changing order as y, x, z etc.). The critical state is then determined by the Eq. (2.5.10) when one or several critical compressive members are on the point of buckling. (Here, the critical loads need not assume theoretical values, but may be determined experimentally, for instance using the Southwell plot (see Ref. 15). This determines the critical value (W^0) of the load parameter which follows from the equations (2.5.4) and (2.5.10) i.e.

$$\mathbf{AP}^0 = \boldsymbol{\beta} w^0 \tag{2.11.1}$$

In this state the onset of buckling in the perfect system happens and from now on the lattice distorts in post-buckling.

Equilibrium of the distorted and buckled lattice is established by simply equating the load vector ($\delta\mathbf{W}$) of the incremental external loads, compiled according to the initially established convention of numbering the joints and the members in the lattice, to the vector of *internal* incremental load changes (in which the distortions have been accounted for by Eq. (2.10.19)) at each node strictly observing the same numbering sequences of the nodes and members and entering these changes in the order of the (x, y, z) directions of the global coordinate system, as explained in detail before. Thus, for the equilibrium in any post-buckling state the following condition must hold.

$$\delta \mathbf{W} = \boldsymbol{\beta}' \, \delta w = \delta \mathbf{F} = \sum_{\alpha=1}^{N} \delta \mathbf{F}_\alpha; \qquad \alpha = 1, 2, \ldots, n = N \tag{2.11.2}$$

where the summation is on the index (α) according to the defined sequence of member numbering. Substituting next for $(\delta \mathbf{F}_\alpha)$ from Eq. (2.10.19), we get

$$\begin{aligned}\delta \mathbf{F} = \sum_{\alpha=1}^{N} \delta \mathbf{F}_\alpha &= \sum_{\alpha=1}^{N} \mathbf{C}_\alpha^{aa}\, \delta \mathbf{P} + \sum_{\alpha=1}^{N} (\mathbf{C}_\alpha^{aa} \mathbf{T}' \mathbf{C}_\alpha^{aaT})\mathbf{u} - \sum_{\alpha=1}^{n} \left(\frac{P_\alpha^0}{L_\alpha} \mathbf{I}_\alpha^{AA} \right)\mathbf{u} \\ &= (\mathbf{C}_1^{aa} + \mathbf{C}_2^{aa} + \cdots + \mathbf{C}_N^{aa})\, \delta \mathbf{P} \\ &\quad + (\mathbf{C}_1^{aa}\mathbf{T}'\mathbf{C}_1^{aaT} + \mathbf{C}_2^{aa}\mathbf{T}'\mathbf{C}_2^{aaT} + \cdots + \mathbf{C}_N^{aa}\mathbf{T}'\mathbf{C}_N^{aaT})\mathbf{u} \\ &\quad - \left(\frac{P_1^0}{L_1} \mathbf{I}_1^{AA} + \frac{P_2^0}{L_2} \mathbf{I}_2^{AA} + \cdots + \frac{P_N^0}{L_N} \mathbf{I}_N^{AA} \right)\mathbf{u}. \end{aligned} \tag{2.11.3}$$

Now, if the displacement vector $(\mathbf{u})$ were zero, for instance, and the lattice undistorted, but the internal forces were increased by the vector $(\delta \mathbf{P})$ one easily recognizes that all except the first term in this equation would be zero and that the sum of the terms in this term, i.e.

$$\sum_{\alpha=1}^{N} \mathbf{C}_\alpha^{aa} = \mathbf{C}_1^{aa} + \mathbf{C}_2^{aa} + \cdots + \mathbf{C}_n^{aa} = \mathbf{A} = \mathbf{C}^T \tag{2.11.4}$$

represents the statical matrix $(\mathbf{A})$ of the initial equilibrium equations of the undistorted lattice, which is equal to the kinematic matrix $(\mathbf{C})$ transposed. In that hypothetical case, the equilibrium change would be expressed by the equation

$$\mathbf{A}\, \delta \mathbf{P} = \delta \mathbf{W} \tag{2.11.5}$$

The sum of the matrix products in the second term in Eq. (2.11.3) may be evaluated as follows,

$$\begin{aligned} &\mathbf{C}_1^{aa}\mathbf{T}'\mathbf{C}_1^{aaT} + \mathbf{C}_2^{aa}\mathbf{T}'\mathbf{C}_2^{aaT} + \cdots \\ &\quad = \mathbf{C}_1^{aa}\mathbf{T}'(\mathbf{C}_1^{aaT} + \mathbf{C}_2^{aaT} + \mathbf{C}_3^{aaT} + \cdots + \mathbf{C}_N^{aaT}) \\ &\quad + \mathbf{C}_2^{aa}\mathbf{T}'(\mathbf{C}_1^{aaT} + \mathbf{C}_2^{aaT} + \cdots + \mathbf{C}_N^{aaT}) \\ &\quad + \cdots \\ &\quad + \mathbf{C}_N^{aa}\mathbf{T}'(\mathbf{C}_1^{aaT} + \mathbf{C}_2^{aaT} + \cdots + \mathbf{C}_N^{aaT}) \end{aligned} \tag{2.11.6}$$

This is so, since

$$\mathbf{C}_j^{aa}\mathbf{T}'\mathbf{C}_k^{aaT} = \mathbf{0}; \quad \text{if } j \neq k.$$

We verify this result for say $(j = 3)$ and $(k = 9)$. Then

$$\mathbf{C}_3^{aa}\mathbf{T}'\mathbf{C}_9^{aaT} = \underbrace{\left[\begin{array}{c|c|c} \mathbf{0} & \overbrace{\mathbf{0}}^{\text{3rd Col.}} & \mathbf{0} \\ \hline \mathbf{0} & \mathbf{C}_{3i} & \mathbf{0} \\ \hline \mathbf{0} & \mathbf{0} & \mathbf{0} \\ \hline \mathbf{0} & -\mathbf{C}_{3i} & \mathbf{0} \\ \hline \mathbf{0} & \mathbf{0} & \mathbf{0} \end{array}\right]}_{=\mathbf{C}_3^{aa}} \mathbf{T}' \underbrace{\left(\begin{array}{c|c|c|c|c} \mathbf{0}^T & \mathbf{0}^T & \mathbf{0}^T & \mathbf{0}^T & \mathbf{0}^T \\ \hline \mathbf{0} & \mathbf{C}_{9l}^T & \mathbf{0} & -\mathbf{C}_{9l}^T & \mathbf{0} \\ \hline \mathbf{0} & \mathbf{0} & \mathbf{0} & \mathbf{0} & \mathbf{0} \end{array}\right)}_{=\mathbf{C}_9^{aaT}} \begin{matrix} \\ \leftarrow \text{9th Row} \\ \\ \end{matrix} = \mathbf{0} \tag{2.11.7}$$

where member 3 is supposed to be connected to the joints (i) and (j), and member 9 to joints (l) and (m), say. The operation ($\mathbf{C}_3^{aa}\mathbf{T}'$) has the effect of multiplying the 3rd column of ($\mathbf{C}_3^{aa}$) by ($P_3^0 L_3^{-1}$) and the product of this matrix and the matrix ($\mathbf{C}_9^{aaT}$) is clearly a zero matrix as long as the column number ($j = 3$) and the row number ($k = 9$) in the transposed matrix are two different numbers.

Therefore, the form of Eq. (2.11.6) is correct. Now clearly, the sums in the parentheses in Eq. (2.11.6) represent the statical matrix ($\mathbf{A}^T$), which follows from Eq. (2.11.4). So, Eq. (2.11.6) may be written simply

$$\sum_{\alpha=1}^{N} \mathbf{C}_\alpha^{aa}\mathbf{T}'\mathbf{C}_\alpha^{aaT} = (\mathbf{C}_1^{aa} + \mathbf{C}_2^{aa} + \cdots + \mathbf{C}_N^{aa})\mathbf{T}'\mathbf{A}^T = \mathbf{A}\mathbf{T}'\mathbf{A}^T \tag{2.11.8}$$

But, ($\mathbf{A}^T = \mathbf{C}$), by (2.5.6), where ($\mathbf{C}$) is the kinematic matrix, and the last result becomes

$$\sum_{\alpha=1}^{N} \mathbf{C}_\alpha^{aa}\mathbf{T}'\mathbf{C}_\alpha^{aaT} = \mathbf{C}^T\mathbf{T}'\mathbf{C} \tag{2.11.9}$$

Thus the expression for the vector ($\delta\mathbf{F}$) in Eq. (2.11.3) can be written

$$\delta\mathbf{F} = \sum_{\alpha=1}^{N} \delta\mathbf{F}_\alpha = \mathbf{A}\,\delta\mathbf{P} + \mathbf{A}\mathbf{T}'\mathbf{A}^T\mathbf{u} - \mathbf{P}_L^0\,\mathbf{u} \tag{2.11.10}$$

where ($\mathbf{P}_L^0$) is defined as the sum of the twice augmented unit matrices multiplied by the factors (P_α^0/L_α; $\alpha = 1, \ldots, n = N$), which appears in the third term in Eq. (2.11.3), i.e.

$$\mathbf{P}_L^0 = \frac{P_1^0}{L_1}\mathbf{I}_1^{AA} + \frac{P_2^0}{L_2}\mathbf{I}_2^{AA} + \cdots + \frac{P_N^0}{L_N}\mathbf{I}_N^{AA} \tag{2.11.11}$$

Then, the matrix ($\mathbf{P}_L^0$) is constructed by first constructing the twice augmented ($\mathbf{I}_\alpha^{AA}$) matrices for each member (α), relative to the numbers of the nodes to which it is attached, multiplying these augmented matrices by (P_α^0/L_α) for each (α) and adding them for ($\alpha = 1$ to N).

The result of Eq. (2.11.10) may be written alternatively

$$\delta\mathbf{F} = \sum_{\alpha=1}^{N} \delta\mathbf{F}_\alpha = \mathbf{A}\,\delta\mathbf{P} + (\mathbf{A}\mathbf{T}'\mathbf{A}^T - \mathbf{P}_L^0)\mathbf{u}$$

or

$$\delta\mathbf{F} = \mathbf{A}\,\delta\mathbf{P} + \mathbf{R}\mathbf{u} \tag{2.11.12}$$

where the matrix

$$\mathbf{R} = \mathbf{A}\mathbf{T}'\mathbf{A}^T - \mathbf{P}_L^0 \tag{2.11.13}$$

represents the contribution of the nodal displacement field to the changes in the internal incremental forces due to the rotations of the members in the post-buckling mechanism in the lattice.

The post-buckling equilibrium equations in their general form are then obtained by equating the vectors ($\delta\mathbf{F}$) of the internal load changes at all nodes in the three directions of the global coordinate system (x, y, z) to the vector ($\delta\mathbf{W}$) of the external load changes at these nodes in the same directions, which is accomplished by the result of Eq. (2.11.2) which, in view of the results of Eq. (2.11.10) and the subsequent two equations now yield,

$$\delta\mathbf{F} = \sum_{\alpha=1}^{n=N} \delta\mathbf{F}_\alpha = \mathbf{A}\,\delta\mathbf{P} + \mathbf{A}\mathbf{T}'\mathbf{A}^T\mathbf{u} - \mathbf{P}_L^0\,\mathbf{u} = \delta\mathbf{W} = \boldsymbol{\beta}'\,\delta w \tag{2.11.14}$$

and hence

$$\mathbf{A}\,\delta\mathbf{P} + \mathbf{A}\mathbf{T}'\mathbf{A}^T\mathbf{u} - \mathbf{P}_L^0\,\mathbf{u} = \boldsymbol{\beta}'\,\delta w \tag{2.11.14a}$$

If ($\mathbf{A}$) is replaced by ($\mathbf{C}^T$) where ($\mathbf{A} = \mathbf{C}^T$) we get the general form of equilibrium equations for an isostatic buckled lattice

$$\boxed{\mathbf{C}^T\,\delta\mathbf{P} + (\mathbf{C}^T\mathbf{T}'\mathbf{C} - \mathbf{P}_L^0)\mathbf{u} = \boldsymbol{\beta}'\,\delta w} \tag{2.11.15}$$

where ($\mathbf{C}$) is the kinematic matrix of this lattice and ($\boldsymbol{\beta}'$) the postcritical vector of the load-proportionality constants which may be generally different from the vector ($\boldsymbol{\beta}$), operative in the subcritical range of loading.

2.12 Reduction of the General Equilibrium Equations and Their Solution

The general equilibrium equations (2.11.15) for the buckled lattice valid for any mode of deformation caused by the flexural distortions of its members, which we derived in the last section, can be uncoupled into two independent smaller sets of reduced equilibrium equations. Then, following the method originally used by **Britvec** and **Davister** (Ref. 23) and **Davister** (Ref. 27) to solve such equilibrium equations for complex hyperstatic lattices (discussed later), we apply this method of solution to the set of equations (2.11.15).

The first step in this solution consists in regrouping the elements in the vector ($\delta\mathbf{P}$) according to the incremental axial loads related to the critical members, which are on the point of buckling, and those which never shorten flexurally. We designate the first group by the vector ($\delta\mathbf{P}_c$) and the second by ($\delta\mathbf{P}_s$) where ($c + s = n = N$). These elements are best entered in the ascending order of the original member numbering sequence starting with the member with the lowest number in the leading position of the vector ($\delta\mathbf{P}_c$) and continuing downwards with higher member indices. The same should be done with the elements of the vector ($\delta\mathbf{P}_s$). Then, the first term in Eq. (2.11.15), i.e.

$$\boxed{\mathbf{C}^T\,\delta\mathbf{P}} + (\mathbf{C}^T\mathbf{T}'\mathbf{C} - \mathbf{P}_L^0)\mathbf{u} = \boldsymbol{\beta}'\,\delta w \qquad \text{(2.11.15)bis}$$

may be partitioned, as follows

$$\mathbf{C}^T\,\delta\mathbf{P} = (\mathbf{C}_c^T \mid \mathbf{C}_s^T)\begin{pmatrix}\delta\mathbf{P}_c\\ \hline \delta\mathbf{P}_s\end{pmatrix} \tag{2.12.1}$$

where $(\mathbf{C}_c^T)$ and $(\mathbf{C}_s^T)$ are the corresponding submatrices of the transposed kinematic matrix $(\mathbf{C})$.

The regrouping of the elements of the vector $(\delta\mathbf{P})$ into $(\delta\mathbf{P}_c)$ and $(\delta\mathbf{P}_s)$ gives rise to the corresponding regrouping of the columns of the matrix $(\mathbf{C}^T)$, and that has the consequence that also the rows of the transposed matrix $(\mathbf{C})$ are regrouped *correspondingly*. This means that also the vector $(\mathbf{e})$ is regrouped according to the sub-vectors $(\mathbf{e}_c)$ and $(\mathbf{e}_s)$ which now appear in the same sequence and in the ascending numbering order of the shortenings of the individual critical and nonbuckling members, as do the corresponding axial load increments in the vectors $(\delta\mathbf{P}_c)$ and $(\delta\mathbf{P}_s)$ respectively. Therefore, the vector $(\mathbf{e})$ is regrouped using the kinematic relation (2.3.6),

$$\mathbf{e} = \mathbf{C}\mathbf{u} \tag*{(2.3.6) bis}$$

i.e.

$$\mathbf{e} = \begin{pmatrix}\mathbf{e}_c\\ \hline \mathbf{e}_s\end{pmatrix} = \begin{pmatrix}\mathbf{C}_c\\ \hline \mathbf{C}_s\end{pmatrix}\mathbf{u} \tag{2.12.2}$$

where the submatrices $(\mathbf{C}_c)$ and $(\mathbf{C}_s)$ in both Eq. (2.12.1) and (2.12.2) are the same, only in Eq. (2.12.1) in the transposed order. We note that the sequence of the columns in the matrix $(\mathbf{C})$ and, therefore, also in the submatrices $(\mathbf{C}_c)$ and $(\mathbf{C}_s)$ is not affected by the regrouping of the rows in these matrices, which are the columns of the transposed matrices $(\mathbf{C}^T)$ and $(\mathbf{C}_c^T)$ viz. $(\mathbf{C}_s^T)$, so that the rows of the transposed matrices which, in turn, are the columns of the original matrices $(\mathbf{C})$, $(\mathbf{C}_c)$ and $(\mathbf{C}_s)$ are unaffected by this regrouping. This means that also the sequence of the (x, y, z) nodal displacement components in the vector $(\mathbf{u})$ is unaffected by this partitioning. Therefore, if the matrices $(\mathbf{C})$ and $(\mathbf{C}^T)$ in the second term in Eq. (2.11.15) are partitioned in the same way, the vector $(\mathbf{u})$ multiplying that term is not affected by this partitioning. However, the diagonal elements of the matrix $(\mathbf{T}')$, which appears in this second term, must be regrouped, so that they multiply the corresponding elements of the submatrices $(\mathbf{C})$ and $(\mathbf{C}^T)$ in the product $(\mathbf{C}^T\mathbf{T}'\mathbf{C})$. Then, the second term in Eq. (2.11.15) may be rewritten as follows

$$\mathbf{C}^T\mathbf{T}'\mathbf{C} = (\mathbf{C}_c^T \mid \mathbf{C}_s^T)\left(\begin{array}{c|c}\mathbf{T}_c' & \mathbf{0}\\ \hline \mathbf{0} & T_s'\end{array}\right)\begin{pmatrix}\mathbf{C}_c\\ \hline \mathbf{C}_s\end{pmatrix} \tag{2.12.3}$$

and the matrix product $(\mathbf{T}'\mathbf{C})$, as

$$\mathbf{T}'\mathbf{C} = \left(\begin{array}{c|c}\mathbf{T}_c' & \mathbf{0}\\ \hline \mathbf{0} & \mathbf{T}_s'\end{array}\right)\begin{pmatrix}\mathbf{C}_c\\ \hline \mathbf{C}_s\end{pmatrix} \tag{2.12.4}$$

To solve the equilibrium Eqs. (2.11.15) we premultiply this set of equations by the inverse matrix $(\mathbf{C}^T)^{-1}$. The kinematic matrix $(\mathbf{C})$ is obviously by definition a nonsingular matrix, if for a given set of the contractions $(\mathbf{e})$ there is a unique displacement field of the nodes $(\mathbf{u})$. Then from (2.3.6) it follows that

$$\mathbf{u} = \mathbf{C}^{-1}\mathbf{e} = \mathbf{C}_1\mathbf{e} \tag{2.6.3)bis}$$

as already shown in Section 2.6, Eq. (2.6.3). Premultiplying (2.11.15) by $(\mathbf{C}^T)^{-1}$ gives

$$\mathbf{I}\,\delta\mathbf{P} + [\mathbf{T}'\mathbf{C} - (\mathbf{C}^T)^{-1}\mathbf{P}_L^0]\mathbf{u} = (\mathbf{C}^T)^{-1}\boldsymbol{\beta}'\,\delta w \tag{2.12.5}$$

Since, by Eq. (2.6.3),

$$(\mathbf{C}^T)^{-1} = (\mathbf{C}^{-1})^T = \mathbf{C}_1^T$$

the last result becomes

$$\delta\mathbf{P} + (\mathbf{T}'\mathbf{C} - \mathbf{C}_1^T\mathbf{P}_L^0)\mathbf{u} = \mathbf{C}_1^T\boldsymbol{\beta}'\,\delta w \tag{2.12.6}$$

Regrouping now the rows of the matrices $(\mathbf{C})$ and $(\mathbf{T}')$, as explained before, and regrouping also the columns of the inverse matrix $(\mathbf{C}^{-1} = \mathbf{C}_1)$, in the same sequence as the elements of the vector $(\mathbf{e})$ into the subvectors $(\mathbf{e}_c)$ and $(\mathbf{e}_s)$, then, according to the relation (2.6.3)bis, we get

$$\mathbf{u} = \mathbf{C}^{-1}\mathbf{e} = \mathbf{C}_1\mathbf{e} = (\mathbf{C}_{1c} \mid \mathbf{C}_{1s})\begin{pmatrix}\mathbf{e}_c\\ \hline \mathbf{e}_s\end{pmatrix} = \mathbf{C}_{1c}\mathbf{e}_c \tag{2.12.7}$$

since $(\mathbf{e}_s \equiv 0)$.

With these modifications and the kinematic relation (2.3.6) viz. (2.6.3), Eq. (2.12.6) becomes

$$\delta\mathbf{P} + \mathbf{T}'\mathbf{e} - \mathbf{C}_1^T\mathbf{P}_L^0\mathbf{C}_1\mathbf{e} = \mathbf{C}_1^T\boldsymbol{\beta}'\,\delta w \tag{2.12.8}$$

Partitioning now according to the subvectors $(\mathbf{e}_c)$ and $(\mathbf{e}_s)$

$$\begin{pmatrix}\delta\mathbf{P}_c\\ \hline \delta\mathbf{P}_s\end{pmatrix} + \begin{pmatrix}\mathbf{T}'_c & \mathbf{0}\\ \mathbf{0} & \mathbf{T}'_s\end{pmatrix}\begin{pmatrix}\mathbf{e}_c\\ \hline \mathbf{e}_s\end{pmatrix} - \begin{pmatrix}\mathbf{C}_{1c}^T\\ \hline \mathbf{C}_{1s}^T\end{pmatrix}\mathbf{P}_L^0(\mathbf{C}_{1c} \mid \mathbf{C}_{1s})\begin{pmatrix}\mathbf{e}_c\\ \hline \mathbf{e}_s\end{pmatrix} = \begin{pmatrix}\mathbf{C}_{1c}^T\\ \hline \mathbf{C}_{1s}^T\end{pmatrix}\boldsymbol{\beta}'\,\delta w$$

where $(\mathbf{e}_s \equiv 0)$.

After carrying out the matrix multiplications we get

$$\begin{pmatrix}\delta\mathbf{P}_c\\ \hline \delta\mathbf{P}_s\end{pmatrix} + \begin{pmatrix}\mathbf{T}'_c\mathbf{e}_c\\ \hline \mathbf{T}'_s\mathbf{e}_s\end{pmatrix} - \begin{pmatrix}\mathbf{C}_{1c}^T\\ \hline \mathbf{C}_{1s}^T\end{pmatrix}\mathbf{P}_L^0(\mathbf{C}_{1c}\mathbf{e}_c + \mathbf{C}_{1s}\mathbf{e}_s) = \begin{pmatrix}\mathbf{C}_{1c}^T\\ \hline \mathbf{C}_{1s}^T\end{pmatrix}\boldsymbol{\beta}'\,\delta w \tag{2.12.9}$$

Here the matrix $(\mathbf{C}_{1c}^T)$ is of order $(c \times N)$ and the matrix $(\mathbf{C}_{1s}^T)$ of order $(s \times N)$. $(\mathbf{P}_L^0)$ is obviously a square $(N \times N)$ matrix. Putting $(\mathbf{e}_s = 0)$, the above set reduces to

$$\begin{pmatrix}\delta\mathbf{P}_c\\ \hline \delta\mathbf{P}_s\end{pmatrix} + \begin{pmatrix}\mathbf{T}'_c\mathbf{e}_c\\ \hline \mathbf{0}\end{pmatrix} - \begin{pmatrix}\mathbf{C}_{1c}^T\mathbf{P}_L^0\mathbf{C}_{1c}\mathbf{e}_c\\ \hline \mathbf{C}_{1s}^T\mathbf{P}_L^0\mathbf{C}_{1c}\mathbf{e}_c\end{pmatrix} = \begin{pmatrix}\mathbf{C}_{1c}^T\boldsymbol{\beta}'\,\delta w\\ \hline \mathbf{C}_{1s}^T\boldsymbol{\beta}'\,\delta w\end{pmatrix} \tag{2.12.10}$$

These equations can obviously be uncoupled, so that the reduced equilibrium equations become

$$\delta \mathbf{P}_c + \mathbf{T}_c' \mathbf{e}_c - \mathbf{C}_{1c}^T \mathbf{P}_L^0 \mathbf{C}_{1c} \mathbf{e}_c = \mathbf{C}_{1c}^T \boldsymbol{\beta}' \, \delta w \tag{2.12.11}$$

and

$$\delta \mathbf{P}_s - \mathbf{C}_{1s}^T \mathbf{P}_L^0 \mathbf{C}_{1c} \mathbf{e}_c = \mathbf{C}_{1s}^T \boldsymbol{\beta}' \, \delta w \tag{2.12.12}$$

Using now the general law in (2.1.1) Eq. (2.12.11) may be written

$$(\mathbf{K}_c \mathbf{T}_c' + \mathbf{T}_c' - \mathbf{C}_{1c}^T \mathbf{P}_L^0 \mathbf{C}_{1c}) \mathbf{e}_c = \mathbf{C}_{1c}^T \boldsymbol{\beta}' \, \delta w \tag{2.12.13}$$

Putting

$$\mathbf{E}_c = (\mathbf{K}_c + \mathbf{I}_c) \mathbf{T}_c' - \mathbf{C}_{1c}^T \mathbf{P}_L^0 \mathbf{C}_{1c}$$

or

$$\mathbf{E}_c = (\mathbf{I} + \mathbf{K}_c^{-1}) \mathbf{T}_c - \mathbf{C}_{1c}^T \mathbf{P}_L^0 \mathbf{C}_{1c} \tag{2.12.13a}$$

where $(\mathbf{E}_c)$ is referred to as the *post-buckling stiffness matrix* with respect to the critical flexural contractions $(\mathbf{e}_c)$. We get finally

$$\boxed{\mathbf{E}_e \mathbf{e}_c = \mathbf{C}_{1c}^T \boldsymbol{\beta}' \, \delta w} \tag{2.12.14}$$

This set of equations can be solved provided all the critical members buckle. The post-buckling stiffness matrix $(\mathbf{E}_c)$ plays an important part in the discussion of the stability of the critical state, as we shall see later.

If fewer members buckle, let us say only (b) out of (c) critical members buckle in a post-buckling equilibrium mode governed by the reduced equilibrium equations (2.12.14), then the straight or nonbuckling critical members are absorbed in the vectors $(\mathbf{e}_s)$, which for this purpose and for the purpose of this regrouping, its elements now containing also the zero shortenings of the unloading nonbuckling critical members, may be designated by $(\mathbf{e}_{s'})$, where now $(b + s' = N)$, as explained before. Then clearly, the reduced governing equilibrium equations (2.12.14) assume the form

$$\mathbf{E}_b \mathbf{e}_b = \mathbf{C}_{1b}^T \boldsymbol{\beta}' \, \delta w; \qquad b = 1, 2, 3, \ldots, c \tag{2.12.15}$$

The remaining equations (2.12.12) now become

$$\delta \mathbf{P}_{s'} - \mathbf{C}_{1s'}^T \mathbf{P}_L^0 \mathbf{C}_{1b} \mathbf{e}_b = \mathbf{C}_{1s'}^T \boldsymbol{\beta}' \, \delta w \tag{2.12.16}$$

The solution of these equilibrium equations is best obtained on a digital computer. The formalism is as follows: First the post-buckling stiffness matrix $(\mathbf{E}_b)$ is inverted. This matrix is usually much smaller than the matrix $(\mathbf{S})$ in Eq. (2.7.16) or the matrix $(\mathbf{E})$ in Eq. (2.7.18), used in the equilibrium approach in Section 2.7, as (c) is usually much smaller than N. The solution of Eqs.

(2.12.15) results in the vector of flexural shortenings ($\mathbf{e}_b$) in a particular buckling mode (b), i.e.

$$\mathbf{e}_b = \mathbf{E}_b^{-1}\mathbf{C}_{1b}^T\boldsymbol{\beta}'\,\delta w \tag{2.12.17}$$

Finally from the set (2.12.16), the changes in the axial loads in the straight members are obtained in the form

$$\delta\mathbf{P}_{s'} = (\mathbf{C}_{1s'}^T\mathbf{P}_L^0\mathbf{C}_{1b}\mathbf{E}_b^{-1}\mathbf{C}_{1b}^T + \mathbf{C}_{1s'}^T)\boldsymbol{\beta}'\,\delta w \tag{2.12.18}$$

For this buckling mode to be statically admissible under the load vector ($\boldsymbol{\beta}'$), the axial load changes ($\delta\mathbf{P}_{s'}$) in the ($c-b$) initially critical members, must be all *negative* for *one* sign of the load increment (δw) under proportionate loading, since any initially critically stressed compressive perfect member must unload in post-buckling in order to remain straight. This condition must be tested for each mode. If this condition is not satisfied, the particular mode is not statically admissible and it cannot occur under the load vector ($\boldsymbol{\beta}'$). The only way in which such a mode could develop is to stiffen the straight initially critical members in which the axial load changes in this mode increase, so as to prevent them from buckling.

The axial load changes in the (b) buckling members can be found using the equation for the general law in Eq. (2.9.9). Partitioning again the vector ($\mathbf{e}$) into ($\mathbf{e}_b$) and ($\mathbf{e}_{s'}$), we get

$$\delta\mathbf{P} = \begin{pmatrix}\delta\mathbf{P}_b\\ \hline \delta\mathbf{P}_{s'}\end{pmatrix} = \left(\begin{array}{c|c}\mathbf{K}_b & \mathbf{0}\\ \hline \mathbf{0} & \mathbf{K}_{s'}\end{array}\right)\left(\begin{array}{c|c}\mathbf{T}_b' & \mathbf{0}\\ \hline \mathbf{0} & \mathbf{T}_{s'}'\end{array}\right)\begin{pmatrix}\mathbf{e}_b\\ \hline \mathbf{e}_{s'}\end{pmatrix}$$

or

$$\delta\mathbf{P} = \left(\begin{array}{c|c}\mathbf{K}_b & \mathbf{0}\\ \hline \mathbf{0} & \mathbf{K}_{s'}\end{array}\right)\begin{pmatrix}\mathbf{T}_b'\mathbf{e}_b\\ \hline \mathbf{T}_{s'}'\mathbf{e}_{s'}\end{pmatrix} = \begin{pmatrix}\mathbf{K}_b\mathbf{T}_b'\mathbf{e}_b\\ \hline \mathbf{K}_{s'}\mathbf{T}_{s'}'\mathbf{e}_{s'}\end{pmatrix} \tag{2.12.19}$$

or

$$\delta\mathbf{P}_b = \mathbf{K}_b\mathbf{T}_b'\mathbf{e}_b \tag{2.12.20}$$

If all the members are prismatic

$$\mathbf{K}_b = \tfrac{1}{2}\mathbf{I}_b \tag{2.12.20a}$$

and

$$\delta\mathbf{P}_b = \tfrac{1}{2}\mathbf{T}_b'\mathbf{e}_b \tag{2.12.21}$$

substituting for ($\mathbf{e}_b$) from (2.12.17),

$$\delta\mathbf{P}_b = \tfrac{1}{2}\mathbf{T}_b'\mathbf{E}_b^{-1}\mathbf{C}_{1b}^T\boldsymbol{\beta}'\,\delta w \tag{2.12.22}$$

The displacements of the nodes may be found from Eq. (2.8.3), where the subscript (b) instead of (c) is now used, i.e.

$$\mathbf{u} = \mathbf{C}_{1b}\mathbf{e}_b = \mathbf{C}_{1b}\mathbf{E}_b^{-1}\mathbf{C}_{1b}^T\boldsymbol{\beta}'\,\delta w \tag{2.12.23}$$

Here, as before, use was made also of Eq. (2.12.17).

2.13 Some Applications of the Theory to Simple Space Structures Made of Flexible Elastic Members

The theory developed in the preceding sections may be applied to any isostatic space lattice of arbitrary configuration. Very common configurations are reticulated beams, such as that shown in Fig. 2.13.1, which may be used in a variety of applications, e.g. as parts of cranes, as light beams in systems deployed in outer space to support an antenna or a nuclear reactor attached to a satellite etc. Regardless of their size and shape, such beams are subjected to transverse and axial loads applied at their nodes, which may have very different magnitudes and directions. Reticulated structures, in general, are only useful if loaded centrally at the nodes, so that loads are led into the lattice as axial loads in the members, which are usually light and flexible. Flexible members, in general, are weak when loaded transversely, but if they are *braced*, as in a lattice, they offer high resistance to axial loads. On the other hand, their failure in such a braced lattice is not so much governed by their material strength, but rather by their resistance to buckling under compressive axial loads. We have seen, while discussing the theory in previous sections, that material failure in space lattices follows the onset of buckling which takes place in equilibrium or in unstable motion. At first, there is a change of a physical state in the structure under the applied loading. This change of physical state *precedes* the material failure, which happens subsequently. It is of interest, therefore, to analyze the states of post-buckling equilibrium and of motion in practical structures of this type, which are caused by the critical intensities of axial loads. What concerns us most is the location of the most severely affected members in such a system under a particular set of external loads. We then try to isolate the worst cases of such deformation, which tend to degrade most severely the elasto-mechanical response of the system. The dangerous levels of loading are present from the onset of visible flexural deformations in the members, which happen at the near-critical and the critical levels of external loading intensities. As we saw in the preceding sections, the purpose of this analysis is to determine these critical loading levels for different loading systems and to limit the cycles of loading to non-dangerous levels.

In the next section we discuss the effects of initial small geometrical imperfections, which can severely affect these limits and how these effects can be analyzed *quantitatively*. It is shown there that the response of the perfect or imperfection-free structure is essential for this quantification. Therefore, we examine a few sample cases to illustrate the response of such flexible lattices under ideal conditions. As the first example we select a space beam, such as that in Figs. 2.13.1 and 2.13.1a or in 2.13.1b, consisting of twenty bays. (See also, Refs. 24 and 25.) But, a shorter space beam consisting of fewer bays is equally acceptable. The geometrical and material data for this simple structure are shown for a truncated space beam of five bays in Fig. 2.13.1c. The longitudinal members are called longerons, the transverse members forming equilateral triangles, the battens and the remaining

Figure 2.13.1
The control of flexible structures—60-meter deployable flight-mast—(NASA COFS Program, Ref. 24)

Figure 2.13.1(a)
Flight-mast under deployment from space-shuttle in outer space (NASA Ease/Access Post-Mission Report 1992)

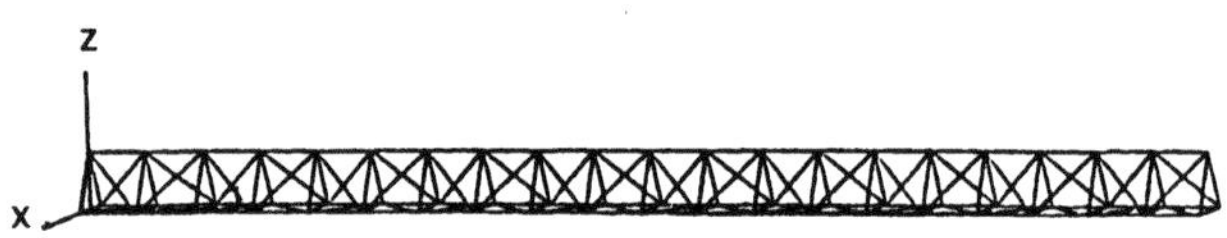

Figure 2.13.1(b)
20-bay reticulated mast-beam (NASA COFS Program, Ref. 24)

members are the diagonal members. The lengths, the cross-sectional areas and the moments of inertia of these areas about the principal axes are shown in the table in Fig. 2.13.1c. Also given are the critical loads based on the modulus for steel ($E = 29 \times 10^6$ psi). These data are based on the design values specified by NASA for one of its space beams[3]. The lengths of the members in meters were as follows:

[3] See NASA (COFS) Technology Program, February 7th 1986.

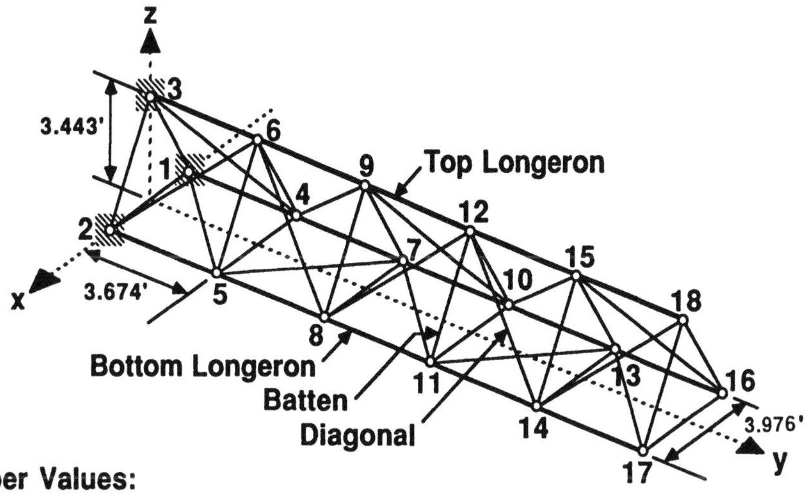

Member Values:

Type	Area (in.^{2})	I (in.^{4})	P_{crit} (lb.)	L (ft.)
Bottom Longerons	.2190	.012110	1,836.64	3.674
Top Longerons	.2800	.016870	2,558.52	3.674
Battens	.0352	.000373	48.35	3.976
Diagonals	.0649	.000335	23.40	5.417
Elastic Modulus	E = 29,000,000 psi			

Figure 2.13.1(c)
Geometry of the pin-jointed 5-bay space beam. (NASA COFS Program, Refs. 24, 25)

Bay length (meters)	1.12
Batten length (meters)	1.212
Diagonal length (meter)	1.651
Beam mass per unit length (Kg/meter)	2.65
Longeron tubes ID/OD (mm)	13.97/19.37 (2 bottom longerons)
	13.97/20.63 (1 top longeron)
Batten tubes ID/OD (mm)	6.35/8.32
Diagonal Rods OD (mm)	7.30

The data in the table in Fig. 2.13.1c are based on these figures converted into in., ft., lb. units. In Figs. 2.13.1c and 2.13.2 the 5-bay beam is shown fixed to a rigid foundation at one end. (Such a foundation may be the end attached to the space shuttle orbiter, if the beam were deployed from such a vehicle. Or, it may be any fixed foundation in a different situation.) The first two loading cases A and B that we consider are shown in Fig. 2.13.2. These loads are assumed to be applied gradually i.e. quasi-statically and they consist of three equal and parallel loads of intensity (W). They are applied in the plane normal to the plane of symmetry at the opposite end, which is free. We note that the top longeron tube is somewhat stiffer than the two bottom longerons and this difference in stiffness makes the (y, z) plane, indicated in Fig. 2.13.2, also the plane of symmetry in this case.

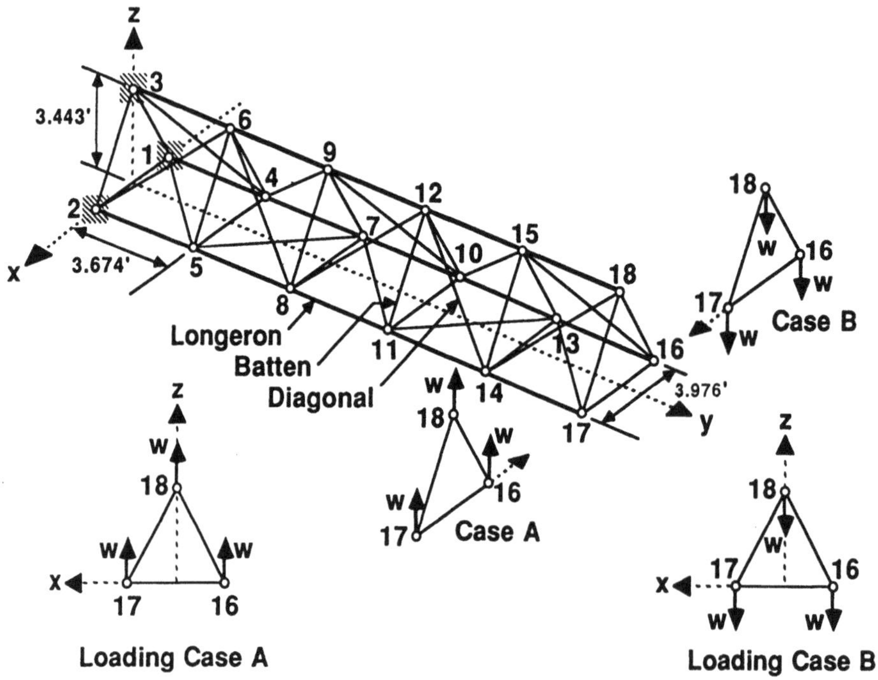

Figure 2.13.2
Loadings on the 5-bay space beam (Refs. 24, 25)

In Case A the three loads (**W**) are applied "upwards" and in Case B "downwards" i.e. in the reversed direction.

Case A

The theory expounded in the last section was applied to this space beam using a digital computer. The joints and the members were numbered as shown in Fig.

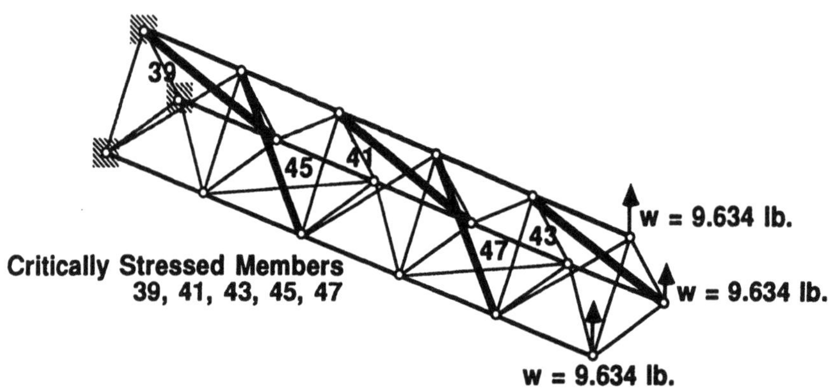

Figure 2.13.3(a)
Case A: Critical state occurs at ($W = 9.634$ lb.)

Member	Relative Flexural Shortening e
39	1.0
41	0.0
43	0.0
45	0.0
47	0.0

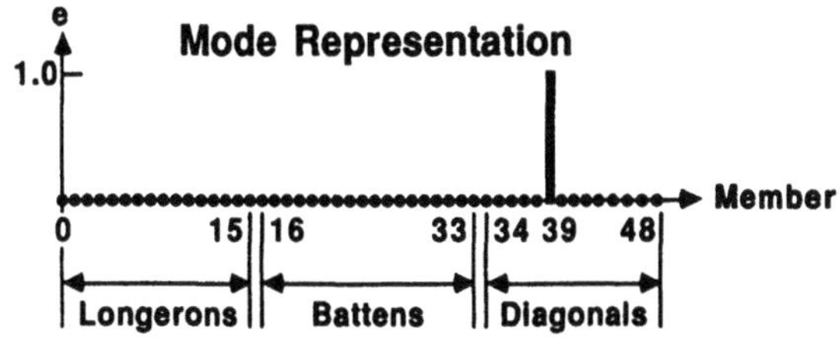

Figure 2.13.3(b)
Case A: The most-degrading post-buckling equilibrium mode. Bar diagram

2.13.2 and 2.13.3. The relations of the members to the joints at each end are as follows:

Member	*Joints*	*Member*	*Joints*
1	3–6	25	10–11
2	6–9	26	11–12
3	9–12	27	12–10
4	12–15	28	13–14
5	15–18	29	14–15
6	1–4	30	15–13
7	4–7	31	16–17
8	7–10	32	17–18
9	10–13	33	18–16
10	13–16	34	1–5
11	2–5	35	5–7
12	5–8	36	7–11
13	8–11	37	11–13
14	11–14	38	13–17
15	14–17	39	3–4
16	1–2	40	4–9
17	2–3	41	9–10
18	3–1	42	10–15
19	4–5	43	15–16
20	5–6	44	2–6
21	6–4	45	6–8
22	7–8	46	8–12
23	8–9	47	12–14
24	9–7	48	14–18

Using the numbering sequence shown in the figure, the basic matrices were derived and coded as shown in Section 2.3, i.e., the kinematic matrix (**C**) and its transpose ($\mathbf{C}^T = \mathbf{A}$), which is the statical matrix. The load proportionality vectors ($\boldsymbol{\beta}$) and ($\boldsymbol{\beta}'$) were identical in this case. The post-buckling stiffness matrix ($\mathbf{E}_c$) and the corresponding vector ($\mathbf{C}_{1c}^T \boldsymbol{\beta}'$) were derived and programmed as explained earlier. The reduced equilibrium equations were formulated for this case according to Eq. (2.12.15) and then were solved on a VAX 11/780 digital computer using the standard numerical techniques. Here, only the essential results of these computations are reproduced numerically. The critically stressed members were found to be the diagonal members 39, 41, 43, 45 and 47. The critical loads (W), at which the critical state occurs, were found to be equal to ($W^0 = 9.634$ lb.), as indicated in Fig. 2.13.3a. Several post-buckling equilibrium modes from this state are possible, however, of most practical interest is that mode which is likely to occur in a test in a similar model structure. Experiments show that invariably that mode develops which is associated with the largest decrease in the total potential energy of the system, as this energy is dissipated in a form of kinetic energy. If (δW_i) represents the increment in the critical load (W_i^0) and (δs_i) its corresponding displacement, assuming that this load remains parallel to its original direction, the expression[4]

$$v = \frac{1}{2} \sum_{i=1,2,3,\ldots} \delta W_i \, \delta s_i$$

represents this energy change. So the mode in which this expression has the largest negative value is called the most degrading mode. In this case this mode occurs when only one member, marked 39, buckles. In Fig. 2.13.4 the expression for (v) in this mode is calculated when ($\delta W = -1.0$ lb.). We get that in this case the displacements of the applied loads (W) are given in reference to the coordinate system in Fig. 2.13.2, as follows,

$$\delta s_1 = u_{16} = -1.263 \text{ in.}; \qquad \delta s_2 = u_{17} = 0.000; \qquad \delta s_3 = u_{18} = -0.632 \text{ in.}$$

[4] For a detailed discussion of this and the related expressions the reader is referred to Ref. 15, Chapters 1, 3, 6 and 7.

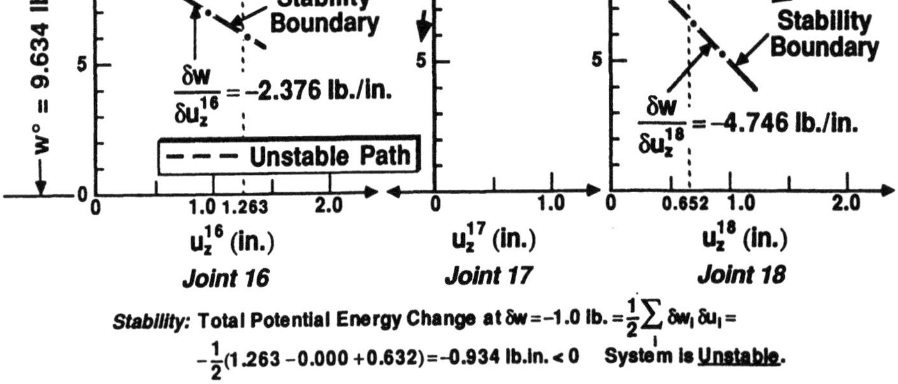

Figure 2.13.4
Case A: Primary unstable equilibrium paths at the critical state and the stability test

Then, the sum

$$v = \frac{1}{2} \sum_{i=1}^{3} \delta W_i \, \delta s_i = -\frac{1}{2}(1.263 - 0.000 + 0.632) = -0.934 \text{ lb. in.}$$

Since this expression is negative, the total potential energy of the system is released as kinetic energy in an onset of unstable motion if the system is not artificially stabilized. (We discuss these aspects in more detail in Section 2.15.) The distorted configuration of the structure may be represented in a bar-graph, as shown in Fig. 2.13.3b. On the abscissa the members are plotted according to their sequencing numbers and the ordinates represent normalized relative flexural contractions, the largest shortening in this case being set equal to unity.

The equilibrium paths represent plots of each applied load against its corresponding displacement, where it is assumed that the load and displacement directions do not change. The three equilibrium paths are shown plotted in Fig. 2.13.4. Notice that the gradients of these paths at joints 16 and 18, which are labelled in Fig. 2.13.2 and which are on the same side as the buckling member 39, are finite and that the displacement of the lower left joint 17 is practically zero, giving an infinite gradient to this path. (Actually, the displacement of joint 17 shows a very small negative value below the three decimal points and it is therefore equated to zero to the accuracy of this analysis.) The decreasing equilibrium paths are statically unstable and are therefore shown dashed.

The chain-dotted lines in the diagrams in Fig. 2.13.4 represent the stability boundaries‡ in this case, on which the imperfect equilibrium paths peak. The perfect paths are modified by the influence of initial imperfections such as small initial curvatures in the members, eccentricities of loading etc. We discuss the forms of these modified or imperfect paths in the next section.

Fig. 2.13.5 shows the computer print of the distorted 5-bay space beam viewed from the free end. Here, the zero displacement of the lower left joint 12 is quite visible. The distortion corresponds to a change in ($\delta W = -1.0$ lb.), as mentioned earlier. The graph in the bottom of the figure shows the dependence of the load (W) on the angle of twist (θ) of the end-section. The path is clearly decreasing, confirming the unstable nature of this *torsional* deformation of the space-beam under the three *symmetric* upward loads (W).

Case B

In this case the same structure is loaded by three equal loads (W) only applied downward, i.e. in the *opposite* direction to that in Case A, Fig. 2.13.6(a). The critical state is reached at the same load intensity of ($W = W^0 = 9.634$ lb.). But, now

‡ We show later, that the gradients of these boundaries are equal to approximately *three times* the gradients of the perfect paths in the same modes.

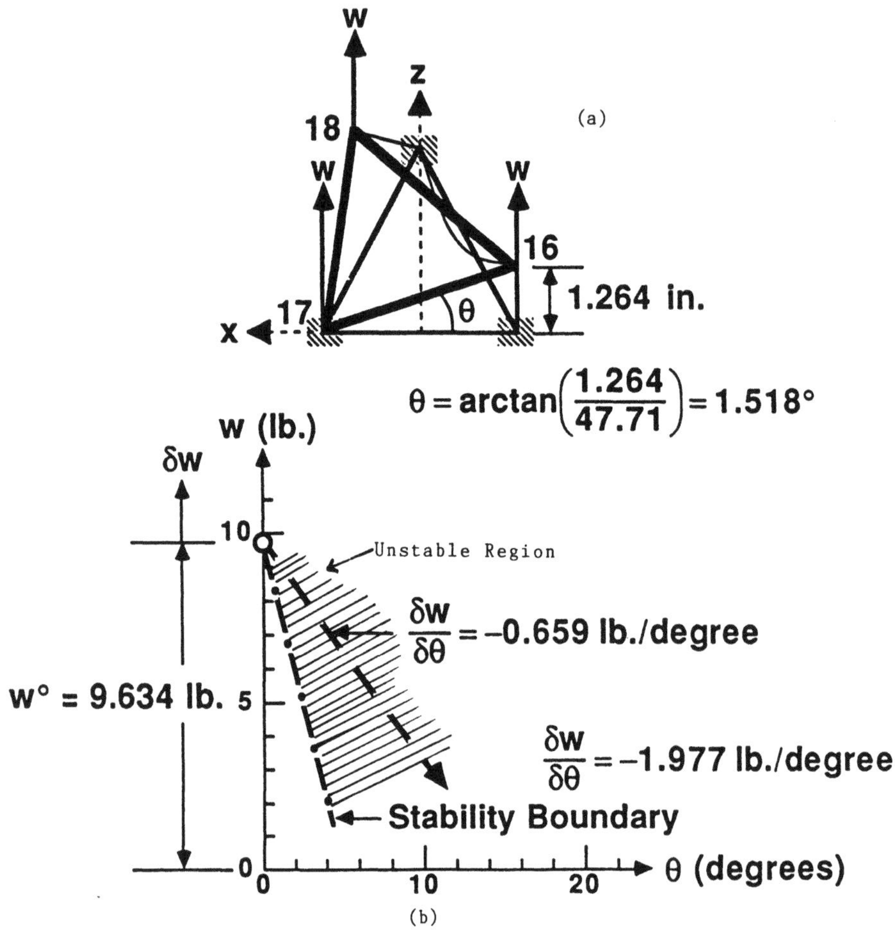

Figure 2.13.5
(a) Case A: Rotation of the loaded end-section of the 5-bay beam in post-buckling equilibrium. (b) Case A: Load-parameter end-rotation diagram of the 5-bay space-beam in unstable equilibrium

entirely different members become critically stressed, as it can be seen from the table in Fig. 2.13.6(b). The most degrading post-buckling mode which develops in this case is also different. Now three members 44, 46 and 48 buckle. The bar-graph for this mode is shown in the same figure in which the relative flexural shortenings are shown normalized with respect to the flexural shortening of member 48, which is arbitrarily set equal to unity. This member is the most severely distorted member and its location is now at the free end of the space beam, diametrically opposite to the member 39, which was the most critical member in Case A.

The equilibrium paths for the three loads and their corresponding displacements are now rising paths and they are shown in Fig. 2.13.7. However, this does not mean that these equilibrium paths are necessarily stable. This is why they are shown dashed. (The reasons for this will be discussed later. See also, Case C, Fig. 2.13.13,

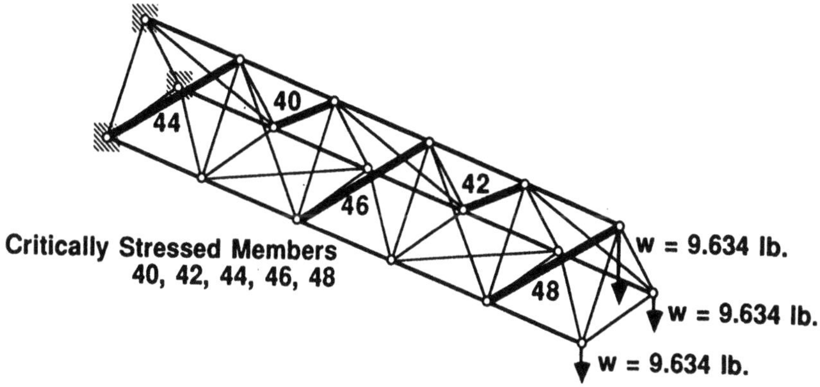

Figure 2.13.6(a)
Case B: Critical state occurs at ($W = 9.634$ lb.)

Member	Relative Flexural Shortening e
40	0.000
42	0.000
44	0.217
46	0.356
48	1.000

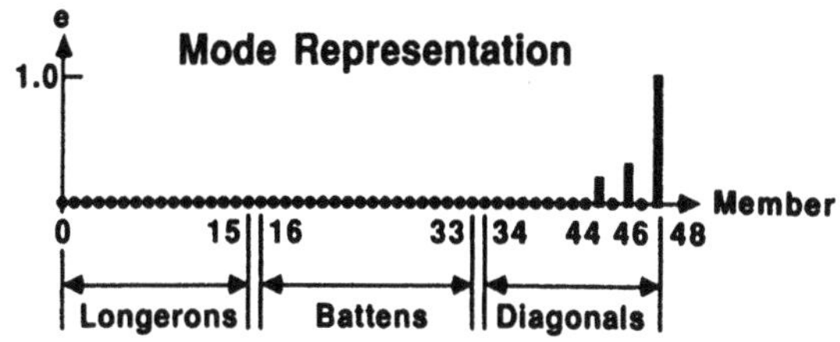

Figure 2.13.6(b)
Case B: The most-degrading post-buckling equilibrium mode. Bar diagram

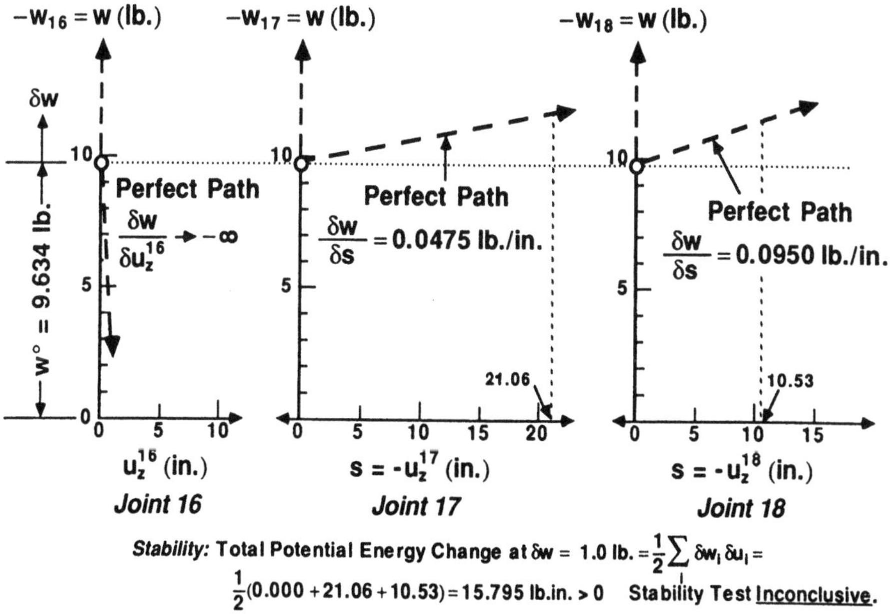

Figure 2.13.7
Case B: Primary equilibrium paths at the critical state

page 122.) The total potential energy change (v) in this mode is positive as shown in Fig. 2.13.7 and its absolute value at ($\delta W = 1.00$ lb.) is now much higher. The meaning of a rising unstable path is best explained if the system is released from an artificially stabilized equilibrium state at some point corresponding to the same value of (δW) on the graphs in Fig. 2.13.7. If this release results in a motion to another mode at lower energy levels, then the original equilibrium state is certainly unstable. However, this does not necessarily mean that the initial or critical state is unstable. In fact, it may be shown similarly as in case C, Fig. 2.13.13, using the method discussed in Section 2.15 that in this case the initial critical state is stable. Therefore, the structure if released from an equilibrium state at higher energies will return to the initial state in this case.

Fig. 2.13.8(a) shows the end-view of this structure under the distortions in the same mode. The distorted view reveals a considerable twist of the flight mast and a totally different distortion pattern from that encountered in Case A. It is now the lower right joint 16 (see also Fig. 2.13.2) which is practically undisplaced and the left lower joint 17 suffers the largest downward displacement. The angle of twist (θ) and the whole distorted configuration corresponds to an increase in the loads ($\mathbf{W} = \mathbf{W}^0$) by the amount ($\delta W = 1.00$ lb.). The dependence of each load ($\mathbf{W}$) on the angle of twist in this mode is also shown in the graph in Fig. 2.13.8(b).

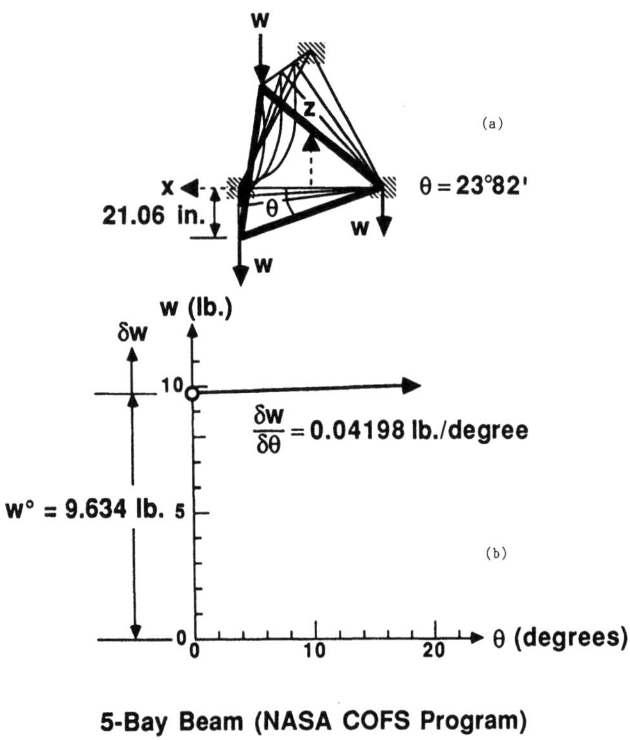

Figure 2.13.8
(a) Case B: Rotation of the loaded end-section of the 5-bay beam in post-buckling equilibrium. (b) Case B: Load-parameter end-section rotation diagram of the 5-bay beam

Case C

The same structure, shown earlier in Fig. 2.13.1(c), is now subjected to a torque applied at the free end by means of two equal and opposite forces (**W**) as shown in Fig. 2.13.9(a). Under the counter-clockwise torque the diagonal members in alternate bays buckle starting with the bay nearest to the supported end. Altogether nine members labeled 34, 36, 38, 39, 41, 43, 44, 46 and 48 buckle and shorten by *equal* amounts, which are shown normalized in the bar-graph in Fig. 2.13.9(b). The critical state occurs when (**W**) reaches the value of ($W^0 = 14.45$ lb.). This is a very low value, since the mast can be crippled by a critical torque (T^0) which is equal to

$$T^0 = 14.45 \times 3.976 = 57.45 \text{ lb. ft.}$$

where the lever arm is the batten length which equals only 3.976 ft. Thc computed distorted configuration of the 5-bay mast is shown in Fig. 2.13.10 in which the distortions are shown enlarged. The end-view, corresponding to this configuration, is shown in Fig. 2.13.11. In Fig. 2.13.12 the twisted mast is shown

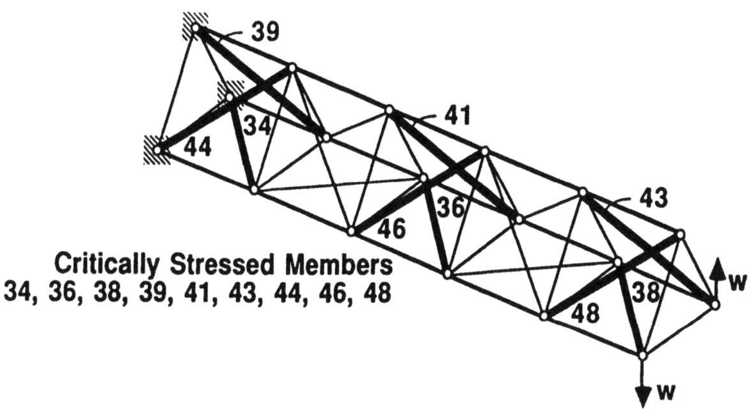

Figure 2.13.9(a)
Case C: Critical state occurs at ($W = 14.45$ lb.)

Member	Relative Flexural Shortening e
34	1.0
36	1.0
38	1.0
39	1.0
41	1.0
43	1.0
44	1.0
46	1.0
48	1.0

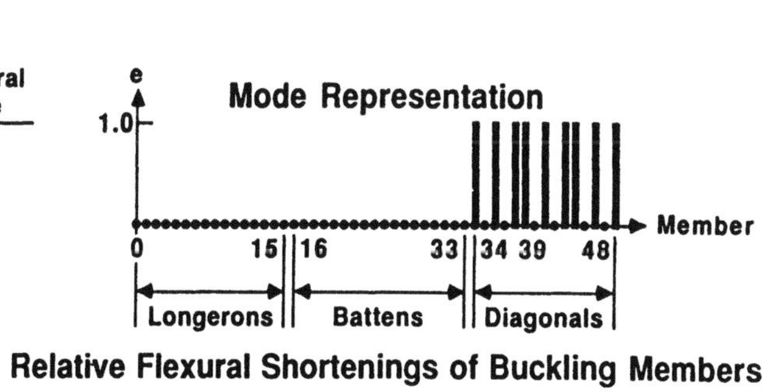

Figure 2.13.9(b)
Case C: The most-degrading post-buckling equilibrium mode. Bar diagram

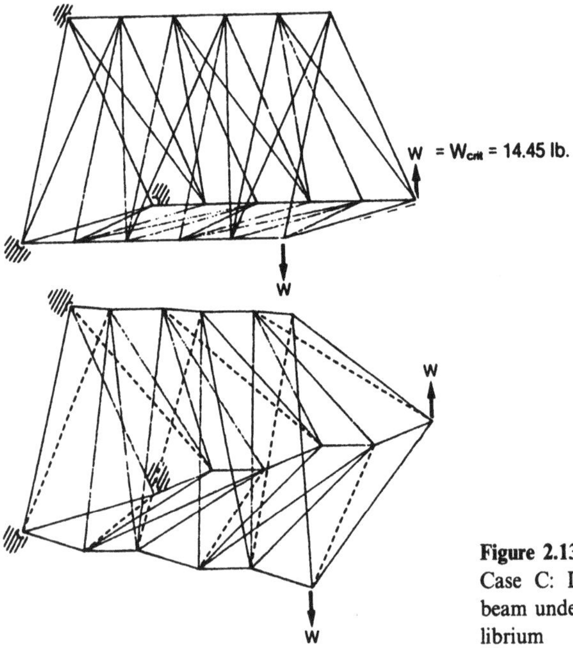

Figure 2.13.10
Case C: Distorted shape of a 5-bay space beam under end-torque in post-buckling equilibrium

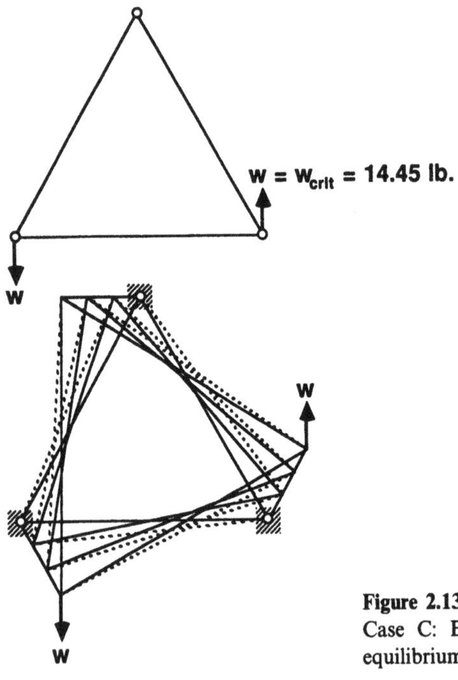

Figure 2.13.11
Case C: End-view of the buckled 5-bay space beam in equilibrium under the post-critical end-torque

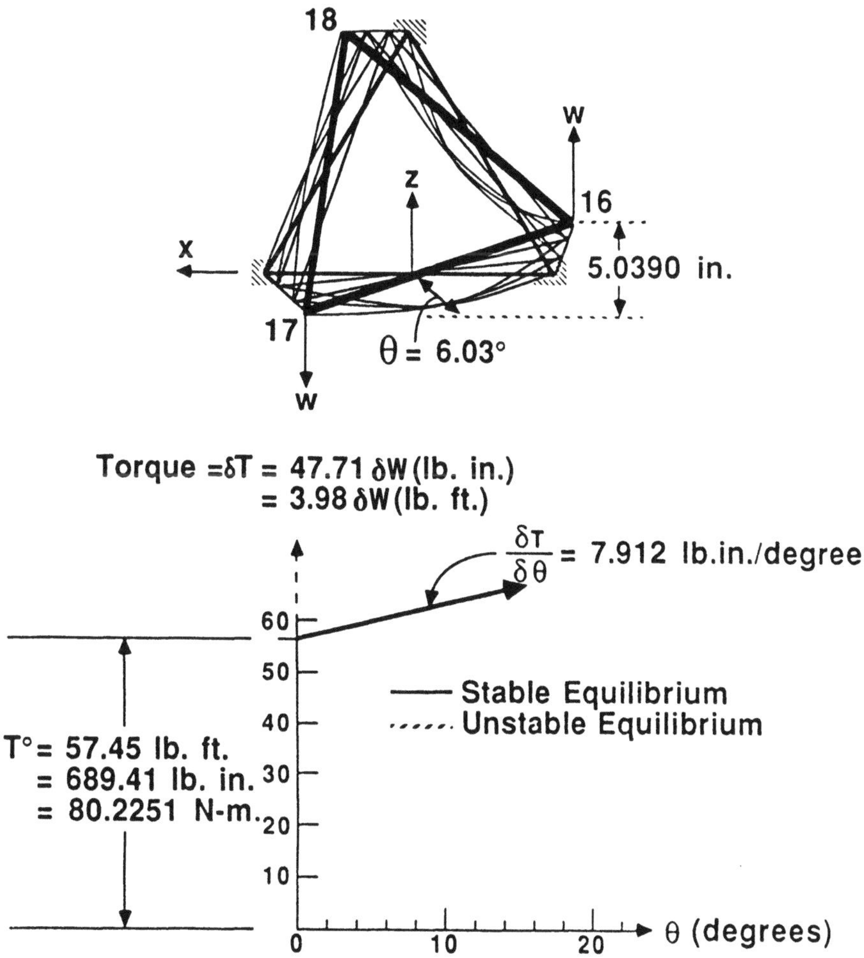

Figure 2.13.12
(a) Case C: Rotation of the loaded end-section in the most-degrading post-buckling equilibrium mode. (b) Case C: Load-parameter end-rotation diagram of the 5-bay beam

under an angle of twist ($\theta = 6.03°$), and the torque-twist relationship is displayed in the bottom diagram in the figure. The increase in the torque, corresponding to ($\theta = 6.03°$), is given by the increment ($\delta T = 47.71$ lb. in.). This value and the value of the angle of twist correspond to an increase in each load intensity (W) from (W^0) by ($\delta W = 1.00$ lb.).

The post-buckling stiffness matrix ($\mathbf{E}_c$) involving all the critical members is shown in Fig. 2.13.13. This matrix plays an important part in the discussion of the stability of the system in the critical state. We treat this problem in Section 2.15 in detail. There, it is shown that the change (v) in the total potential energy of the system from its critical state is given by the quadratic form

$$v = \tfrac{1}{2}\mathbf{e}_c^T \mathbf{E}_c \mathbf{e}_c$$

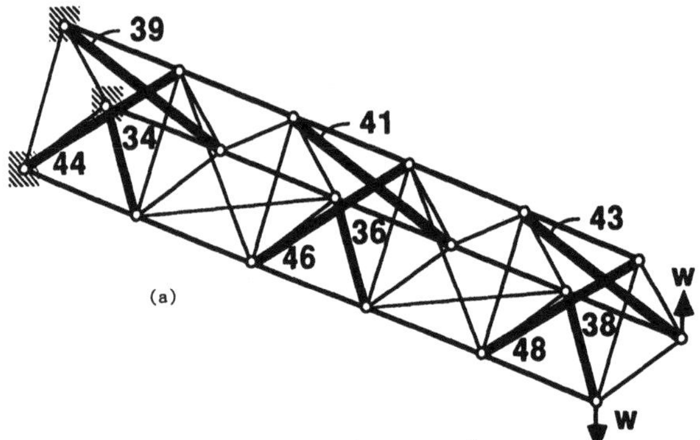

$\mathbf{e}_c^T = [\, e_{34} \;\; e_{36} \;\; e_{38} \;\; e_{39} \;\; e_{41} \;\; e_{43} \;\; e_{44} \;\; e_{46} \;\; e_{48} \,]$

$\mathbf{E}_c =$

e_{34}	e_{36}	e_{38}	e_{39}	e_{41}	e_{43}	e_{44}	e_{46}	e_{48}
0.021856	0	0	0.015561	0	0	0.015561	0	0
.	0.021856	0	0	0.015561	0	0	0.015561	0
.	.	0.021856	0	0	0.015561	0	0	0.015561
.	.	.	0.021856	0	0	0.015561	0	0
.	.	.	.	0.021856	0	0	0.015561	0
.	.	.	.	.	0.021856	0	0	0.015561
.	.	*symmetrical*	.	.	.	0.021856	0	0
.	.	.	.	.	.	.	0.021856	0
.	.	.	.	.	.	.	.	0.021856

(b)

Figure 2.13.13
Case C: The stability of the system. (a) The critical members of the 5-bay space beam under end-torque. (b) The vector $(\mathbf{e}_c)$ of flexural shortenings of the critical members and the stability matrix $(\mathbf{E}_c)$ of the system. The total potential energy change (v) at the critical state is given by $(v = \frac{1}{2}\mathbf{e}_c^T \mathbf{E}_c \mathbf{e}_c)$, where $(\mathbf{E}_c)$ is positive definite in this case. Therefore, the system is stable in the critical state.

where, $(\mathbf{e}_c > \mathbf{0})$, is a positive vector of flexural contractions of *all* the critically stressed members. If this energy change is negative in at least one adjacent configuration defined by $(\mathbf{e}_c > \mathbf{0})$, then the system is *unstable* in the critical state. However, if it can be shown that (v) is *positive* for *all* values of the positive vector $(\mathbf{e}_c)$, then the equilibrium in the critical state is stable and the system does not collapse from that state in an unstable motion. In the Case C, under consideration, it can be verified that this is so, since all the non-zero elements of $(\mathbf{E}_c)$ are positive numbers. Thus, for $(\mathbf{e}_c > \mathbf{0})$ the above quadratic form is always positive in the positive orthant of the coordinate space of $(\mathbf{e}_c)$.

It can be shown that this equilibrium mode is associated with the *lowest* energy levels and that, therefore, it is expected to develop in a practical case.

Case D

The flight mast shown in Fig. 2.13.14 is the same as that in Fig. 2.13.1(c), except that the longitudinal scale has been shortened to display more visibly the distorted form. Now, the entire mast of twenty bays is fixed at one end and

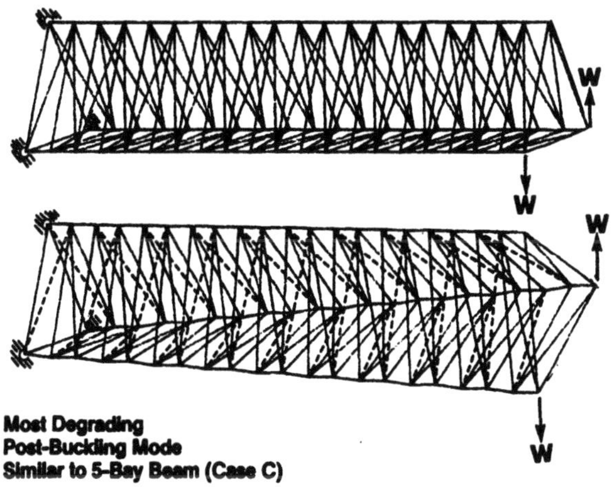

Figure 2.13.14
Case D: The distorted shape of a 20-bay space beam under the end-torque in the most-degrading mode in post-buckling equilibrium is similar to that of the 5-bay beam in Case C. The system is statically stable in the critical state (NASA COFS Program, Ref. 24)

subjected to a counter-clockwise torque at the opposite end by means of two equal and opposite forces (**W**), applied at the two bottom end-joints as shown. The dimensions of each bay of the mast and of the members are the same as given in the table of Fig. 2.13.1c. The critical members in this case are again the diagonal members in alternate bays starting with the bay next to the opposite fixed end in the view of Fig. 2.13.14. The critical state is again reached when ($W = W^0 = 14.45$ lb.), as in Case C. Also, the diagonal members shorten by *equal* amounts in the torsional equilibrium mode which is stable and which is associated with the lowest energy levels, similarly as in Case C.

The end-view of the twisted mast under this torque, which is the same as in Case C, is shown in the computed print in Fig. 2.13.15. The twist is now uniform and not irregular, as in the earlier loading Cases A and B. However, the critical torque sufficient to cripple the 20-bay mast is the same as in Case C i.e. ($T^0 = 57.45$ lb. ft.). This torque is relatively low and it is questionable if it can be tolerated in a practical situation.

It may be shown in this case, as in the previous case, that under a *reversed* torque different diagonal members buckle which are not affected in the Cases C and D. Thus, under the reversed torque diagonal members in the alternate bays buckle, starting with the bay which is *next but one* from the fixed end, so that also the diagonal members in the last bay adjacent to the free end become critically stressed under the reversed critical torque ($T^0 = -57.45$ lb. ft.).

Loading cases, such as these may be very common in practice and it is, therefore, of interest to examine the resistance of the flexible mast in each case and to limit the design tolerances, so that the design is safe from the stability point of

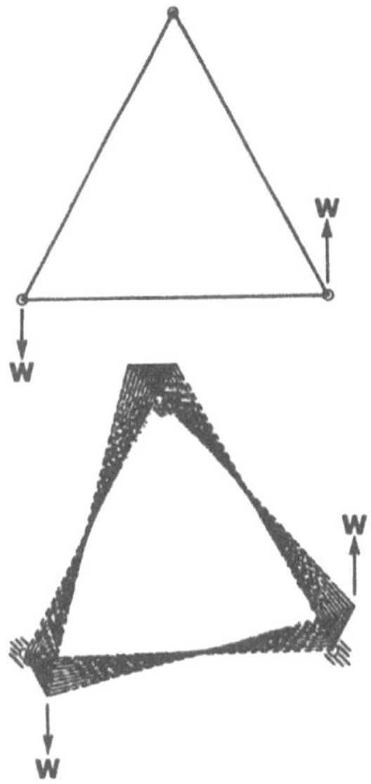

Figure 2.13.15
Case D: The end-view of the buckled 20-bay space beam under end-torque in post-buckling equilibrium (NASA COFS Program, Ref. 24)

view and that it does not result in a catastrophic failure at dangerous loading levels.

2.14 Influence of Initial Imperfections on the Post-Buckling Equilibrium Paths of Pin-Connected Lattices Composed of Flexible Members

Buckling of geometrically imperfect elastic systems and their equilibrium paths were discussed previously in the context of a general theory. This theory is based on the so called geometrical perturbation method described in Ref. 15, Chapter 1, Section 11. There, the notion of an *imperfection parameter* was introduced, by which the totality of small initial, mainly geometrical imperfections may be characterized in a small perturbation of the initial geometry or of the initial geometrical or physical properties of the original perfect system. Experience has confirmed (see also Ref. 15, Chapters 3, 4, 5 and 7) that the so called strong perturbation parameter, which we denote by the symbol (α), does meaningfully reflect the tendencies revealed by the experimental observations of the load-displacement curves of real structures which converge to the equilibrium paths of the equivalent ideal or perfect systems. The value of this parameter is not known and it must be estimated from tests on the prototype or on a scaled-down model in each case, provided that the fabrication and the expected imperfections are "similar" in the model structure to be tested and in the prototype for which this characterization is desired. In fact, several tests may be required to estimate this parameter adequately.

It is possible to introduce into the analysis of imperfection-sensitive structures additional parameters to characterize their response more accurately. Some such attempts have been made, e.g. Ref. 43, but only minor theoretical improvements have been achieved by this means. We confine ourselves here to the simplified analysis of structures composed of flexible members with small initial geometrical imperfections, using a single perturbation parameter (α). Moreover, in this Chapter we are only interested in pin-connected or quasi pin-jointed elastic systems, such as space lattices made of fairly straight flexible members which can buckle under axial loads and bow out in arbitrary direction. This means that if (θ_α^0) represents the end-rotation of some member (α) in the lattice, the flexural shortening, which is proportional to $(\theta_\alpha^0)^2$, of this, or any other member in the lattice, will not depend on the sign of (θ_α^0). We shall assume that this independence also holds for the change in the total potential energy of the perfect system in the coordinate region close to the critical state of the system in which at least one of the compressive members in the lattice is on the point of buckling. We illustrate this approach from first principles to make the theory more understandable to the reader.

Consider, therefore, a system composed of a structure of flexible pin-connected members having small intrinsic imperfections and of a system of external proportionate loads, depending on a parameter (P), which are applied as centrally as possible at its nodes. However, these loads are assumed to remain parallel to one another so that the system remains conservative. Then, in the context of the general theory developed in Ref. 15, the total potential energy (V) of the system, measured from some known datum, is a function of the generalized coordinates (Q_i) of deformation of the system, which are generally (m) in number, and of this load parameter (P). Moreover, if the imperfections, or their effect on the change in this energy from a fundamental (critical) state can be characterized by a single major perturbation parameter (α), then the change (v) in (V) also depends on (α). We denote by (Q_i) the "engineering" or physical generalized coordinates of the system in this case. Their changes from some initial (critical) state are denoted by (q_i). The coordinate change (q_i) may denote the end-rotation (θ_i) of a buckling member denoted by the index[5] (i). There, in the vicinity of the initial state, characterized by $[V = V^0 = V(Q_i^0, P^0)]$, this energy is

$$V = V^0 + v = V(Q_i + q_i, P^0 + p, \alpha) \tag{2.14.1}$$

where (v) is a small but finite change in (V). Here, (q_i) represent some physical coordinate changes or quantities of deformation, such as the end-rotations (θ_i^0) of the buckling members, for instance.

If the change (v) is to be independent of the sign of the coordinate changes (q_i), required by the physical characteristics of a pinned connector, then also (V)

[5] We interchangeably use latin and greek indices in this section such as (i, j, k) to denote members, but not (α), as the perturbation parameter is commonly denoted by this symbol.

must be a symmetrical or even function in (q_i; $i = 1, 2, \ldots, m$). Moreover, if the members are flexible or very slender, the axial elongations or contractions caused by Hooke's law are negligible in the critical state, compared to the subsequent flexural contractions causing the nodes and the applied loads at these nodes to be displaced. If the axial deformations are considered in the prebuckling stage of loading, they can be superimposed and this simply shifts the critical or initial state linearly in the coordinate frame, as we explained in the preceding sections. However, axial extensibility can be definitely neglected in the post-buckling analysis and it also does not affect appreciably the critical level of loading (this effect being less than 1% of the critical value of the load, see Ref. 15, Ch. 2, p. 190.). This means that, initially, in the post-buckling analysis of systems composed of flexible members we can set

$$Q_i^0 = 0 \quad \text{and} \quad V^0 = 0 \tag{2.14.2}$$

and Eq. (1) reduces to

$$v = v(q_i, P^0 + p, \alpha) \tag{2.14.3}$$

If this expression is symmetrical with respect to the coordinate changes (q_i), then, for the ideal or perfect unperturbed system to be symmetrical, in which ($\alpha = 0$), the following condition must hold

$$v(+q_i) = v(-q_i); \qquad i = 1, 2, \ldots, m \tag{2.14.4}$$

where (m) is now the total number of the members. (Note that in an hyperstatic lattice (m) may be greater than (N)). Because of this symmetry, pin-connected systems are sometimes called "symmetrical systems", (see Ref. 15) where, obviously, this symmetry is *not* an optical symmetry of the initial geometry of the structure.

To carry out this analysis, we proceed as in Ref. 15 and in Ref. 18, and expand the total potential energy change (v) in a Taylor series at ($Q_i = Q_i^0 = 0$). But, Eq. (2.14.4) requires that the coefficients of all the terms in this expansion, which depend on odd powers of (q_i) and relate to the perfect system, must vanish on account of this symmetry, so that these terms are struck out from this series expansion. Carrying out this expansion in the usual manner, using the index (tensor) notation in which a comma followed by a subscript, say (i), imples partial differentiation with respect to the coordinate (q_i) or another quantity denoted by the same subscript, e.g. ($V_{,p} = \partial V/\partial p$) or ($V_{,\alpha} = \partial V/\partial \alpha$) or ($V_{,i} = \partial V/\partial q_i$), or ($V_{,ij} = \partial^2 V/\partial q_i \partial q_j$) etc., where repeated indices in the factors in a product imply summation from 1 to m, as in standard tensor notation. We get, similarly as in Eq. (2.11.4), Ch. 1-11 in Ref. 15,

$$\begin{aligned} v = {} & V_{,p}p + V_{,\alpha}\alpha + V_{,i}q_i \\ & + \frac{1}{2!}(V_{,pp}p^2 + V_{,\alpha\alpha}\alpha^2 + V_{,ij}q_iq_j + 2V_{,p\alpha}p\alpha + 2V_{,\alpha i}\alpha q_i + 2V_{,pi}pq_i) \\ & + \frac{1}{3!}(V_{,ppp}p^3 + V_{,\alpha\alpha\alpha}\alpha^3 + 3V_{,pp\alpha}p^2\alpha + 3V_{,\alpha\alpha p}\alpha^2 p + 3V_{,ppi}p^2 q_i \end{aligned}$$

$$+3V_{,pij}pq_iq_j + V_{,ijk}q_iq_jq_k + 3V_{,pii}pq_i^2 + 3V_{,\alpha ij}\alpha q_iq_j + 6V_{,\alpha pi}\alpha pq_i$$

$$+3V_{,\alpha\alpha i}\alpha^2 q_i) + \frac{1}{4!}(V_{,iiii}q_i^4 + 3V_{,iijj}q_i^2q_j^2 + 6V_{,\alpha\alpha ij}\alpha^2 q_iq_j + V_{,pppp}p^4$$

$$+12V_{,p\alpha ij}p\alpha q_iq_j + 6V_{,ppii}p^2q_i^2) + \frac{1}{5!}(\cdots \tag{2.14.5}$$

sum on $i, j = 1, 2, \ldots, m$
no sum on α, p

Here, also all the quartic terms must be considered and these appear only in the even powers of (q_i) if the differentiation does *not* involve the parameter (α). (The quartic terms whose coefficients vanish on account of the symmetry of (v) have been omitted in this last expression. Also, the cubic terms with the coefficients like $(V_{,ijk})$ vanish because of symmetry.) Because of the symmetry of the perfect system the coordinate changes (q_i) appear as complete squares i.e. as (q_i^2) in all terms not involving (α). Let us emphasize that there is no requirement of symmetry in (v) in the perturbed system with respect to the imperfection parameter (α) nor with respect to the terms involving the parameter (α). This means, that a perturbed perfect symmetrical system need not be a symmetrical system with respect to the distribution of geometrical perturbations (imperfections) under a small parameter (α). The expression for (v) reduces to that for the equivalent perfect system upon setting $(\alpha = 0)$.

Next, equilibrium conditions for a generic perturbed system are obtained by partially differentiating Eq. (2.14.5) with respect to any coordinate (q_γ), for $(\gamma = 1, \ldots, m)$ and equating the result of this differentiation to zero. The first thing we notice here is that at $(\alpha = 0)$ and $(q_i = 0; i = 1, 2, \ldots, m)$ also, $(V_{,\gamma})$ must be zero for symmetry and for equilibrium in the initial state, i.e.

$$v_{,\gamma} = V_{,\gamma} = 0; \qquad j = 1, 2, \ldots, m$$

This means that the equivalent perfect system is symmetrical with respect to the coordinates (q_i), and that $(v_{,\gamma})$ vanishes on account of this symmetry as $(V_{,\gamma} = 0; \gamma = 1, \ldots, m)$. This also means that such a perfect symmetrical system is *necessarily* in statical equilibrium in the initial configuration in which $(q_i = 0)$. We also note that, as a consequence of this symmetry and of this initial equilibrium, the largest terms in the expression for (v) for the equivalent perfect system, when $(\alpha = 0)$, are given by the quadratic form

$$V_{,ii}q_i^2; \qquad \text{sum on } i = 1, 2, \ldots, m$$

which determines the sign of (v) in this region. We know from previous discussions (see Ref. 15, Ch. 1, etc.) that the sign of this sum determines the stability of the perfect system in the initial (critical) state. The coefficients $(V_{,ii})$ are called the *stability coefficients* for the system. We shall see later that in a completely symmetrical generic system such as this, all the stability coefficient are zero.

Also, the coefficients $(V_{,pi})$, $(V_{,ppi})$ etc. vanish on account of this symmetry. We observe that, on setting $(\alpha = 0)$, $(q_i = 0)$ for all (i), the expression for (v) becomes

$$v = pV_{,p} + \frac{1}{2!}p^2V_{,pp} + \frac{1}{3!}p^3V_{,ppp} + \cdots \tag{2.14.6}$$

This expression represents the total potential energy change in the fundamental (initial) state due to a change in the load parameter (p). As this change is zero, if (p) increases at no deformations, the coefficients $(V_{,p}, V_{,pp})$ etc. vanish in this state. A state or point in which these conditions hold, is called a *multiple branching point*. (For more details of this theory, the reader is referred to Ref. 15, Ch. 1, Sections 1-10, 1-11 and 1-12.) Thus, in the equilibrium states of the slightly imperfect or perturbed system in the vicinity of the initial (critical) state of the equivalent perfect system, we get on differentiating Eq. (2.14.5), partially with respect to a coordinate (q_γ),

$$\begin{aligned} v_{,\gamma} = {} & V_{,\gamma} + V_{,\alpha\gamma}\alpha + V_{,\gamma\gamma}q_\gamma + V_{,p\gamma\gamma}pq_\gamma + V_{,\alpha\gamma i}\alpha q_i \\ & + \frac{1}{2!}V_{,\alpha\alpha\gamma}\alpha^2 + V_{,\alpha p\gamma}p\alpha + \frac{1}{3!}(V_{,\gamma\gamma\gamma\gamma}q_\gamma^3 + 3V_{,\gamma\gamma ii}q_\gamma q_i^2 + 3V_{,\alpha\alpha\gamma i}\alpha^2 q_i \\ & + 3V_{,pp\gamma\gamma}p^2q_\gamma + 6V_{,\alpha p\gamma i}\alpha p q_i) + \cdots = 0; \end{aligned} \tag{2.14.7}$$

$\gamma = 1, 2, \ldots, m$
sum on $i = 1, \ldots, m$
no sum on α, p

Then, if the symmetry and initial equilibrium conditions are considered, these simplified equilibrium equations hold for a generic symmetrical system with nonsymmetrical small perturbation or initial imperfection characteristics, characterized by the perturbation parameter (α). We get from (2.14.7) the following (m) simplified equations,

$$\begin{aligned} v_{,\gamma} = {} & V_{,\gamma\gamma}q_\gamma + V_{,\alpha\gamma}\alpha + V_{,p\gamma\gamma}pq_\gamma + V_{,\alpha\gamma i}\alpha q_i + \frac{1}{2!}V_{,\alpha\alpha\gamma}\alpha^2 + V_{,\alpha p\gamma}p\alpha \\ & + \frac{1}{3!}(V_{,\gamma\gamma\gamma\gamma}q_\gamma^3 + 3V_{,\gamma\gamma ii}q_\gamma q_i^2 + 3V_{,\alpha\alpha\gamma i}\alpha^2 q_i \\ & + 3V_{,pp\gamma\gamma}p^2q_\gamma + 6V_{,\alpha p\gamma i}\alpha p q_i) + \cdots = 0; \end{aligned} \tag{2.14.7a}$$

$\gamma = 1, 2, \ldots, m$
sum on $i = 1, 2, \ldots, m$
no sum on α, p

Perfect Systems as a Special Case

The equilibrium equations (2.14.7a) must also hold in the special case when $(\alpha = 0)$ and the generic system is perfect. In that case one finds,

$$V_{,\gamma\gamma}q_\gamma + V_{,p\gamma\gamma}pq_\gamma + \frac{1}{3!}(V_{,\gamma\gamma\gamma\gamma}q_\gamma^3 + 3V_{,\gamma\gamma ii}q_\gamma q_i^2 + 3V_{,pp\gamma\gamma}p^2q_\gamma) + \cdots = 0;$$

$\gamma = 1, 2, \ldots, m$ (2.14.8)
sum on $i = 1, 2, \ldots, m$

Now, (q_γ) is a factor and we can write

$$q_\gamma[V_{,\gamma\gamma} + V_{,p\gamma\gamma}p + \frac{1}{3!}(V_{,\gamma\gamma\gamma\gamma}q_\gamma^2 + 3V_{,\gamma\gamma ii}q_i^2 + 3V_{,pp\gamma\gamma}p^2) + \cdots] = 0;$$

$$\gamma = 1, 2, \ldots, m \quad \text{sum on } i = 1, 2, \ldots, m \qquad (2.14.8a)$$

This admits the trivial equilibrium solution in which the coordinates of deformation and therefore, the distortions of the equivalent *perfect* system are identically zero, i.e. the load-parameter (P) axis is an equilibrium path, for

$$q_\gamma \equiv 0; \qquad \gamma = 1, 2, \ldots, m \qquad (2.14.8b)$$

The nontrivial equilibrium paths of the perfect system are given by the remaining (m) equations

$$V_{,\gamma\gamma} + V_{,p\gamma\gamma}p + \frac{1}{3!}(V_{,\gamma\gamma\gamma\gamma}q_\gamma^2 + 3V_{,\gamma\gamma ii}q_i^2 + 3V_{,pp\gamma\gamma}p^2) + \cdots = 0;$$

$$\gamma = 1, 2, 3, \ldots, m \quad \text{sum on } i = 1, 2, 3, \ldots, m \qquad (2.14.8c)$$

In particular, these equations must hold in the critical state or point in which $(p = 0)$ and $(q_i = 0; i = 1, 2, \ldots, m)$. But, this requires that also the second partial derivatives of the total potential energy function in the critical state vanish, i.e.

$$V_{,\gamma\gamma} = 0; \qquad \gamma = 1, 2, \ldots, m \qquad (2.14.9)$$

Therefore, all the *stability coefficients* $(V_{,jj}, j = 1, 2, \ldots, m)$, for the perfect symmetrical system vanish in the critical state, as mentioned earlier. This means that, in the context of the general stability theory expounded in Ref. 15, all the coordinates (q_i) are *critical*. These critical coordinates correspond to the end-rotations, which we denoted earlier by (θ_j^0), of the critically stressed members within the lattice, which ultimately govern the post-buckling mechanisms, which can develop from this state. By virtue of this last result, the *equilibrium conditions* for a generic imperfect symmetrical system, with a nonsymmetrical distribution of imperfections, characterized by a single major parameter (α), simplify further to

$$V_{,\alpha\gamma}\alpha + V_{,p\gamma\gamma}pq_\gamma + V_{,\alpha\gamma i}\alpha q_i + \frac{1}{2}V_{,\alpha\alpha\gamma}\alpha^2 + V_{,\alpha p\gamma}p\alpha + \frac{1}{3!}(V_{,\gamma\gamma\gamma\gamma}q_\gamma^3$$
$$+ 3V_{,\gamma\gamma ii}q_\gamma q_i^2 + 3V_{,\alpha\alpha\gamma i}a^2 q_i + 3V_{,pp\gamma\gamma}p^2 q_\gamma + 6V_{,\alpha p\gamma i}\alpha p q_i) + \cdots = 0;$$

$$\gamma = 1, 2, \ldots, m \quad \text{sum on } i = 1, 2, \ldots, m \qquad (2.14.10)$$

Notice that the coefficients $(V_{,\alpha\gamma})$, $(V_{,\alpha\alpha\gamma})$ and $(V_{,\alpha p\gamma})$ etc. do *not* belong to the original perfect symmetrical system, but to the *perturbed* imperfect system and are, therefore, nonzero. We first solve these equations for the case $(\alpha = 0)$, i.e. when the generic symmetrical system is perfect. The terms involving (p^2) are small compared to the terms involving (p) and the (p^2) terms depend on the order of (q_i^4). We neglect those terms in a first approximation and obtain the simplified nontrivial equations after dividing by (q_γ) i.e.

$$V_{,p\gamma\gamma}p+\frac{1}{3!}(V_{,\gamma\gamma\gamma\gamma}q_\gamma^2+3V_{,\gamma\gamma ii}q_i^2)+\cdots=0; \quad \gamma=1,2,\ldots,m \quad \text{sum on } i=1,2,\ldots,m \tag{2.14.11}$$

These equations are linear in the coordinates $(q_\gamma^2; i=1,2,\ldots,m)$ and their solution has the general form

$$q_i^2=\frac{1}{C_i}p; \qquad i=1,2,\ldots,m \tag{2.14.12}$$

where the constants (C_i) can be computed, if the partial derivatives of the total potential energy function (V) i.e. $(V_{,p\gamma\gamma})$, $(V_{,\gamma\gamma\gamma\gamma})$ and $(V_{,\gamma\gamma ii})$ are known numerically. *Alternatively*, the coefficients (C_i) can be computed using the theory expounded in Chapter 1, Section 1.2, or using the solutions of the Eqs. 2.12.15 in Chapter 2, in which $(\delta w\equiv p)$. The coordinates (q_i) represent the end rotations of the buckling members and these are directly related to the flexural contractions (e_γ) by the relations $(e_\gamma/L_\gamma=q_\gamma^2/4)$, Ref. 15, Ch. 2. The coefficients (C_i) must be all positive or negative, since (p) is of one sign in the case of a statically admissible buckling mode.

In the relation (2.14.12) (p) is treated as the independent variable. If, on the other hand, one of the generalized coordinates of deformation e.g. (q_i) or (q_r) is chosen as the independent or reference variable, then, putting $(p=p_{\text{pf.}})$ for the perfect system, we get

$$p_{\text{pf.}}=C_rq_r^2 \tag{2.14.13}$$

where (C_r) is a positive *or* a negative constant contained in the set of $(C_i; i=1,\ldots,m=N)$, (N) being the number of members in an isostatic system in this case. The remaining coordinates are calculated from

$$q_1^2=\frac{C_r}{C_i}q_r^2=b_iq_r^2; \qquad i=1,2,\ldots,m \quad i\neq r \tag{2.14.14}$$

or

$$\frac{q_i}{q_r}=\sqrt{\frac{C_r}{C_i}}=\sqrt{b_i}$$

where

$$b_i=\frac{C_r}{C_i}>0 \tag{2.14.14a}$$

Clearly, (b_i) are always positive constants, for the solution to be real. Writing, $(p=P-P^0)$, where (P) is the total value of the load parameter, the equilibrium paths of an equivalent perfect system may be now rewritten as

$$P=P^0+C_iq_i^2 \tag{2.14.15}$$

where the constants $(C_i; i=1,2,\ldots,m)$ are all positive or negative i.e. of one sign, for the mode to be statically admissible. The forms of these equilibrium paths are shown in Fig 2.14.1. It is obvious that in the first approximation these

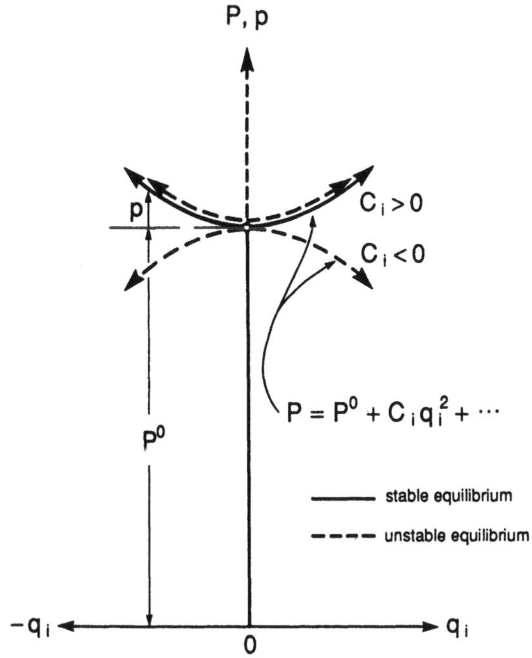

Figure 2.14.1
Typical post-buckling equilibrium paths of perfect completely symmetrical pin-jointed systems.—Plot of the load-parameter (P) against a deformation coordinate (q_i) representing the end-tangent rotation of a buckling member

paths are parabolic. The rising equilibrium paths may be statically stable or unstable (unstable paths are shown dashed). The falling paths are always statically unstable. The load parameter axis itself constitutes an equilibrium path, which loses its stability in the critical state in which ($P = P^0$).

Imperfect or Perturbed Systems

In order to find an approximate solution of the equilibrium equations (2.14.10) for the perturbed imperfect system, we neglect in the first approximation the higher order terms involving ($p^2 q_\gamma$), ($\alpha^2 q_\gamma$), and ($\alpha p q_\gamma$). Thus

$$\begin{aligned} &V_{,\alpha\gamma}\alpha + V_{,p\gamma\gamma}pq_\gamma + V_{,\alpha p\gamma}p\alpha + V_{,\alpha\gamma i}\alpha q_i + \frac{1}{2}V_{,\alpha\alpha\gamma}\alpha^2 \\ &+ \frac{1}{3!}(V_{,\gamma\gamma\gamma\gamma}q_\gamma^3 + 3V_{,\gamma\gamma ii}q_\gamma q_i^2) \\ &\underbrace{+\frac{1}{2}V_{,\alpha\alpha\gamma i}\alpha^2 q_i + \frac{1}{2}V_{,pp\gamma\gamma}p^2 q_\gamma + V_{,\alpha p\gamma i}\alpha p q_i + \cdots}_{\text{higher order terms}} = 0; \end{aligned}$$

$$\gamma = 1, 2, \ldots, m$$
$$\text{sum on } i = 1, 2, \ldots, m \qquad (2.14.16)$$

where ($p = P - P^0$) stands for the change in the load parameter of the perturbed or imperfect system, measured from the critical value of this parameter in the critical state of the equivalent perfect system. Then, eliminating the load parameter change (p) for the imperfect system, we get

$$p_{\text{impf.}}(V_{,p\gamma\gamma}q_\gamma + V_{,\alpha p\gamma}\alpha) = -V_{,\alpha\gamma}\alpha - V_{,\alpha\gamma i}\alpha q_i - \frac{1}{2}V_{,\alpha\alpha\gamma}\alpha^2 \underbrace{-\frac{1}{3!}(V_{,\gamma\gamma\gamma\gamma}q_\gamma^2 + 3V_{,\gamma\gamma ii}q_i^2)q_\gamma + \cdots}_{= -V_{,p\gamma\gamma}p_{\text{pf.}}} \tag{2.14.17}$$

where, by definition, ($p_{\text{impf.}} = P_{\text{impf.}} - P^0$) and ($P^0$) is the critical value of the load parameter (P) of the equivalent perfect system.

We recognize that the last term on the right side represents the load parameter change ($p_{\text{pf.}}$) of the equivalent *perfect* system given by Eqs. (2.14.11) and (2.14.13) when ($\alpha = 0$). Using Eq. (2.14.11) and writing in the last result for ($p = p_{\text{impf.}}$), to distinguish the load parameter change for the perturbed imperfect system from ($p_{\text{pf.}}$) for the perfect system, this result becomes

$$p_{\text{impf.}}(V_{,p\gamma\gamma}q_\gamma + V_{,\alpha p\gamma}\alpha) = -V_{,\alpha\gamma}\alpha - V_{,\alpha\gamma i}\alpha q_i - \tfrac{1}{2}V_{,\alpha\alpha\gamma}\alpha^2 + V_{,p\gamma\gamma}q_\gamma p_{\text{pf.}} + \cdots; \quad \gamma = 1, 2, \ldots, m \tag{2.14.18}$$

If ($p_{\text{impf.}}$) is eliminated from anyone of these *uncoupled* equations, we get the dependence of ($p_{\text{impf.}}$) on only one coordinate change (q_γ) i.e.

$$p_{\text{impf.}} = -\frac{V_{,\alpha\gamma}\alpha + V_{,\alpha\gamma i}\alpha q_i + \frac{1}{2}V_{,\alpha\alpha\gamma}\alpha^2}{V_{,p\gamma\gamma}q_\gamma + V_{,\alpha p\gamma}\alpha} + \frac{V_{,p\gamma\gamma}q_\gamma}{V_{,p\gamma\gamma}q_\gamma + V_{,\alpha p\gamma}\alpha}p_{\text{pf.}}; \quad \gamma = 1, 2, \ldots, m \tag{2.14.19}$$

Here, the coordinates (q_i) may be approximately expressed by the coordinate (q_γ) using the initial solution of the equivalent perfect system (2.14.13) viz. (2.14.14). Writing

$$V_{,\alpha\gamma i}q_i = V_{,\alpha\gamma i}\sqrt{\frac{C_\gamma}{C_i}}\,q_\gamma = V'_{,\alpha\gamma\gamma}q_\gamma; \qquad \text{sum on } i = 1, 2, \ldots, m \tag{2.14.19a}$$

where

$$V'_{,\alpha\gamma\gamma} = V_{,\alpha\gamma i}\sqrt{\frac{C_\gamma}{C_i}}; \qquad \text{sum on } i = 1, 2, 3, \ldots, m$$

the expression in Eq. (2.14.19) may be simplified by this identity. From this expression we see that the change in the load parameter for the perturbed system is in a sense governed by the parameter change of the equivalent perfect system. If the coordinate (q_γ) becomes relatively large compared to the perturbation or imperfection parameter (α), then

$$\lim p_{\text{impf.}} = -\frac{V'_{,\alpha\gamma\gamma}}{V_{,p\gamma\gamma}}\alpha + p_{\text{pf.}} \qquad \text{for } q_\gamma \text{ large} \tag{2.14.20}$$

and the *gradients* of the two paths are approximately the same regardless of the choice of the coordinate of deformation (q_γ). The difference ($p_{\text{pf.}} - p_{\text{impf.}}$) depends on the first term which is proportional to (α), so that the two paths are always distinct. Since the imperfect path has obviously a solution at ($P_{\text{impf.}} = 0$), we see that it must also have a maximum or peak value, if the gradient of the perfect path is negative i.e. if the perfect system is unstable in the critical state. We see also that now (q_γ) is not a factor in Eq. (2.14.18) and, therefore, the imperfect paths do not coincide with the load-parameter axis at any point, as this was the case for the generic perfect systems.

To find the stationary values of the imperfect paths we differentiate Eq. (2.14.18) and (2.14.19a) with respect to any coordinate of deformation (q_γ; $\gamma = 1, 2, \ldots, m$) and equate the result of this differentiation to zero. This gives

$$\frac{dp_{\text{impf.}}}{dq_\gamma}(V_{,p\gamma\gamma}q_\gamma + V_{,\alpha p\gamma}\alpha) + p_{\text{impf.}}V_{,p\gamma\gamma}$$

$$= -V'_{,\alpha\gamma\gamma}\alpha + V_{,p\gamma\gamma}p_{\text{pf.}} + V_{,p\gamma\gamma}q_\gamma\frac{dp_{\text{pf.}}}{dq_\gamma} \qquad (2.14.21)$$

Putting

$$\frac{dp_{\text{impf.}}}{dq_\gamma} = 0; \qquad \gamma = 1, 2, \ldots, m$$

one obtains the condition for the stationary value of the equilibrium path of the slightly perturbed system i.e.,

$$V_{,p\gamma\gamma}p_{\text{impf.}} = -V'_{,\alpha\gamma\gamma}\alpha + V_{,p\gamma\gamma}q_\gamma\frac{dp_{\text{pf.}}}{dq_\gamma} + V_{,p\gamma\gamma}p_{\text{pf.}} \qquad (2.14.22)$$

But, from (2.14.12) we deduce that ($dp_{\text{pf.}}/dq_\gamma = 2C_\gamma q_\gamma$), where now ($q_\gamma$) is any one coordinate of deformation. Substitution for ($p_{\text{pf.}}$) from Eq. (2.14.13) into the relation (2.14.22), defines the coordinates ($q_\gamma = q_\gamma^*$) and ($p_{\text{impf.}} = p^*_{\text{impf.}}$) of the *locus* of the stationary values of the equilibrium paths of the perturbed system for different values of (α), i.e.

$$p^*_{\text{impf.}} = -\frac{V'_{,\alpha\gamma\gamma}}{V_{,p\gamma\gamma}}\alpha + 3C_\gamma q_\gamma^{*2} = -\frac{V'_{,\alpha\gamma\gamma}}{V_{,p\gamma\gamma}}\alpha + 3p^*_{\text{pf.}} \qquad (2.14.23)$$

This locus is a parabola having a curvature ($6C_\gamma$) in the (p^*, q_γ^*) coordinate plane and a stationary value at the origin ($p^* = q_\gamma^* = 0$). Differentiating the last equation with respect (q_γ^*) yields

$$\frac{d^2p^*_{\text{impf.}}}{dq_\gamma^{*2}} = 6C_\gamma = 3\frac{d^2p^*_{\text{pf.}}}{dq_\gamma^{*2}}: \qquad \gamma = 1, 2, \ldots, m \qquad (2.14.24)$$

where ($p^*_{\text{pf.}} = C_\gamma q_\gamma^{*2}$) is the plot of the perfect path in the same coordinate plane. Since ($P^*_{\text{impf.}} = P^0 + p^*_{\text{impf.}}$) represents the ordinate of this locus measured from the ($q_\gamma = q_\gamma^*$) coordinate axes at ($P = P_{\text{impf.}} = 0$), then, if ($C_\gamma < 0$), the locus of the stationary values of the imperfect paths is a decreasing function of the coordinates ($q_\gamma = q_\gamma^*$), its rate of decrease being 3 *times* the negative gradient of the unstable perfect path. This suggests, that a branch of the *imperfect* paths is an initially rising function of the coordinates of deformation (q_γ) that has a *maximum below* the loading levels of the perfect path. This locus, therefore, represents a *stability boundary* where the imperfect equilibrium paths change their statical stability from *stable* to *unstable* and vice versa. The curvature of this stability boundary equals *three times* the curvature of the perfect path of the equivalent perfect system in the same mode of elastic deformation.

Alternatively, the proof of a maximum of the imperfect path for ($C_\gamma < 0$) may be accomplished on differentiating Eq. (2.14.21) with respect to (q_γ) and evaluating the second derivative at the point of its stationary value on the path on setting

$$\frac{dp_{\text{impf.}}}{dq_\gamma} = 0$$

Thus

$$\frac{d^2p_{\text{impf.}}}{dq_\gamma^2}(V_{,p\gamma\gamma}q_\gamma + V_{,\alpha p\gamma}\alpha) + \frac{dp_{\text{impf.}}}{dq_\gamma}V_{,p\gamma\gamma} + \frac{dp_{\text{impf.}}}{dq_\gamma}V_{,p\gamma\gamma}$$
$$= V_{,p\gamma\gamma}\frac{dp_{\text{pf.}}}{dq_\gamma} + V_{,p\gamma\gamma}\frac{dp_{\text{pf.}}}{dq_\gamma} + V_{,p\gamma\gamma}q_\gamma\frac{d^2p_{\text{pf.}}}{dq_\gamma^2}$$

At the stationary point where ($q_\gamma = q_\gamma^*$) and ($dp_{\text{impf.}}/dq_\gamma = 0$),

$$\frac{d^2p_{\text{impf.}}}{dq_\gamma^2}(V_{,p\gamma\gamma}q_\gamma^* + V_{,\alpha p\gamma}\alpha) = 2V_{,p\gamma\gamma}\frac{dp_{\text{pf.}}}{dq_\gamma} + V_{,p\gamma\gamma}q_\gamma^*\frac{d^2p_{\text{pf.}}}{dq_\gamma^2}$$

Substituting for ($dp_{\text{pf.}}/dq_\gamma$) from (2.14.13), gives on differentiating

$$\frac{d^2p_{\text{impf.}}}{dq_\gamma^2}(V_{,p\gamma\gamma}q_\gamma^* + V_{,\alpha p\gamma}\alpha) = 6V_{,p\gamma\gamma}C_\gamma q_\gamma^*$$

or

$$\frac{d^2p_{\text{impf.}}}{dq_\gamma^2} = \frac{6V_{,p\gamma\gamma}C_\gamma q_\gamma^*}{V_{,p\gamma\gamma}q_\gamma^* + V_{,\alpha p\gamma}\alpha}$$

which may be written

$$\frac{d^2p_{\text{impf.}}}{dq_\gamma^2} = 6C_\gamma\frac{1}{1 + \dfrac{V_{,\alpha p\gamma}}{V_{,p\gamma\gamma}}\dfrac{\alpha}{q_\gamma^*}}$$

It is shown in Eq. (2.14.28), that at the stationary point ($q_\gamma = q_\gamma^*$), approximately

$$q_\gamma^* = \left(-\frac{V_{,\alpha\gamma}}{2C_\gamma V_{,p\gamma\gamma}}\right)^{1/3}\alpha^{1/3} \tag{2.14.28}$$

Thus

$$\frac{\alpha}{q_\gamma^*} = \left(-\frac{V_{,\alpha\gamma}}{2C_\gamma V_{,p\gamma\gamma}}\right)^{-1/3}\alpha^{2/3}$$

and the second derivative

$$\frac{d^2p_{\text{impf.}}}{dq_\gamma^2} = 6C_\gamma \frac{1}{1 + \dfrac{V_{,\alpha p\gamma}}{V_{,p\gamma\gamma}}\left(-\dfrac{V_{,\alpha\gamma}}{2C_\gamma V_{,p\gamma\gamma}}\right)^{-1/3}\alpha^{2/3}}$$

$$= 6C_\gamma \frac{1}{1 + \left(-\dfrac{2C_\gamma V^3_{,\alpha p\gamma}}{V^2_{,p\gamma\gamma}V_{,\alpha\gamma}}\right)^{1/3}\alpha^{2/3}}$$

Since the last term in the denominator is usually small compared to unity the sign of the second derivative in the last equation is determined by the sign of the coefficient (C_γ). Then, for ($C_\gamma < 0$) the path has a maximum at the stationary point. On the other hand, if ($C_\gamma > 0$) the perfect path is a rising path and the branch of the imperfect path may have a minimum at the stationary point, this branch of the imperfect path approaching the perfect parabolic rising path from *above* the critical point, similarly as shown for the unstable case in Fig. 2.14.2. However, the descending imperfect path in the case of a rising perfect equilibrium parabolic path is of little practical interest, as in that case another *rising* imperfect path (branch) exists which develops from the unloaded state of the system in which ($p = P - P^0 = -P^0$) or ($P = 0$).

Therefore, the locus given by Eq. (2.14.23), defines a *stability boundary* within a band-width proportional to $[(V'_{,\alpha\gamma\gamma}/V_{,p\gamma\gamma})\alpha]$ on which all the rising imperfect paths of the perturbed unstable system experience a loss of statical stability, regardless of the magnitude of (α), since beyond that locus these *paths decrease*. Eq. (2.14.23) may be written in a more conspicuous form on recognizing from (2.14.13) that

$$C_\gamma q_\gamma^{*2} = p^*_{\text{pf.}}$$

which is the value of the load parameter change of the perfect system at the coordinate value ($q_\gamma = q_\gamma^*$), as

$$p^*_{\text{impf.}} = -\frac{V'_{,\alpha\gamma\gamma}}{V_{,p\gamma\gamma}}\alpha + 3p^*_{\text{pf.}} \tag{2.14.25}$$

In words, the gradient of the locus of points representing the peak values ($P^*_{\text{impf.}}$) of the "imperfect" paths at the coordinate change ($q_\gamma = q_\gamma^*$) is proportional to approximately 3 times the gradient of the path ($P^*_{\text{pf.}}$) of the equivalent "perfect" system in the same mode of deformation and at the same coordinate value (q_γ^*). Substituting this into Eq. (2.14.19) for ($p_{\text{impf.}}$), one gets,

$$-\frac{V'_{,\alpha\gamma\gamma}}{V_{,p\gamma\gamma}}\alpha + 3p^*_{\text{pf.}} = -\frac{V_{,\alpha\gamma}\alpha + V'_{,\alpha\gamma\gamma}\alpha q_\gamma^* + \frac{1}{2}V_{,\alpha\alpha\gamma}\alpha^2}{V_{,p\gamma\gamma}q_\gamma^* + V_{,\alpha p\gamma}\alpha}$$

$$+ p^*_{\text{pf.}}\frac{V_{,p\gamma\gamma}}{V_{,p\gamma\gamma}q_\gamma^* + V_{,\alpha p\gamma}\alpha}q_\gamma^*$$

After some rearrangement of the terms, this yields

$$V_{,\alpha\gamma}\alpha + \left(\frac{1}{2}V_{,\alpha\alpha\gamma} - \frac{V'_{,\alpha\gamma\gamma}}{V_{,p\gamma\gamma}}V_{,\alpha p\gamma}\right)\alpha^2 + 2C_\gamma V_{,p\gamma\gamma}q_\gamma^{*3} + 3C_\gamma V_{,\alpha p\gamma}\alpha q_\gamma^{*2} = 0 \tag{2.14.26}$$

This is a cubic equation in (q_γ^*). However, the first approximate solution may be found on retaining the largest *independent terms* i.e.

$$V_{,\alpha\gamma}\alpha + 2C_\gamma V_{,p\gamma\gamma}q_\gamma^{*3} = 0; \qquad \gamma = 1, 2, \ldots, m \tag{2.14.27}$$

(Notice that now the coefficient $(V'_{,\alpha\gamma\gamma})$ does not appear in these equations.) These equations are uncoupled and yield the value of (q_γ^*) i.e.

$$q_\gamma^* = \left(-\frac{V_{,\alpha\gamma}}{2C_\gamma V_{,p\gamma\gamma}}\right)^{1/3}\alpha^{1/3} \tag{2.14.28}$$

Thus, in the vicinity of the stationary point this solution requires that the coordinates (q_γ^*) are fairly large in relation to (α). Substituting this result in (14.23) gives

$$p_{\text{impf.}}^* = -\frac{V'_{,\alpha\gamma\gamma}}{V_{,p\gamma\gamma}}\alpha + 3C_\gamma\left(-\frac{V_{,\alpha\gamma}}{2C_\gamma V_{,p\gamma\gamma}}\right)^{2/3}\alpha^{2/3} \tag{2.14.29}$$

To the first order of small quantities, this expression reduces to

$$p_{\text{impf.}}^* = 3C_\gamma\left(-\frac{V_{,\alpha\gamma}}{2C_\gamma V_{,p\gamma\gamma}}\right)^{2/3}\alpha^{2/3} + \text{higher order terms in } \alpha \tag{2.14.30}$$

or, by (2.14.28), approximately,

$$p_{\text{impf.}}^* = 3C_\gamma q_\gamma^{*2} = 3p_{\text{pf.}}^* \tag{2.14.31}$$

when $(q_\gamma = q_\gamma^*)$.

Now, $(p_{\text{impf.}}^*)$ denotes the equation of the locus of the stability boundary in the coordinate plane (p, q_γ). We have seen that the flexural shortening (λ_γ)† of a buckling pin-jointed member (γ) is directly proportional to the end-tangent rotation (α_γ^2), which is identical to (q_γ^2) in this case. This means that any nodal displacement (s_i) at some joint (i) of a pin-connected lattice is proportional to (q_r^2), where (q_r) is a reference coordinate of deformation for this system in a particular mode of post-buckling deformation. If, instead of the coordinate (q_γ^2) or (q_r^2) a reference nodal displacement (s_r) is used to plot an equilibrium path of a perfect and the slightly perturbed imperfect system, then we conclude from Eq. (2.14.31) that the gradient $(dp_{\text{impf.}}/ds_r)_{s_r = s_r^*}$ equals 3 times the gradient of the perfect equilibrium path in the same post-buckling mode of deformation represented in the same coordinate plane (p, s_r), for then it follows directly from Eq. (2.14.31) that

$$\left.\frac{dp_{\text{impf.}}}{ds_r}\right|_{s_r = s_r^*} = 3\frac{dp_{\text{pf.}}}{ds_r} \tag{2.14.32}$$

This statement is also true at the critical state itself, where (q_r) and (q_r^*) are zero.

† Ref. 15, Ch. 2.

More accurately, if the α-term in (2.14.29) is retained,

$$p^*_{\text{impf.}} = -\frac{V'_{,\alpha\gamma\gamma}}{V_{,p\gamma\gamma}}\alpha + 3p^*_{\text{pf.}} \tag{2.14.33}$$

where the coefficient $(-V'_{,\alpha\gamma\gamma}/V_{,p\gamma\gamma})$, as well as the parameter (α), may be positive or negative. Thus, for relatively large values of the reference coordinate of deformation compared with the perturbation parameter (α), the locus of the stability boundary represents a narrow *band* whose width is proportional to the first term in Eq. (2.14.33) which is of the order of magnitude of the perturbation parameter (α). Thus, if $(p_{\text{pf.}})$ follows a parabolic curve the stability boundary is a parabola of band-width proportional to (α) and of 3 times the curvature of this path. If the perfect path is a line in a coordinate plane, the stability boundary is also a line of 3 times the gradient of the perfect path in the same plane, having a band-width equal to $[(-V'_{,\alpha\gamma\gamma}/V_{,p\gamma\gamma})\alpha]$. These results are summarized in Figs. 2.14.2 and 2.14.3.

We now return to Eq. (2.14.19) to find the simplified or the first approximate mathematical solution for the parameter change $(p_{\text{impf.}})$, which is accurate in the vicinity of the stationary point of the path of the perturbed system, i.e. near the stability boundary of this system for a particular mode of deformation. The domain of the stability boundary falls into the most sensitive region of an unstable perfect system where slight perturbations of the original system, characterized by the perturbation parameter (α), are the most sensitive or critical and where these perturbations or imperfections can cause a catastrophic collapse of the system, by causing the perturbed system to lose its statical stability in a sudden onset of unstable motion under the influence of intrinsic imperfections. We see from Eq. (2.14.19) that approximately, *in the vicinity of the stationary* point $(q_\gamma = q^*_\gamma)$, which is proportional to $(\alpha^{1/3})$, as $(\alpha^{1/3} \gg \alpha)$ for small (α).

$$p_{\text{impf.}} = -\frac{V_{,\alpha\gamma}}{V_{,p\gamma\gamma}}\frac{\alpha}{q_\gamma} + p_{\text{pf.}} + \text{higher order terms};$$

$$\gamma = 1, 2, 3, \ldots, m \tag{2.14.34}$$

This result is very simple and it holds with respect to any coordinate of deformation (q_γ). The constant

$$-\frac{V_{,\alpha\gamma}}{V_{,p\gamma\gamma}} = A_{\alpha p\gamma} \tag{2.14.34a}$$

depends on partial derivatives of the total potential energy of the system with respect to the coordinate (q_γ), the imperfection parameter (α) and the loading parameter $(P = P^0)$ applied to the original system. This constant has to be determined in each case experimentally from a test, since its theoretical evaluation is at best difficult. We write in a simplified way $(A_{\alpha p\gamma} = A)$, bearing in mind the dependence of (A) on the above coordinates and parameters and Eq. (2.14.34) becomes more simply,

$$p_{\text{impf.}} = \frac{A\alpha}{q_\gamma} + p_{\text{pf.}} \tag{2.14.35}$$

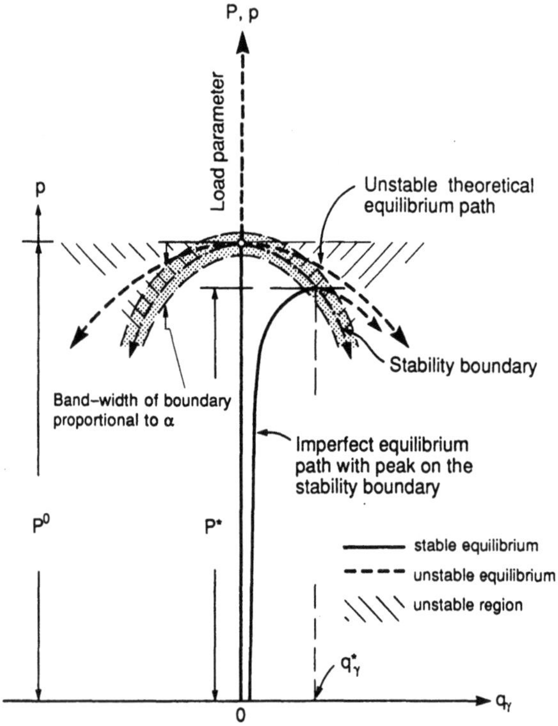

Figure 2.14.2
A typical rising-falling equilibrium path of an unstable imperfect pin-jointed system with its stability boundary.—Plot of the load-parameter (P) against a typical coordinate of deformation (q_γ) representing the end-tangent rotation of a buckling member

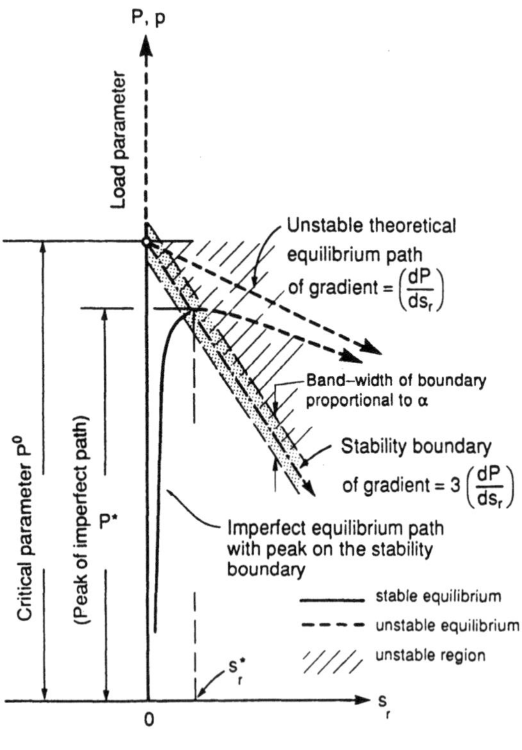

Figure 2.14.3
A typical rising-falling equilibrium path of an unstable imperfect pin-jointed system with its stability boundary.—Plot of the load-parameter (P) against a typical reference joint-displacement (s_r). The equivalent perfect system is said to be completely symmetrical in this case (see also Ref. 15)

Suppose that now we change the variables and introduce a nodal reference displacement ($s = s_r$) as the coordinate to replace (q_γ). As already mentioned, it can be shown that in a kinematically and statically admissible mode of buckling deformation (see Ref. 15, Chapters 2 and 3) [$q_\gamma = 2(e_\gamma / L_\gamma)^{1/2}$], where ($e_\gamma$) is the flexural shortening of the same member, and (L_γ) its initial length. Any nodal reference displacement (s_r) is in the first approximation linearly proportional to the flexural shortenings of the buckling members which, in a system, are in turn proportional to a reference shortening (e_r). So, (q_γ) in Eq. (2.14.35) can be replaced by ($\sqrt{s_r}$), and a constant of proportionality. This means if ($\sqrt{s_r}$) is used instead of (q_γ) in (2.14.35), the constant (A) is modified and replaced by a constant (B). Then, under the same perturbation parameter (α), the equation (2.14.35) for the imperfect path in the vicinity of the stationary point is modified to

$$p_{\text{impf.}} = \frac{B\alpha}{\sqrt{s_r}} + p_{\text{pf.}} \tag{2.14.36}$$

This equation represents an equilibrium path which has a maximum in the (p, s_r) coordinate plane, when

$$\frac{dp_{\text{impf.}}}{ds_r} = -\frac{1}{2} B\alpha s_r^{-3/2} + \frac{dp_{\text{pf.}}}{ds_r} = 0 \tag{2.14.37}$$

Now, as ($p_{\text{pf.}}$) is linearly related to (s_r),

$$\frac{dp_{\text{pf.}}}{ds_r} = C_{\text{pf.}}$$

where ($C_{\text{pf.}}$) represents the theoretical slope (gradient) of the load parameter (p) with respect to the reference displacement (s_r). For example, in Case (i) in Test 1, Section 1.2, page 20, ($C_{\text{pf.}} = -7\sqrt{3}W^0/6L$). (The constant $C_{\text{pf.}}$ is not to be confused with the constants (C_γ) or (C_r) in Eqs. (2.14.12) or (2.14.13)!) This constant may also represent the rate of change of an external load applied at some reference node (r), with respect to its corresponding displacement (s_r) at the same node, or the gradient of an equilibrium path of a load applied at some node (r) and another displacement, etc. So, ($C_{\text{pf.}}$) can be readily found using the theory expounded in Section 1.2 and in Sections 1 to 12 in this chapter, provided the system is isostatic. [However, we show in Ch. 3 using a somewhat modified theory that ($C_{\text{pf.}}$) can be readily found theoretically for any *hyperstatic* pin-jointed lattice.] Solving Eq. (2.14.37) for the displacement ($s_r = s_r^*$) where the path has a peak, we find

$$B\alpha = 2C_{\text{pf.}} s_r^{*3/2} \tag{2.14.37a}$$

and Eq. (2.14.36) for the imperfect equilibrium path becomes

$$p_{\text{impf.}} = P_{\text{impf.}} - P^0 = \pm 2C_{\text{pf.}} \frac{s_r^{*3/2}}{\sqrt{s_r}} + C_{\text{pf.}} s_r \tag{2.14.38}$$

where the second term may have a plus or minus sign. Finally,

$$\boxed{P_{\text{impf.}} = P^0 \pm 2C_{\text{pf.}} \frac{s_r^{*3/2}}{\sqrt{s_r}} + C_{\text{pf.}} s_r} \tag{2.14.39}$$

where ($C_{pf.}$) may be a positive or a negative constant. If ($C_{pf.} > 0$), that is, if the perfect path is a rising equilibrium path, the positive sign denotes the descending-ascending upper branch of the imperfect paths of the perturbed system shown in Fig. 2.14.4a, which has a minimum at ($s_r = s_r^*$) when

$$P_{impf.} = P^0 + 3C_{pf.}s_r^* \tag{2.14.39a}$$

which approaches the perfect rising path asymptotically *from above*, starting at positive infinity at ($s_r = 0$). Such paths have been experimentally observed to exist, but only under controlled laboratory conditions (see, for example, Ref. 15, Ch. 5).

In that case the negative sign at the second term represents the lower rising imperfect path with no stationary point in Fig. 2.14.4a. Simiarly, if ($C_{pf.} < 0$), i.e. the perfect path is a descending unstable path, the positive sign of the second term represents the lower stable-unstable path of the imperfect system, which has a maximum at the point ($s_r = s_r^*$) where

$$P_{impf.} = P^0 - 3C_{pf.}s_r^* \tag{2.14.39b}$$

The negative sign in the last case represents the upper descending unstable branch shown in Fig. 2.14.4b.

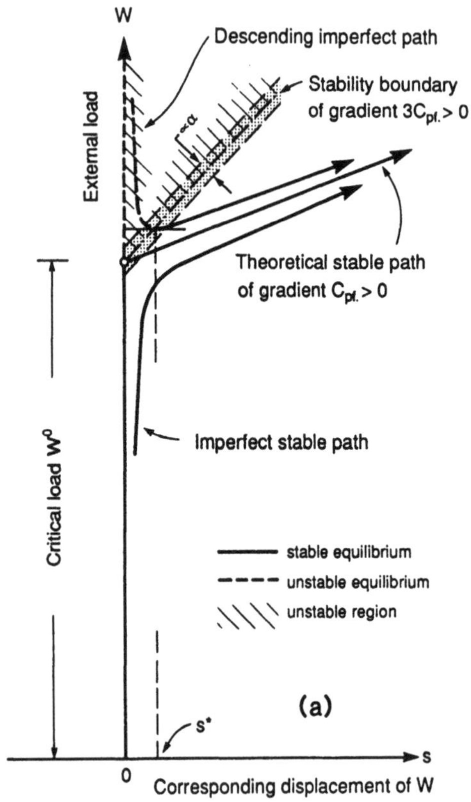

Figure 2.14.4(a)
A typical rising stable equilibrium path of an imperfect pin-jointed system plotted against the corresponding displacement (s) of an applied load (W) and a descending-ascending unstable-stable equilibrium path of this system with its stability boundary.—Notice that the stability boundary is situated above the theoretical stable path of the equivalent perfect system. The descending-ascending path may be obtained from a suitable model test under controlled laboratory conditions. In this case, the stable path represents the "natural" path of the system

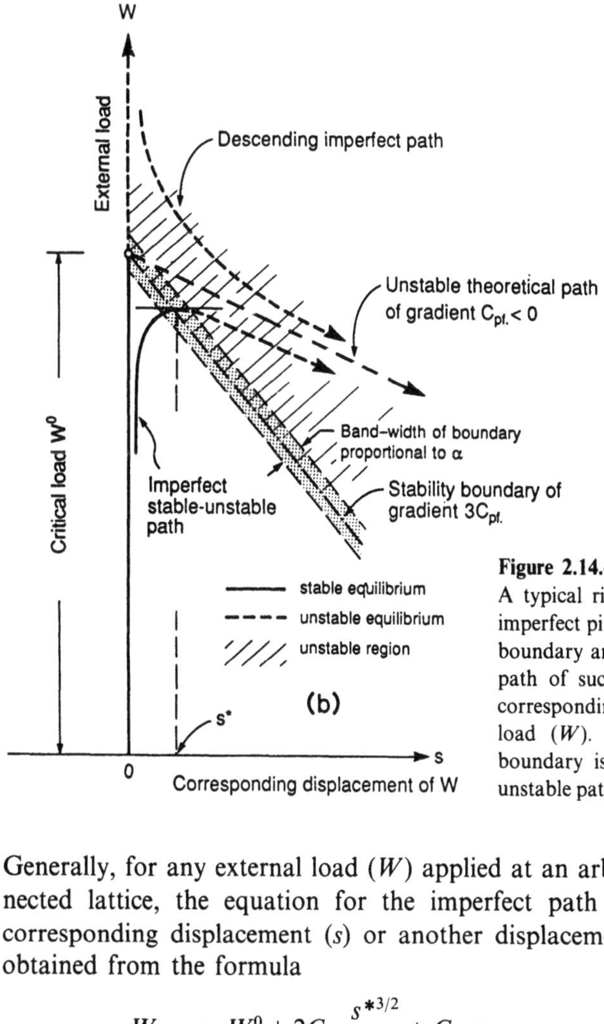

Figure 2.14.4(b)
A typical rising-falling path of an unstable imperfect pin-jointed system with its stability boundary and a typical descending unstable path of such a system plotted against the corresponding displacement (s) of an applied load (W). Notice that now the stability boundary is situated below the theoretical unstable path of the equivalent perfect system

Generally, for any external load (W) applied at an arbitrary node of a pin-connected lattice, the equation for the imperfect path relating this load to its corresponding displacement (s) or another displacement or coordinate is then obtained from the formula

$$W_{\text{impf.}} = W^0 \pm 2C_{\text{pf.}} \frac{s^{*3/2}}{\sqrt{s}} + C_{\text{pf.}} s \tag{2.14.40}$$

where the constants ($C_{\text{pf.}}$) are *different* depending on which displacement (s) they are related to.

Of most practical interest are the two lower branches of equilibrium paths given by Eq. (2.14.40), shown in Figs. 2.14.4(a) and (b), which agree with the experimental observations (Ref. 15) of the modified primary equilibrium paths in the pre- and post-buckling ranges of symmetrical or quasi pin-connected imperfect plane and space model structures. The value of (s^*) in each case must be determined *experimentally* from a test.

The results of Eq. (2.14.40) are summarized graphically in Fig. 2.14.4(a) and (b). There, also the descending unstable-stable and unstable paths for each case are indicated, for each of the two signs in the second term in Eq. (2.14.40). But, of most practical interest are the typical modified or imperfect primary ascending stable paths, shown in Fig. 2.14.4a, and the modified or imperfect ascending-descending stable-unstable paths shown in Fig. 2.14.4b.

Example

We apply the theory developed in this section to the 5-bay space beam studied in Section 5 of this chapter. The equation (2.14.40) derived earlier for an arbitrary load-displacement relationship under the perturbed or imperfect geometry, is now applied to each load (**W**) and its corresponding displacement (**u**). The general form of this equation

$$W_{\text{impf.}} = W^0 + 2C_{\text{pf.}} \frac{s^{*3/2}}{\sqrt{s}} + C_{\text{pf.}} s \qquad (2.14.40)\text{bis}$$

may be written for each load (W), where now (W) represents the load of the slightly imperfect structure i.e. ($W = W_{\text{impf.}}$) and the displacement (s) represents the corresponding displacement (u) of each load (W). In this case, (s) assumes the values of the three displacements (u_z^{16}), (u_z^{17}), and (u_z^{18}), while ($W_{\text{impf.}}$) represents the corresponding loads of the imperfect structure, applied at the joints 16, 17 and 18 respectively. Since these displacements are in a fixed ratio in any mode of the equivalent perfect structure, their relationships are not arbitrary. The gradient of the perfect path at joint 16 in the most degrading mode in this case is equal to ($C_{\text{pf.}} = -0.792$ lb./in.) and the gradient of the corresponding stability boundary at this joint, with respect to the same displacement, is [3(-0.792) = -2.376 lb./in.]. If the value of (u_z^{*16}) is arbitrarily set equal to 0.50 in., the path plotted according to Eq. (2.14.40) is as shown in the diagram in Fig. 2.14.5. For a value of ($u_z^{*16} = 0.80$ in.) the path shows a more pronounced nonlinearity *before* reaching the stability boundary. Both paths peak on this boundary at the displacement values of 0.50 and 0.80 in. respectively, as indicated in the figure. The displacements (u_z^{16}) and (u_z^{18}) in the mode of the perfect structure are in the ratio of

$$u_z^{16}/u_z^{18} = 1.26398/0.63199 = 2.00$$

where the changes (δW) at these values are equal to (-1.00 lb.). These displacements are indicated in the graphs for the perfect path at the joints 16 and 18 in Fig. 2.14.5. The imperfect paths at the joint 18 were plotted for the peak value of the load corresponding to the displacements ($u_z^{18} = 0.25$ in.) and ($u_z^{18} = 0.40$ in.). Again, these values were set arbitrarily. In an actual experiment the correponding values of (u_z^{*16}) and (u_z^{*18}) (not necessarily in the same ratio 2, as above) would have to be determined from test data, by plotting these paths for a scaled-down model structure under test. The displacement of joint 17 is practically zero and the imperfection-sensitivity is not considered at the joint. The total potential energy change of (-0.934 lb. in.) applies under perfect conditions, which were discussed earlier in Section 2.13.

One important conclusion follows from this illustration, namely that the sensitive region in which the imperfect paths peak can shift considerably along the

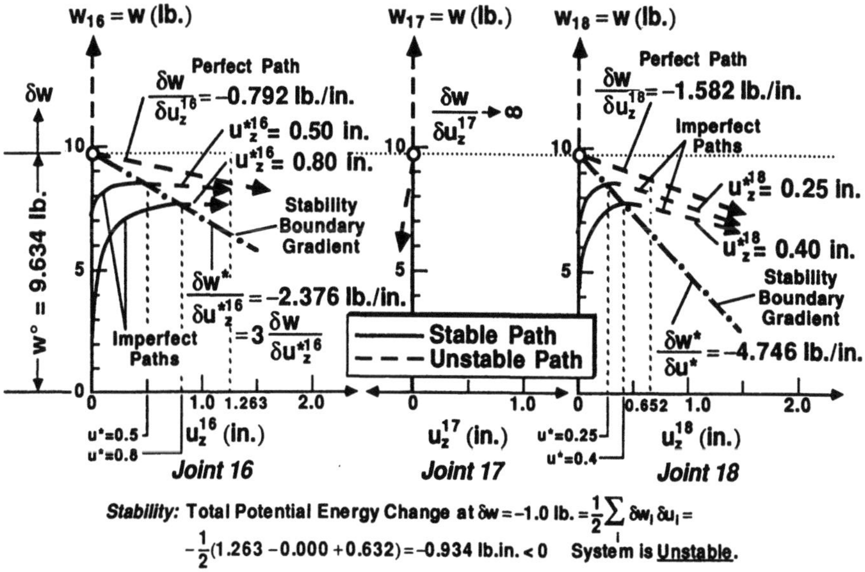

Figure 2.14.5
Case A: Imperfect primary equilibrium paths for the 5-bay beam plotted in a load-displacement representation.—Stability test in the critical state

stability boundary resulting in more and more *nonlinear* forms of the paths *before* this boundary is reached. Moreover, it can be shown by experiments (see Ref. 15, Chapter 6) that repeated loading *degrades* this response so that the two imperfect paths shown in this case could easily result in practice from repeated loading sequences. This means that the response of such systems changes in the process of loading and reloading so that it cannot be uniquely defined for all times, as it is subject to some variations which must be estimated from tests using an approach such as this. This phenomenon is of consequence for the control of the response of such a system intended, for example, to prevent the onset of instability or to control the distortions of the lattice at prescribed loading intensities. This illustration is provided to stress the character of the imperfection-sensitivity of flexible structures and of the regions in which this sensitivity can become pronounced, especially in systems which are highly *unstable* under perfect conditions. In such cases this imperfection-sensitivity depends essentially on two things, the steepness of the unstable paths of the equivalent perfect system and on the initial magnitude of the imperfections. But, under the same average initial imperfections this sensitivity increases as the negative gradients of the unstable equilibrium paths of the equivalent perfect system increase. These negative gradients are detected at the critical eigenvalues of the external loads and they also govern, as we have seen from this illustration, the subsequent sensitivity and the magnitude of the peak loads which are realized in the equivalent imperfect system.

2.15 Stability Analysis of Equilibrium States

The stability of the system in the critical equilibrium state subjected to a set of critical external loads (W_i) is determined by its ability or inability to develop an open or unstable motion from this state under the influence of a small disturbance while these loads remain constant. Such a disturbance need not be physical. It may be geometrical, such as an imperfection causing a slight perturbation of the ideal geometry of the system, or it may be material, causing a slight perturbation of its ideal physical properties. In such a motion, the system develops a characteristic mode or mechanism of post-buckling deformation of its members. It is shown in Ref. 15 that in any such given post-buckling mode the released kinetic energy equals the change in the total potential energy of the system from this state, provided there are no losses and conditions are mainly conservative, and that this energy under *constant* loads can be measured in the *same* distorted configuration, but while the structure is in an equilibrium state, which is achieved by introducing arbitrary virtual (or real) small forces (δW_i) at the nodes of the lattice of the system, such that under the loads ($W_i^0 + \delta W_i$) equilibrium is achieved. Here not all critical loads (W_i^0) need be nonzero, nor are the increments (δW_i) necessarily nonzero at all nodes. The index (i) covers the numbering sequence of the three global coordinate directions at all the nodes. The number of these incremental loads (δW_i) must be sufficient to meet all the degrees of freedom of the lattice and to equilibrate the system of loads which it supports. Then, if also the corresponding displacements (δs_i) in the directions of the loads (δW_i) are introduced, and these are generally directed along the global coordinate axes at each node, it can be shown (see Ref. 15, Ch. 1, 3, 7, etc., see also Section 1.2, p. 12) that the change in the total potential energy (v) can be expressed as

$$v = \frac{1}{2} \sum_{i=1}^{n} \delta W_i \, \delta s_i \tag{2.15.1}$$

where (n) denotes the necessary number of these incremental loads, which is generally also the number of the degrees of freedom of the system. The quantity (v) defines the stability of the system in the critical state in the sense that if (v) is *negative* in *at least one* adjacent position, the equilibrium in the critical state is certainly unstable, as conversion of the total potential into kinetic energy can take place. To show this, let (V) be the total potential energy of the system, measured from some arbitrary initial datum, such that ($V = U + H$), where U is the elastic energy and H the potential energy of the external loads measured with respect to the same datum. Let (T) be the kinetic energy of this system, while it is in some kind of motion. Also (T) is measured with respect to the same datum. Then, if there are no losses (or if these are small, depicted by the term ε) and if (E) is the total initial mechanical energy of the system,

$$E = V + T - \varepsilon \tag{2.15.2}$$

and a change in the total energy under conservative conditions

$$\Delta E = \Delta V + \Delta T - \Delta\varepsilon = 0 \tag{2.15.3}$$

For a conversion into kinetic energy (T) has to be positive and the condition

$$\Delta T = -\Delta V + \Delta\varepsilon > 0 \tag{2.15.4}$$

must hold. If $|\Delta\varepsilon|$ is small compared to (ΔV), then clearly ΔV must be always *negative* for this conversion to take place. Hence, the condition that ($\Delta V = v$) must be negative for the system to collapse by motion, since then, under constant critical loads (constant in direction and magnitude) acting on the structure, the system accelerates rapidly away from the initial critical state and unstable motion takes place, causing it to collapse.

We now return to the Eq. (2.12.14) in Section 2.14 and show how it can be correlated to the expression for the energy change (v), measured from the critical state itself. We showed that equilibrium in any postcritical state of the system is governed by the Eq. (2.12.14), i.e.

$$\mathbf{E}_c \mathbf{e}_c = \mathbf{C}_{1c}^T \boldsymbol{\beta}' \, \delta w \tag{2.12.14)bis}$$

where ($\mathbf{e}_c$) are the flexural contractions of the critical compressive members which are on the point of buckling in the critical state itself. The ($\boldsymbol{\beta}'$) vector of load proportionality constants is *arbitrary* in this equation and it is important to consider it as such. (δw) is the change in the load parameter in adjacent equilibrium states. This suggests that here the flexural contractions ($\mathbf{e}_c$) can be considered as *arbitrary independent* quantities. In other words, for arbitrary ($\mathbf{e}_c$) it is always possible to find a vector ($\boldsymbol{\beta}'$) to satisfy the equilibrium equations of the buckled form of a lattice governed by Eq. (2.12.14). If we premultiply Eq. (2.12.14) by ($\frac{1}{2}\mathbf{e}_c^T$) we get:

$$\tfrac{1}{2}\mathbf{e}_c^T \mathbf{E}_c \mathbf{e}_c = \tfrac{1}{2}\mathbf{e}_c^T \mathbf{C}_{1c}^T \boldsymbol{\beta}' \, \delta w \tag{2.15.5}$$

Now, in Section 2.8, Eq. (2.8.3), we showed that

$$\mathbf{u} = \mathbf{C}_{1c} \mathbf{e}_c \tag{2.8.3)bis}$$

Taking the transpose of this equation we get

$$\mathbf{u}^T = \mathbf{e}_c^T \mathbf{C}_{1c}^T \tag{2.15.6}$$

Also, ($\boldsymbol{\beta}' \, \delta w$) represents here the incremental loads necessary to hold the system in equilibrium in an arbitrary post-buckled configuration defined by the vector ($\mathbf{e}_c$). Then

$$\boldsymbol{\beta}' \, \delta w = \delta \mathbf{W} \tag{2.15.7}$$

Therefore, the right side of the Eq. (2.15.5) can be written in the alternative form

$$\tfrac{1}{2}\mathbf{e}_c^T \mathbf{E}_c \mathbf{e}_c = \tfrac{1}{2}\mathbf{u}^T \delta \mathbf{W} \tag{2.15.8}$$

But, in the notation of Eq. (2.15.1) the right side of (2.15.8) is identical to the right side of Eq. (2.15.1), i.e.

$$\frac{1}{2}\mathbf{u}^T \delta \mathbf{W} = \frac{1}{2}\sum_{i=1}^{n} \delta W_i \, \delta s_i \tag{2.15.9}$$

We thus arrive at the very important result of Eq. (2.15.5) written in the form of Eq. (2.15.1), i.e.

$$v = \frac{1}{2} \sum_{i=1}^{n} \delta W_i \, \delta s_i = \frac{1}{2} \mathbf{e}_c^T \mathbf{E}_c \mathbf{e}_c \tag{2.15.10}$$

Thus, the change (v) in the total potential energy is given by the quadratic form

$$\tfrac{1}{2} \mathbf{e}_c^T \mathbf{E}_c \mathbf{e}_c$$

where the matrix of this form is identical to the post-buckling stiffness matrix ($\mathbf{E}_c$) of the system. Therefore, to ascertain whether a system is stable or unstable in the critical state, Eq. (2.15.10) can be used directly in conjunction with Theorem 1 in Section 1.2 on page 13, or Theorem 5 in Ref. 15. So, if at least one adjacent configuration can be found for which the expression

$$v = \tfrac{1}{2} \mathbf{e}_c^T \mathbf{E}_c \mathbf{e}_c < 0$$

is negative for ($\mathbf{e}_c \geqq 0$) the system in its critical state is certainly *unstable*. The magnitude of (v) in such an adjacent configuration may be indicative of this instability, as it represents the maximum amount of total potential energy (assuming the losses are negligible) which can be converted into kinetic energy and which is a measure of the catastrophic collapse of the system.

If, on the other hand,

$$v = \tfrac{1}{2} \mathbf{e}_c^T \mathbf{E}_c \mathbf{e}_c$$

is found positive in the entire positive orthant of the flexural shortenings (e_i; $i = 1, 2, \ldots, c$) defined by

$$\mathbf{e}_c \geqq 0$$

the system in the critical state is certainly statically *stable*. If displaced from this state, energy has to be added to it. If released from the displaced position the system returns to its initial critical state. This is the meaning of its stability in the sense of Theorem 1 in Section 1.2.

We have used Eq. (2.15.10) in the preceding sections to demonstrate the stability of critical states in particular systems which we investigated. Some additional applications of this theory are addressed in the next section.

2.16 Some Applications of the Stability Theory to Practical Space Lattices and Structures

Case E1

The shortened 5-bay pin-jointed flight mast discussed in Section 2.13, Fig. 2.13.1(c) is shown again in Fig. 2.16.1. The mast is fixed at one end and free at the opposite end. It is now loaded by three upward equal loads (**W**) at the nodes

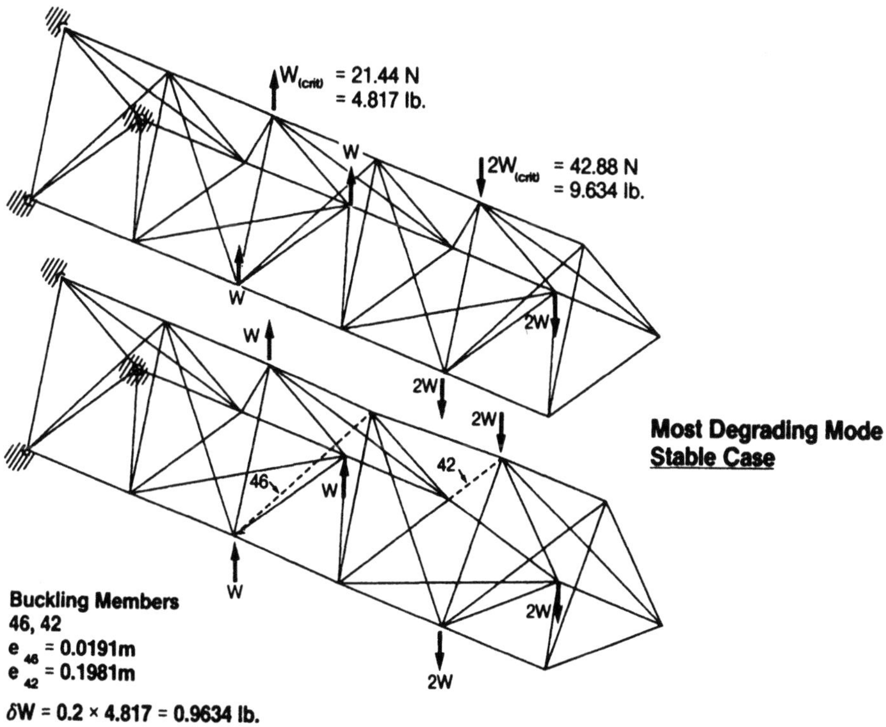

Figure 2.16.1
Case E1: Transversely loaded 5-bay beam (NASA COFS Program, Ref. 24). Distorted shape in the most-degrading post-buckling equilibrium mode. The system is stable in the critical state

at the end of the second bay from the fixed end, as shown in the figure, and with three downward loads of twice the intensity i.e. of $(2W)$ at the beginning of the last bay, counting the bays from the fixed end. The critical state is now reached when $(W = 4.817 \text{ lb.})$ and the members marked 42 and 46 are simultaneously critically stressed, Fig. 2.16.2(a). The most degrading equilibrium mode occurs

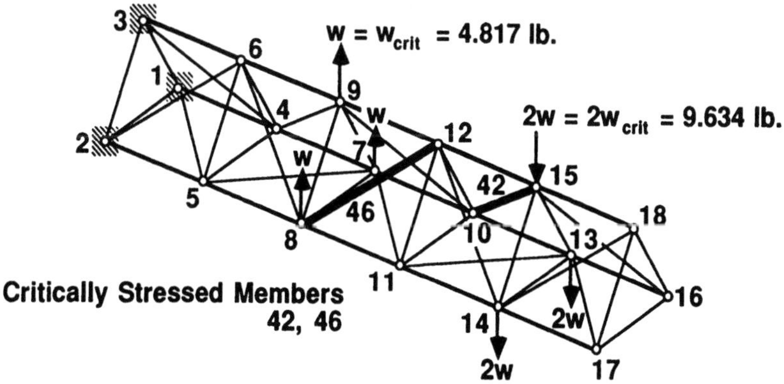

Figure 2.16.2(a)
Case E1: The critical state occurs at $(W = 4.817 \text{ lb.})$

when these two members buckle simultaneously as shown in the bar-graph of Fig. 2.16.2(b). They shorten in the ratio indicated in the table in Fig. 2.16.2(b). An isometric computed view of the 5-bay mast in the buckled configuration is shown in Fig. 2.16.3. There also the $(\mathbf{E}_c)$ matrix for this case is reproduced and it is easily verified that the quadratic form for the total potential energy change

Member	Relative Flexural Shortening e
42	1.0000
46	0.0965

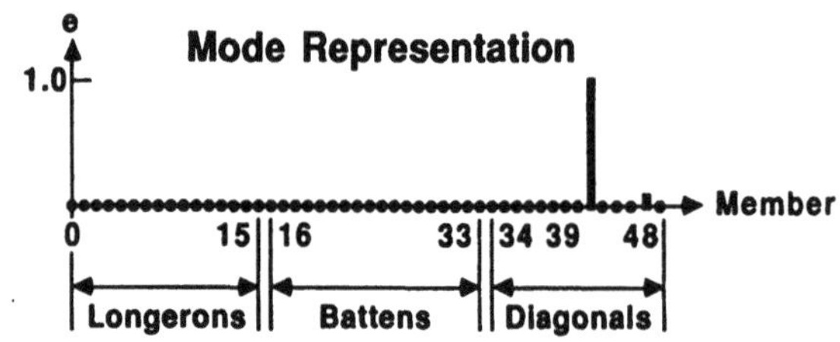

Figure 2.16.2(b)
Case E1: Bar diagram of the most-degrading post-buckling equilibrium mode

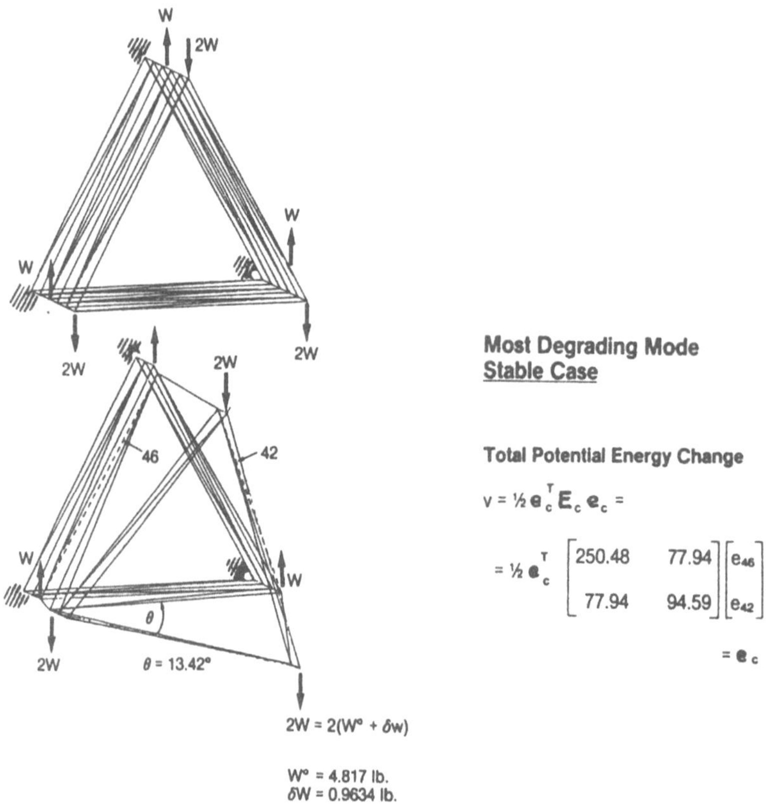

Figure 2.16.3
Case E1: Distorted shape of the transversely loaded 5-bay beam in the most-degrading post-buckling equilibrium mode. Stability test in the critical state. Note that the stability matrix $(\mathbf{E}_c)$ is positive definite (stable system)

(v) is positive definite in the positive quadrant of the vector

$$\mathbf{e}_c = \begin{pmatrix} e_{46} \\ e_{42} \end{pmatrix}$$

as shown in Fig. 2.16.3. The system is therefore *stable* in the critical state. Fig. 2.16.4 shows the graph representing the variation of the load (2**W**) in the most degrading equilibrium mode against the twist (θ) of the end-section under the same load vector ($\boldsymbol{\beta}' = \boldsymbol{\beta}$) corresponding to the system of conservative external loads (**W**) and (2**W**) indicated in Figs. 2.16.1 to 2.16.4.

Case E2

In Fig. 2.16.5 the same structure as in Case E1 is shown loaded by the same loads (**W**) and (2**W**), except that now the load directions are *reversed.* Also, in this case the critical state occurs when (W) reaches the value ($W^0 = 4.817$ lb.). However, the situation regarding the stability of the system is now totally different. Also, different members marked 41 and 47 become critically stressed and they buckle simultaneously in the most degrading equilibrium mode under the same loads but under reversed vector ($\boldsymbol{\beta}' = \boldsymbol{\beta}$), in this case given by three downward loads (**W**) and three upward loads of intensity (2**W**), as shown in Fig. 2.16.5. Figure 2.16.6

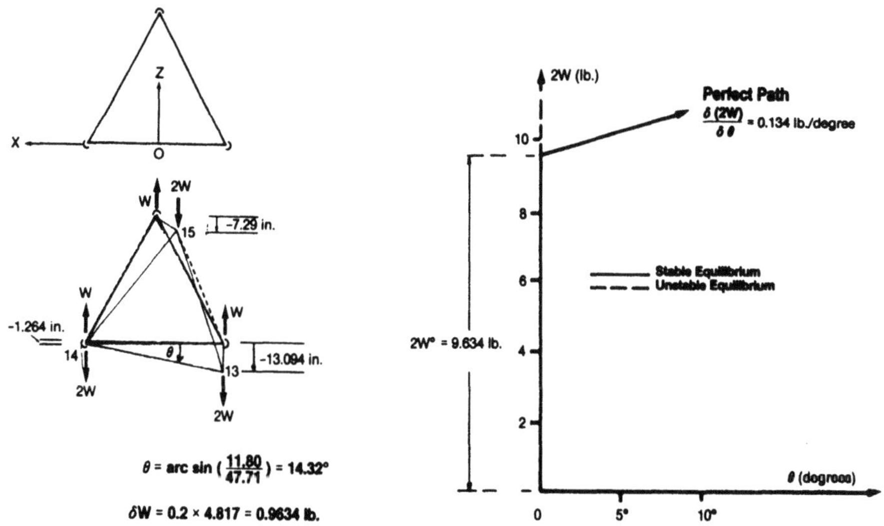

(a) (b)

Figure 2.16.4
(a) Case E1: End-view of the buckled 5-bay space beam under transverse loading in post-buckling equilibrium. Rotation of the loaded end-section. (b) Case E1: End-load end-rotation equilibrium diagram in stable post-buckling

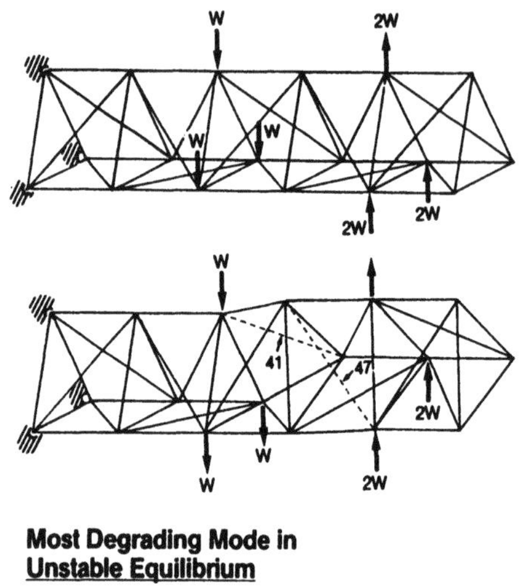

Figure 2.16.5
Case E2: Transversely loaded 5-bay beam.—Inverse loading (NASA COFS Program, Ref. 24). Distorted shape of the space beam in the most-degrading unstable equilibrium mode under the inverse loading with the members 41 and 47 buckling

shows the distorted structure in this mode corresponding to a decrease in the load ($\mathbf{W}$) from ($\mathbf{W}^0$) by the amounts ($\delta W = -0.9634$ lb.). The post-buckling stiffness matrix ($\mathbf{E}_c$) is not positive definite in this case and the change (v) in the total potential energy of the system is given by

$$v = \frac{1}{2}(e_{41} \quad e_{47})\begin{pmatrix} -61.29 & -77.94 \\ -77.94 & 94.59 \end{pmatrix}\begin{pmatrix} e_{41} \\ e_{47} \end{pmatrix}$$

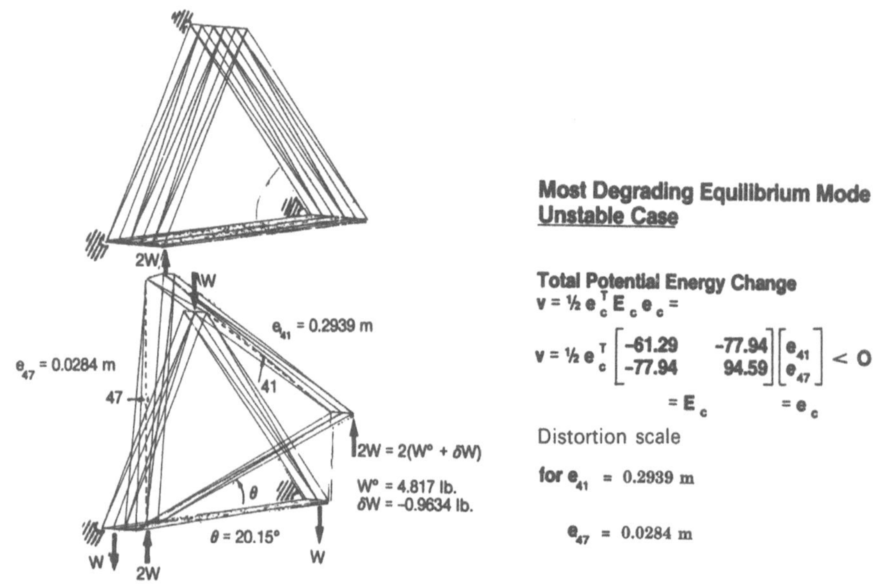

Figure 2.16.6
Case E2: The stability test of the 5-bay beam under the inverse transverse quasi-static loading and the distorted shape of the beam in the most-degrading unstable equilibrium post-buckling mode. End-view

and this expression can be certainly *negative* in the positive quadrant given by

$$\mathbf{e}_c = \begin{pmatrix} e_{41} \\ e_{47} \end{pmatrix} \geqq 0.$$

Therefore, the cantilevered space beam is *unstable* in the critical state under the *reversed* loading and it fails catastrophically under the reversed loads of critical intensity (W^0) and ($2W^0$). From this we conclude, that if the loads of intensity (W) and ($2W$) were applied as inertial loads under some kind of slow oscillation of the cantilevered reticulated beam, the catastrophic failure of the beam can be expected to occur in *a preferred direction*, which corresponds to the same direction discovered to be catastrophic in Case E2, when the loading of the same intensity is applied statically. This similarity or analogy would certainly hold under low frequencies of oscillation of such a system.

Case F1

In this case we illustrate the application of the stability theory to the cantilevered flight mast shown in Fig. 2.13.1(b) consisting of 20 bays. The mast is assumed to be fixed at the left end. Now, the loading represents the inertial loading that is derived from the second mode of a steady state vibration of the mast using a linear theory. This loading is then assumed to act on the structure at its nodes, being applied gradually as static loading. The typical specified loading intensities‡ are shown in Fig. 2.16.7 (top) where, at the mid-span at the top part, this intensity at the maximum amplitude equals (-53.274 Newtons) and at the tip ($+75.621$ Newtons). Similarly, the equivalent intensities at the bottom nodes are indicated in Fig. 2.16.7. These intensities correspond to an upward tip deflection of the mast of 3.33 mm. The equivalent flexural stiffness of the mast for this computation was assumed to be ($EI = 26.5 \cdot 10^6$ Newton-meter^2) and the mass 2.65 kg/meter (NASA-Specifications). The actual loading applied to the real structure was made proportional to these load intensities with the load proportionality parameter being now equal to (w). In Case F1 the inertial loading was distributed to correspond to an upward tip deflection in relation to the mast as drawn in Fig. 2.13.1b (i.e. in the z-direction in Fig. 2.13.1(b)).

Under this loading distribution the critical state is reached when ($w = w^0 = 0.187$). This means that the first diagonal member, marked 4 on the near side in the view of Fig. 2.13.1(b) at the fixed end of the mast, buckles at only 18.7% of the maximum load-amplitudes, which corresponds approximately to an upward tip deflection of only 0.623 mm. In other words, the mast could not complete the first cycle of vibration without buckling. This result is quite revealing as it shows the fallacies of linear theories of vibration being applied to a flexible structure, which obviously no longer apply when near critical states in the structure are reached. Figure 2.16.8 shows an enlarged view of the affected region with the buckling member 4 being shown dashed. In this case, the post-buckling

‡ Ref. 24, Progress Report, Naval Research Laboratory, Wash. D.C., Contract No. N00014-86-K-2014, 1986.

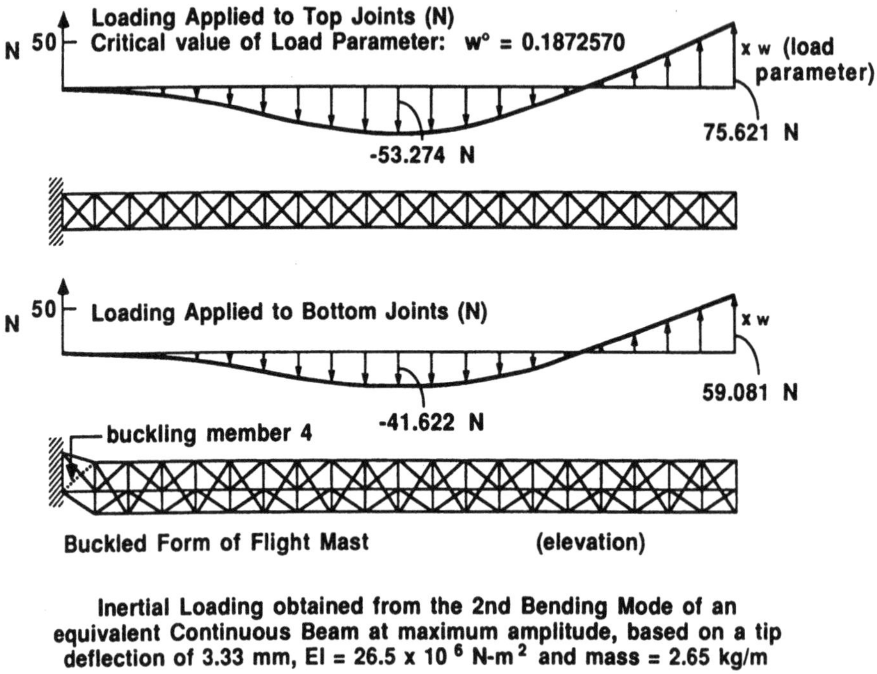

Figure 2.16.7
20-bay flight-mast of length 22.40 m (NASA COFS Program, Ref. 24). The resulting buckled shape of the mast under the maximum transverse inertial quasi-static loading derived from the second normal bending mode of the equivalent continuous mast-beam and the resulting buckled shape of the mast

stiffness matrix ($\mathbf{E}_c$) reduces to a single positive number and, if the structure were held in statical equilibrium in this mode, the post-buckling equilibrium would be governed by the equation

$$168.34 e_4 = 555.76\ \delta w$$

where the units are in Newtons and meters, as indicated in Fig. 2.16.8. The distorted configuration is shown in Fig. 2.16.9 for a value of ($\delta w = 0.09$). The load intensities along the central region of the mast are indicated by arrows pointing in the downward direction. Also, in Fig. 2.16.9 there is a graphical representation of the relationship between the load parameter (w) and the angle of twist (θ) of the unbuckled part of the mast. This gradient is clearly very flat, which means that the mast twists at practically no increase in the loads in the critical and post-critical regions. Since ($\mathbf{E}_c$) is a positive number, the system is theoretically stable in the critical state, but this stability is only slight, as it is very weak in resisting additional loads, beyond the critical state.

Case F2

Here, the same 20-bay pin-jointed mast is analyzed only under the *reversed* inertial loading which is applied statically to the real structure, corresponding to the

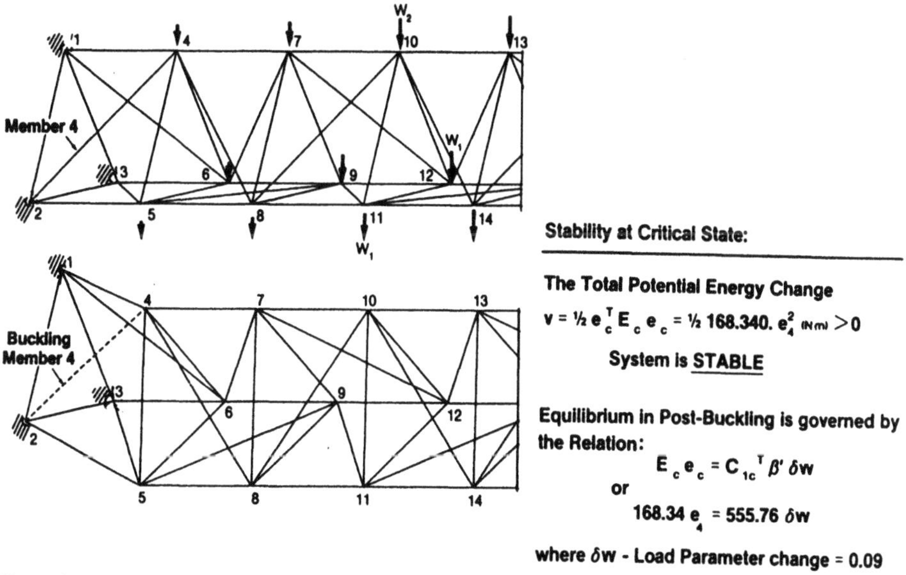

Figure 2.16.8
Enlarged view of the buckled shape of the 20-bay flight-mast under the maximum transverse inertial quasi-static loading derived from the second normal bending mode of the equivalent continuous mast-beam and the stability test in the critical state. Notice that only one member (member 4) contracts flexurally in this mode

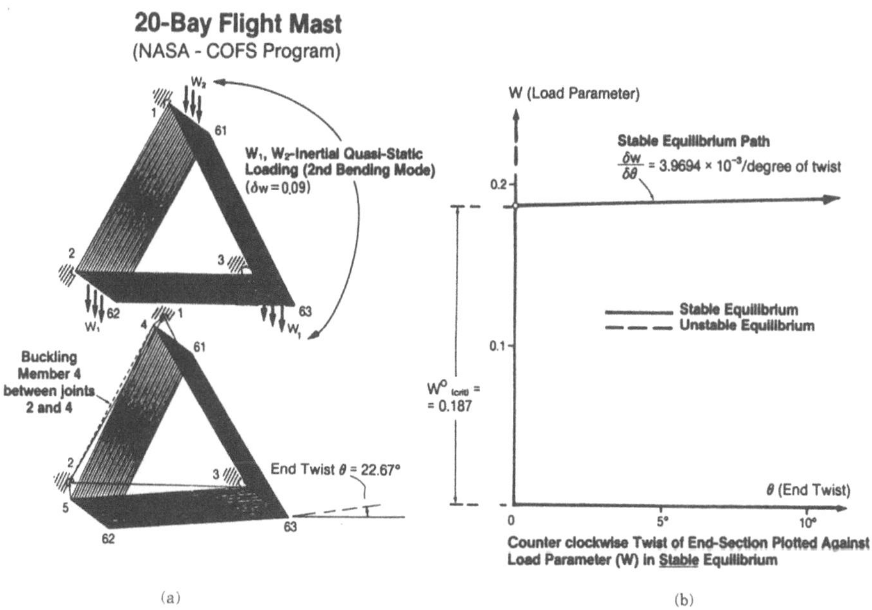

Figure 2.16.9
(a) 20-bay flight-mast (NASA COFS Program, Ref. 24).—Twist of the buckled mast under the maximum inertial transverse quasi-static loading derived from the second normal bending mode of the equivalent continuous mast-beam with only member 4 contracting flexurally. (b) The corresponding load-parameter end-twist slightly stable equilibrium path

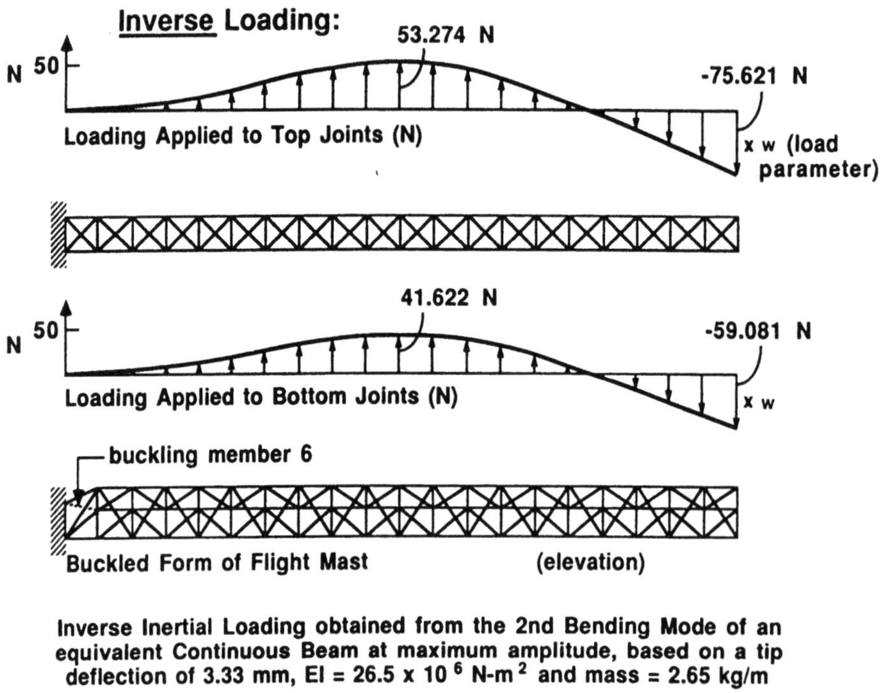

Figure 2.16.10
20-bay flight-mast of length 22.40 m under the inverse maximum transverse inertial quasi-static loading derived from the second normal bending mode of the equivalent continuous mast-beam. The buckled shape of the mast with only one member (member 6) buckling (NASA COFS Program, Ref. 24)

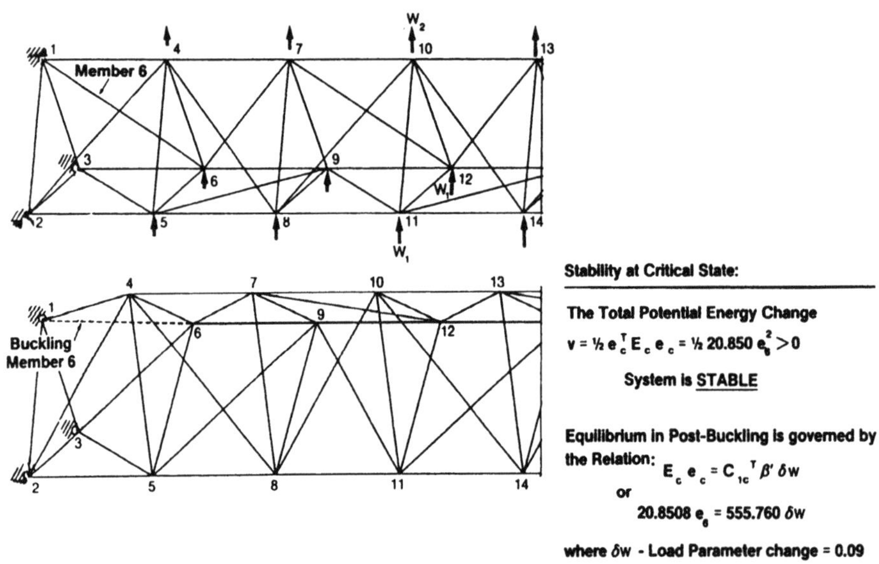

Figure 2.16.11
Enlarged view of the buckled shape of the 20-bay flight-mast under the inverse maximum transverse inertial quasi-static loading derived from the second normal bending mode of the equivalent continuous mast-beam and the stability test in the critical state (NASA COFS Program, Ref. 24)

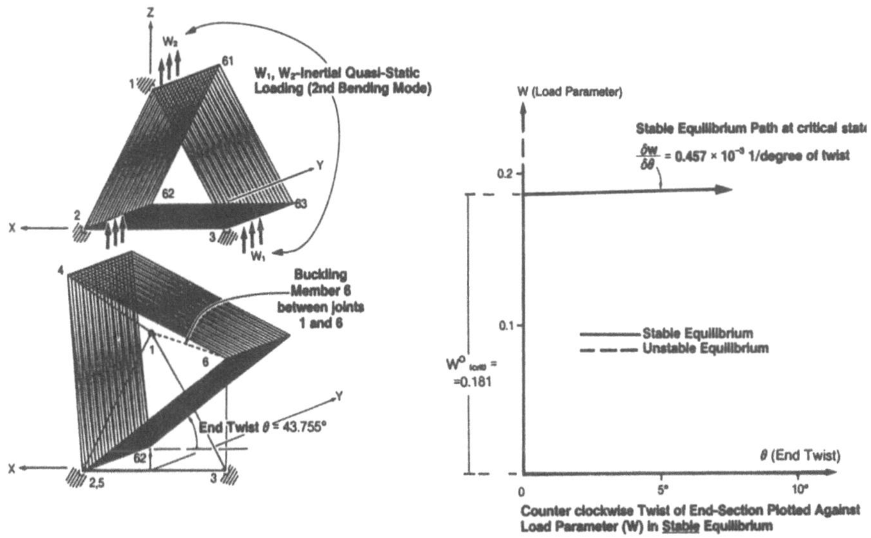

(a) (b)

Figure 2.16.12
(a) Twist of the buckled 20-bay flight-mast under the inverse maximum transverse inertial quasi-static loading derived from the second normal bending mode of the equivalent continuous mast-beam.—Notice that only one member (member 6) contracts flexurally in this mode. End-view. (b) The corresponding load-parameter end-twist slightly stable equilibrium path

downward tip deflection of the mast of 3.33 mm. The load intensities are now clearly the same as in the previous case, only they are of opposite signs, as shown in Fig. 2.16.10. However, now a different diagonal member, marked 6, buckles which is situated on the opposite or far side of the mast in the view of Fig. 2.13.1(b), as also shown enlarged in Fig. 2.16.11. Also in this case (E_c) is a positive number, as the equilibrium is governed by the equation

$$20.8508 e_6 = 555.760\,\delta w.$$

However, the distortions are most pronounced, especially in torsion. The distorted configuration is shown for ($\delta w = 0.09$) in Fig. 2.16.12. The system is, therefore, theoretically stable, but its stiffness in post-buckling, characterized by the gradient of the load-twist equilibrium path ($\delta w/\delta\theta = 0.457 \times 10^{-3}$), is even less than in the previous case, depicted in Fig. 2.16.9, when this gradient was found to be ($\delta w/\delta\theta = 3.9694 \times 10^{-3}$).

This suggests that if this distributed loading were applied slowly in the original and then in the reverse direction, the distortions would involve totally different members and they would be by no means symmetrical, as this may be the case if a linear analysis were used in describing the oscillations of such a beam. The same would be true if these loading patterns were applied in a dynamic process involving slow oscillations at low frequencies. The flexural deformations in the buckling members would dominate the distorted pattern and these would be different depending on the sign and the direction of the inertial loads.

3 Static and Dynamic Buckling of Complex Hyperstatic Pin-Connected Elastic Systems

3.1 Introduction—Postbuckling of Hyperstatic Lattices

In this chapter we consider hyperstatic lattices composed of three-dimensional space elements such as tetrahedrons, octahedrons, combinations of cubes and tetrahedrons etc., in which each element is made of light flexible members jointed together by hinged or quasi-hinged (hinge-like) connectors at the nodes. By giving these members appropriate lengths, the lattice may be shaped into any curved three dimensional array of space elements, as demonstrated in Fig. 3.1.1. Curved three-dimensional arrays of space elements constitute shell-type geometrical configurations. Lattices and, in general, structures of this type are called *reticulated shells.*

Depending on the global geometry of the lattice, we speak of a spherical, a cylindrial or torodial reticulated shell or of a flat reticulated plate, Fig. 3.1.1. If such a curved lattice is used as a ground structure it may be enveloped by a cover shell or membrane supported over concentrated areas at the peripheral nodes of

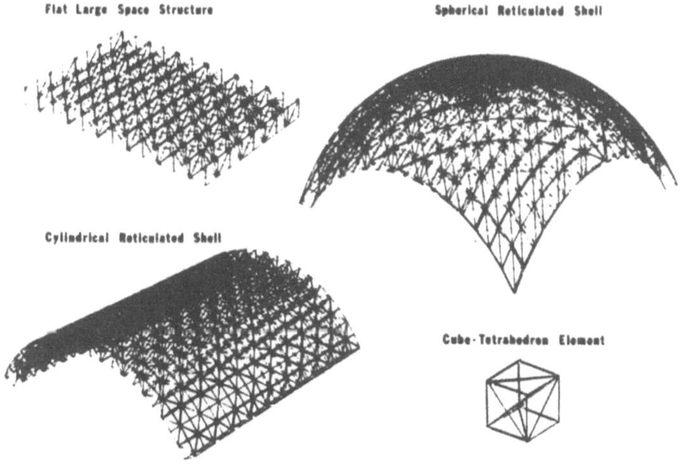

Figure 3.1.1
Different types of space lattices generated from cube-tetrahedron elements

the lattice, while the lattice itself is supported along its boundaries on a rigid foundation. As before, the external loads are applied centrally at the nodes. The loaded joints are usually also peripheral nodes.

In this case the number of members in the lattice is greater than $(3J - 6)$, where (J) is the number of joints and $(3J - 6 = N)$ is the number of members necessary to make the lattice just-stiff or isostatic, as explained in Chapter 2, Section 3. The presence of redundant members accounts for a somewhat different behavior or response of the lattice, as the loads are increased. A hyperstatic lattice composed of flexible slender members also fails by buckling, similarly as an isostatic lattice, except that the external loads, which the hyperstatic lattice can support, may be much higher, this depending on the distribution of the redundant members within the array of three-dimensional elements. Namely, redundant members act as constraints against buckling within the lattice and, therefore, a sufficient number of compressive members, including the redundant ones, must be critically stressed in order that these may contract flexurally thereby permitting the formation of a kinematic buckling mechanism within the lattice, which then results in the displacements of the nodes and in a partial or global deformation mechanism. If the members are highly flexible their axial deformations contribute little to the displacements of the nodes. Therefore, such a lattice can only deform appreciably if the members deform flexurally and that can only happen, if the axial loads in the members approach the critical values at which these members buckle. Thus, the redundant compressive members are always involved in the buckling process of a hyperstatic lattice and they contract flexurally under post-critical axial loads, which then *increase* slightly from their critical values. How many of these redundant members participate in this process, depends on whether the buckling mechanism is global, partial or even local. In a global mechanism all axially stressed redundant members contract flexurally as compressive buckling members, and so participate in the post-buckling collapse mechanism. Thus, on account of the statically redundant members the onset of flexural buckling in the lattice may be delayed, as the load-increases are absorbed by the remaining subcritically stressed bars and the load-carrying capacity of the structure is increased globally since the redundant members can support *additional* external loads. Under such conditions, other (isostatic) members may be restrained from failure, until, in a required combination or sequence, a sufficient number of compressive members—redundant and nonredundant—contract flexurally, thereby permitting the formation of a *kinematic buckling mechanism* within the hyperstatic lattice.

Theoretical studies and earlier tests on such model-lattices show that they can carry surprisingly high loads*[1] compared to their dead-weight. This means that also reticulated shells of this type, enveloped by a cover-shell can support relatively large pressures integrated over their surface in relation to their own weight. This is usually in contrast to conventional monocoque shells, especially

---

[1] *See for example, Refs. 16 and 19; †op.cit.

shallow shells, whose pressure-bearing capacity to weight ratios under imperfection-sensitivity may be much lower and of the order of, say, 10. In comparable enveloped optimized reticulated shells this ratio may be 100 or higher†[1]. This increase in load-bearing capacity compared to dead-weight of such a structure becomes significant at larger spans and especially in the coverage of large areas, where conventional shells or similar metal or concrete structures are less effective.

The post-buckling analysis of hyperstatic lattices and trusses expounded in this Chapter is based on earlier theories published by the author in several papers and books e.g. Refs. 19 to 25 and it is an extension of the theory presented in Chapter 2. The following analysis reveals, that the post-buckling process in such systems usually develops in unstable equilibrium, which cannot be maintained in practice, because the applied loads cannot be adjusted and controlled, so that the structure under its critical or near critical loading develops a dynamic collapse by motion.

The primary design criterion for such structural systems then becomes *the elimination of the onset of this motion*, as the material failure develops *subsequently* in motion. The failure by unstable motion usually occurs in the so called *most degrading mode*, which is responsible for the *largest conversion of the total potential into the kinetic energy of the system*. This conversion, in turn, corresponds to the paths of the largest decreases of the total potential energy from the energy level in the ultimate critical state of the system, where this process originates from rest. Moreover, such a collapse can occur under the simplest conservative loading conditions and it is triggered off by slight *geometrical* disturbances or *initial imperfections* within the system, which deviates from its perfect geometrical shape which is assumed to exist in the theoretical computation of the critical state, its energy decreasing from the corresponding energy levels in the perfect system. Elimination of such a "most degrading" collapse mode obviously leads to an improved design under conservative loading conditions. This elimination is accomplished by means of the stiffening of certain *key* members and it results in a slightly modified structure. This optimization process is primarily concerned with the techniques by which the most degrading buckling modes are mitigated or eliminated altogether. Such a mode is usually associated with the largest decrease in the total potential energy of the system. This can happen either along an equilibrium path, when equilibrium under slightly modified applied proportionate loads may be possible, or in a different mode in unstable motion under constant critical loads. For this reason a direct evaluation of the most degrading modes of both types is of much interest.

One should emphasize that the buckling mechanism representing the most degrading mode in statical *equilibrium* under proportionate applied loads (usually decreasing loads), may be *very different* from the mode taking place in an unstable dynamic collapse by *motion*, in which the system is driven by constant critical loads. Different critically stressed members may buckle in each type of

---

†[1] See, for example, Refs. 16, 19, 20.

mode, the most degrading dynamic mode always resulting in the largest conversion of the total potential into the kinetic energy.

On account of the slenderness of their members, structural systems of this kind behave differently from conventional (usually referred to as linear) structures. Axial deformations of the members are now very small, i.e. smaller or comparable to the second order displàcements of the joints of the lattice resulting from purely flexural distortions. This property is perhaps the most important characteristic that distinguishes light reticulated shells or space lattices from conventional trusses. The deformations of such hyperstatic trusses are, therefore, highly influenced by the flexural distortions of the members subjected to appreciable axial loads, especially the compressive loads, causing them to be sensitive to initial geometrical imperfections and, ultimately, to failure by buckling. Once the most degrading buckling mode is theoretically evaluated, it can be eliminated by stiffening the critical members responsible for its existence (the key members are usually few in number compared to the total number of compressive bars). When this is accomplished and an improved buckling response of the lattice obtained, the procedure is repeated by eliminating the next most degrading mode and so on, until the physical possibilities of the structural lattice are exhausted. The procedure then results in an optimal design for a global statical stability limit under the given loading and the geometrical shape of the lattice.

3.2 Initial Equilibrium Equations—Kinematic Admissibility Conditions at the Ultimate Critical State

The equilibrium equations of the system, written on the initial or undistorted geometry of the lattice, are derived similarly as this was done for statically determinate lattices in Section 2.5 in the last Chapter. The only difference now is in the number (M) of the members in the lattice, which is now larger than the number ($n = N$) of the degrees of freedom (recall that (N) denotes the number of members in an isostatic lattice.) The axial loads in the redundant bars are now *included* in the equations. Thus, we may continue the development of the equilibrium conditions starting with Eq. (2.5.2), written at the node (i) for the (m_i) members meeting at this node, which we label by the index (α), the redundant members included. Thus

$$\sum_{\alpha=1}^{m_i} \mathbf{F}_{\alpha i} = \sum_{\alpha=1}^{m_i} P_\alpha \mathbf{C}_{\alpha i} = \mathbf{W}_i \ddagger; \qquad i = 1, 2, \ldots, J \qquad \text{(2.5.2)bis}$$

Proceeding then in the same way, as in the case of statically determinate systems, we arrive at the equations of the *undistorted statically indeterminate* system (structure and the applied loads) in the form

$$\mathbf{AP} = \mathbf{W} = \boldsymbol{\beta} w \qquad \text{(2.5.4)bis and (3.2.1)}$$

‡ Symbols in bold type denote matrices or vectors.

(**A**) is the statical matrix which is built according to the relations in Eq. (2.5.2), (**P**) the ($M \times 1$) axial load vector, (**W**) the ($N \times 1$) vector of the applied loads, written in the same numbering sequence as in the case of statically determinate systems, (w) is the scalar load parameter and ($\boldsymbol{\beta}$) the vector of the load proportionality constants, which must be prescribed. Here, (M) is the total number of members and ($n = N$) the number of the degrees of freedom of the lattice where ($N \leq M$), so that (**A**) is now a ($N \times M$) rectangular statical matrix.

The equilibrium equations valid after the onset of elastic buckling must be written on the flexurally *distorted* geometry of the lattice, similarly as this was done for statically determinate systems. Compared to the approach in Ref. 15 used for the solution of the post-buckling equilibrium equations in simpler systems, the present method is different in that it consists in changing the set of independent variables from the flexural contractions (**e**) to the incremental axial forces ($\delta\mathbf{P}$) in the members, viz. the displacements (**u**) of the joints. In this form, the equations may be reduced to a much smaller set, involving only particular post-buckling modes, similarly as this was done for statically determinate systems discussed in Chapter 2. So, they can be applied to more complex hyperstatic lattices and solved on a digital computer. The reduced set of the independent variables ultimately involves only the flexural contractions ($\mathbf{e}_c$), related to the incremental forces ($\delta\mathbf{P}_c$) in the critical members by the general law.

Before these equations can be derived and solved, the *ultimate* critical state of the hyperstatic lattice under a system of equilibrated proportionate loads at the nodes has to be computed. In this approach the members, now as before, are considered to be *axially practically inextensible* on account of their slenderness, so that the *linear displacement–compatibility* relations at the nodes are *by no means* valid. These relations are replaced by the *kinematic compatibility conditions* in three-dimensional space derived in Sections 2.3 and 2.4 in Chapter 2, which were already used in the analysis of statically determinate lattices. The only difference now is that the Eq. (2.3.6) represents a rectangular (and *not* a square) matrix. Thus (**C**) is now a ($M \times N$) matrix, (M) being the number of the members. (**e**) and (**P**) are vectors of dimension ($M \times 1$). The displacement vector (**u**) has ($n = N$) elements, which represent the nodal freedoms of the lattice, which are (N) in number. Hence, additional kinematic relations for the redundant members must be derived in the sequel.

In hyperstatic (statically indeterminate) lattices we distinguish between the so called *intermediate critical states* and the *ultimate critical state*, from which a kinematically admissible mechanism, consisting of flexurally shortening members may form. Then, buckling of the system may take place either in equilibrium (usually in statically *unstable* equilibrium states) or in an unstable motion, which leads to the dynamic collapse of the system under constant critical loads. The intermediate critical states occur when, in the loading process, certain compressive members become critically stressed and are, "on the point of buckling". These members may be *constrained* by other redundant or non-redundant bars and, therefore, prevented from buckling. Buckling may develop only then, when a sufficient number of compressive members are simultaneously and in the required sequence on the point of buckling so that a kinematically admissible mechanism may form, in which the buckling members are no longer restrained by the surrounding bars and which may buckle freely, as this was the case in structurally determinate systems. The state from which such an admissible mechanism may form in a hyperstatic system is called the *ultimate critical state.* The ensuing buckling process may take place by the formation of a partial or a global mechanism of this type, this depending on the global geometry of the lattice and on the distribution of the external loads applied to it.

The method adopted here for the determination of the ultimate critical state involves a standard linear programming technique of the simplex method. We adopt here a direct approach based on the physical considerations and, as it turns out, on the simplex method, Ref. 30. The method was first developed by **Britvec** and **Nardini** in Ref. 16 and more comprehensively later by **Britvec** and **Davister** in Refs. 20 and 23 and by **Davister** in Ref. 27.

The critical state is evaluated by applying the statical equilibrium equations to the hyperstatic lattice, similarly as this was done for an isostatic (statically determinate) lattice in Chapter 2. These equations are formulated by satisfying the equilibrium conditions between the internal member forces and the external loads acting at the free or practically free joints in the lattice. The number of the independent joint displacements (freedoms) is denoted by $(n = N)$ and the number of members by (M). In a hyperstatic system $(M > N)$, whereas $(M = N)$, if the system is isostatic, as we have seen in Chapter 2. The equilibrium at the joints between the (M) member forces $(\mathbf{P})$ and the (N) external load components in the vector $(\mathbf{W})$ is then given by eq. (2.5.2)bis, written above. It is assumed that the external loads applied at the joints remain proportionate, while the critical load-parameter is being evaluated, then the $(N \times 1)$ load vector $(\mathbf{W})$ may be replaced by a load vector $(\boldsymbol{\beta})$ multiplied by a load-parameter (w) to give Eq. (2.5.4)bis viz (3.2.1). The $(N \times 1)$ vector $(\boldsymbol{\beta})$ corresponds to a particular loading pattern of conservative external loads acting on the supported lattice centrally at its joints, while the scalar (w) is treated as a variable in the loading process.

During this process it is required that each member force (P) satisfy a pair of inequality conditions or restraints $(-P_y \leq P \leq P^0)$, where (P_y) corresponds to the lower limit established by the tensile yield stress of the particular member and (P^0) to an upper limit, associated with the critical compressive load in the elastic range of deformation, which may be either the Euler buckling load or, more accurately, the experimental critical load for this member, to be determined, for example, from the Southwell plot (see Ref. 15, Chapter 3, p. 231). The upper limit may be also the compressive yield stress, if the particular pin-jointed member is prevented from buckling in the elastic range.

The problem of determining the critical state essentially involves finding the value of (w) in Eq. (3.2.1) which causes a sufficient number of members to buckle, such that a buckling mechanism may develop. In addition to satisfying the upper and the lower limits placed on the magnitude of each member force, the equilibrium equations must be satisfied at each intermediate and at the ultimate critical state. The existence of a buckling mechanism is established at each intermediate and at the ultimate critical state by checking if the *kinematic conditions* are satisfied, which permit the formation of a statically and kinematically admissible mechanism within the pin-jointed lattice.

The kinematic relations, relating the flexural shortenings $(\mathbf{e})$ of the buckling members to the displacements $(\mathbf{u})$ of the joints are derived in the same way as this was shown for isostatic systems in Section 3 in Chapter 2, in the form

$$\mathbf{e} = \mathbf{C}\mathbf{u} \tag{2.3.6)bis}$$

where now $(\mathbf{C})$ represents a rectangular (and not a square) kinematic matrix of a *hyperstatic* lattice. Namely, in these equations also the *redundant* members are included. If the number of the redundant members is denoted by (R), then the total number of members (M)

$$M = N + R \tag{3.2.2}$$

Therefore, $(\mathbf{C})$ is now an $(M \times N)$ matrix, the number of rows being equal to (M). We may partition the vector $(\mathbf{e})$ into two subvectors i.e. $(\mathbf{e}_1)$ and $(\mathbf{e}_2)$, where $(\mathbf{e}_1)$ contains the flexural shortenings of the members in *any* isostatic subsystem, which may be obtained from the given system by rejecting the (R) redundant members, the remaining (N) members, which make the structure just-stiff, being considered as the *isostatic* members. Obviously, there are several choices which can be made in this selection of the redundant members and of the corresponding isostatic subsystems within the lattice. At this stage we consider *any one* of these choices. Then, after partitioning,

$$\mathbf{e} = \begin{pmatrix} \mathbf{e}_1 \\ \hline \mathbf{e}_2 \end{pmatrix} = \begin{pmatrix} \mathbf{C}_1 \\ \hline \mathbf{C}_2 \end{pmatrix} \mathbf{u} = \mathbf{C}\mathbf{u} \tag{3.2.3}$$

where, the kinematic matrix $(\mathbf{C})$ must be partitioned accordingly, i.e.

$$\mathbf{C} = \begin{pmatrix} \mathbf{C}_1 \\ \hline \mathbf{C}_2 \end{pmatrix} \tag{3.2.3a}$$

It follows then, from (3.2.3) that,

$$\mathbf{e}_1 = \mathbf{C}_1 \mathbf{u} \tag{3.2.4}$$

and

$$\mathbf{e}_2 = \mathbf{C}_2 \mathbf{u} \tag{3.2.5}$$

For the isostatic system to be kinematically consistent the $(N \times N)$ $(\mathbf{C}_1)$ matrix must be necessarily *nonsingular*, since the joint displacements $(\mathbf{u})$ must be uniquely defined if the flexural contractions $(\mathbf{e}_1)$ in an isostatic system are prescribed. Therefore, Eq. (3.2.4) may be inverted to give

$$\mathbf{u} = \mathbf{C}_1^{-1} \mathbf{e}_1 \tag{3.2.6}$$

Substituting this in Eq. (3.2.5), we find that

$$\mathbf{e}_2 = \mathbf{C}_2\mathbf{C}_1^{-1}\mathbf{e}_1 \tag{3.2.7}$$

Therefore, the necessary kinematic condition for a kinematic buckling mechanism to exist at a critical state of the system is that the flexural contractions of the isostatic compressive critical members and those of the redundant members which buckle and shorten flexurally must be either positive or that some of these, but *not* all, may be zero. The kinematic conditions may be expressed in the form

$$\mathbf{e}_1 = \mathbf{C}_1\mathbf{u} > \mathbf{0} \tag{3.2.8a}$$

and

$$\mathbf{e}_2 = \mathbf{C}_2\mathbf{C}_1^{-1}\mathbf{e}_1 > \mathbf{0} \tag{3.2.8b}$$

In order to prove the existence of a kinematically admissible mechanism at *each intermediate* and the *ultimate critical* state the conditions (3.2.8) must be used.

If this condition cannot be satisfied at an intermediate critical state, this means that under the proportionate equilibrated external loads, applied to the structure at this state, a kinematic buckling mechanism cannot form and that the load parameter (w) of the proportionate applied loads *may be increased further* until the next intermediate critical state is reached in which sufficient compressive critical members in the new isostatic system are on the point of buckling.

It may be shown now, similarly as this was done in Section 5 in Chapter 2, that in the case of a hyperstatic lattice

$$\mathbf{C}^T = \mathbf{A} \tag{3.2.9}$$

and, therefore, according to (3.2.3a), that

$$\mathbf{C}_1^T = \mathbf{A}_1 \tag{3.2.9a}$$

and

$$\mathbf{C}_2^T = \mathbf{A}_2 \tag{3.2.9b}$$

i.e. that the transpose of the kinematic matrix equals the statical matrix in Eq. (3.2.1).

To determine the ultimate critical state numerically, one proceeds as follows: First, an isostatic subsystem of (N) members is selected from the (M) members by making the structure just-stiff. The other ($R = M - N$) members are thought to be temporarily disconnected, while the external loads are applied as prescribed, these being proportional to the load parameter (w). This hypothetical physical process may be pictured by the equivalent algebraic process in which the (N) out of the (M) columns in the statical matrix ($\mathbf{A}$), corresponding to the first selected isostatic system, are put into the leading (N) positions in the regrouped equilibrium equations. The vector ($\mathbf{P}$) of the internal bar forces must be regrouped accordingly. Let us denote these (N) leading columns in the matrix

($\mathbf{A}$) by the submatrix ($\mathbf{A}_\mathrm{I}$) and the remaining ($M - N = R$) columns by ($\mathbf{A}_2$). Similarly, the vector ($\mathbf{P}$) is regrouped and partitioned accordingly into the subvectors ($\mathbf{P}_\mathrm{I}$) and ($\mathbf{P}_2$). Then it follows from the Eq. (3.2.1) that

$$\mathbf{AP} = (\mathbf{A}_\mathrm{I} \mid \mathbf{A}_2)\begin{pmatrix} \mathbf{P}_\mathrm{I} \\ \hline \mathbf{P}_2 \end{pmatrix} = \mathbf{A}_\mathrm{I}\mathbf{P}_\mathrm{I} + \mathbf{A}_2\mathbf{P}_2 = \boldsymbol{\beta} w \qquad (3.2.1a)$$

Here ($\mathbf{A}_\mathrm{I}$) is the statical submatrix of the selected isostatic subsystem in which all the redundant members have been disconnected. Therefore, we set the axial loads in these members equal to zero, i.e. ($\mathbf{P}_2 = 0$). This gives

$$\mathbf{A}_\mathrm{I}\mathbf{P}_\mathrm{I} = \boldsymbol{\beta} w \qquad (3.2.1b)$$

The matrix ($\mathbf{A}_\mathrm{I}$) is nonsingular and it may be inverted. We note that now, as before ($\mathbf{A}_\mathrm{I} = \mathbf{C}_\mathrm{I}^T$), i.e. the statical matrix of any isostatic subsystem is equal to the transposed kinematic matrix of this system. From the last relation we obtain,

$$\mathbf{P}_\mathrm{I} = (\mathbf{A}_\mathrm{I}^{-1}\boldsymbol{\beta})w \qquad (3.2.1c)$$

As (w) is increased from zero, eventually one or simultaneously several axially compressive members in the isostatic subsystem are critically stressed and on the point of buckling. This is the *first intermediate* critical state. In this state the critically stressed members, when considered as a part of the entire system are *constrained* by the surrounding redundant bars and, therefore, prevented from buckling. Next, the isostatic system is modified by replacing the critically stressed members by other members in their respective locations derived from the (R) redundant bars previously disconnected. The critically stressed compressive bars from the previous isostatic subsystem are now replaced by the constant critical loads in these bars which are retained as external loads acting on the new modified isostatic subsystem. If more than one bar are simultaneously critical at the first intermediate critical state, one bar at a time must be replaced by another redundant bar reconnected in its respective location, forming a new isostatic subsystem, the critical load in the replaced bar being kept constant. Then, a check is performed in the other critical bars to verify if these *unload* and do not violate the restraints ($-\mathbf{P}_y < \mathbf{P}_\mathrm{I} < \mathbf{P}^0$) as ($w$) is increased. If that is the case, (w) can be increased further until a second intermediate critical state is reached in which other compressive isostatic members become critical. If, on the other hand, this is *not* the case and the axial loads in the remaining critical bars are found to *increase* in each modified isostatic subsystem in which always *one* different critical bar force (in the set of the critical bar forces at the first intermediate critical state) is kept constant, *two* critical bars of the original isostatic subsystem must be replaced at a time by *two* redundant replacement bars, which are reconnected in their respective locations. By choosing these bars, care must be taken that always a *just-stiff* isostatic subsystem is formed. The process is continued, similarly as before, checking that the member forces in the new isostatic subsystem associated with the vector ($\mathbf{P}_\mathrm{I}$) in Eq. (3.2.4) satisfy the pair of the inequality restraints ($-\mathbf{P}_y < \mathbf{P}_\mathrm{I} < \mathbf{P}^0$) as the load parameter (w) is increased from its previous intermediate critical value, all the other redundant bars, remaining temporarily disconnected.

If in the continuation of this selective testing process all the critical bars at the first intermediate critical state are found to increase, they must all be replaced by a set of the *same number* of different redundant bars forming a new isostatic subsystem (or bar forces derived from the vector ($\mathbf{P}_2$)), while the critical loads of the replaced bars are kept constant, as (w) is increased beyond the first intermediate critical value (w_1^0). In this way a second intermediate critical state of the modified isostatic subsystem is eventually reached in which additional critical compressive bar forces are identified. Once a second intermediate critical state is found the ongoing procedure is the same as at the first intermediate critical state, except that more critical compressive bars must be replaced by the remaining redundant bars including the critical bars, disconnected at the previous intermediate critical state *which now unload.* The axial loads in these bars must be checked out to make sure that they do not violate the inequality constraints ($-\mathbf{P}_y \leq \mathbf{P}_1 \leq \mathbf{P}^0$), as ($w$) is increased beyond ($w_2^0$). This process is continued until higher intermediate critical states of the load-parameter (w) are reached, ending in the final or lowest *ultimate critical state* in which ($w = w^0$), when the buckling of the structure is no longer constrained by the redundant members. In this loading process the redundant forces ($\mathbf{P}_2$) become either equal to one of the limit loads ($\mathbf{P}^0$) or ($\mathbf{P}_y$), or they are equal to zero, if the bars are unused. If an ultimate critical state is reached at which a *partial* and *not a global* kinematic buckling mechanism may form, the forces in some redundant bars may be equal to zero if these bars are not used. In that case the unused bars perform rigid body motions in the post-buckling process within the structure. In a global buckling mechanism, generally, all the redundant members are critically stressed by the limiting loads, when the compressive members within the *last isostatic subsystem* become critical. The redundant bars then shorten flexurally, as unrestrained buckling takes place. **Davister** (Ref. 27) has shown that the numerical procedure in this method of computing the ultimate critical state can be formulated as a linear programming problem, using a direct approach based on the simplex method, this resulting in substantial savings of the computer storage requirements and in a numerically more efficient computation. We illustrate this approach in the continuation of this analysis.

At each intermediate critical state, the kinematic conditions (3.2.8) must be used to see if a partial or a global buckling mechanism may form. Then the conditions (3.2.8) represent the necessary and sufficient *kinematic admissibility* conditions which permit the formation of a kinematically admissible mechanism within the pin-jointed system.

To answer the question whether such a mechanism is also *statically* admissible under the applied loading, additional *statical* admissibility conditions must hold, which we consider later. As long as the replacement redundant members have not been exhausted to form new isostatic subsystems and these subsystems can be loaded by increasing the parameter (w), then, in the continuation of this loading process also the statical equilibrium conditions are ensured.

Therefore, the criterion used to determine the existence of a kinematically admissible mechanism at a particular stage of loading is necessarily accompanied by the criterion used for checking whether (w) may be increased further by activating additional redundant bars in the manner described before and the problem may be formulated as follows:

Maximize (w), subject to the condition that

$$\mathbf{AP} = \boldsymbol{\beta} w \tag{3.2.1)bis}$$

where the vector of axial loads ($\mathbf{P}$) satisfies the inequality condition

$$-\mathbf{P}_y < \mathbf{P} < \mathbf{P}^0$$

The problem stated above is a linear programming problem, which could be solved using the simplex method but, as mentioned before, more conveniently a direct method, which differs from the standard simplex method, in that "slack variables" are not added to the list of variables, may be used instead. This method also results in a saving of the computer storage requirements.

The sequence of isostatic subsystems used in this computation is not unique as different isostatic subsystems may be formed at each intermediate critical state. Then, also the *last* isostatic subsystem, before the *ultimate critical* state is reached from which a kinematically admissible buckling mechanism may form, is *not* the *same* in each sequence. However, the value of the ultimate critical load parameter, at which this state is reached, as well as the *total* number of critical compressive members is the *same*, regardless of the sequence of basic isostatic subsystems adopted in the solution. The procedure to be carried out in the evaluation of the ultimate critical state of a hyperstatic system consisting of inextensible bars, marked by the ultimate value (w^0) of the load parameter and the formation of a kinematically (and statically) admissible post-buckling mechanism may be stated in the following

Summary

Initially, ($\mathbf{P} = \mathbf{0}$) and ($w = 0$). In the first and each subsequent incremental load step the equilibrium equations of the hyperstatic lattice are written according to (3.2.1a), in the form,

$$\mathbf{A}_1 \mathbf{P}_1 = \boldsymbol{\beta} w - \mathbf{A}_2 \mathbf{P}_2 \tag{3.2.10}$$

and, since ($\mathbf{A}_1$) is nonsingular, in the inverted form

$$\mathbf{P}_1 = \mathbf{A}_1^{-1} \boldsymbol{\beta} w - \mathbf{A}_1^{-1} \mathbf{A}_2 \mathbf{P}_2 \tag{3.2.10a}$$

in which at least some of the redundant axial loads ($\mathbf{P}_2$) are zero while others remain constant at their critical or limit values. Thus a change ($\Delta \mathbf{P}_2$) in the vector ($\mathbf{P}_2$) is zero between any two intermediate states. Writing Eq. (3.2.10) with respect to the incremental load step (Δw) gives

$$\Delta \mathbf{P}_1 = \mathbf{A}_1^{-1} \boldsymbol{\beta}\, \Delta w \tag{3.2.10b}$$

The load step (Δw) in Eq. (3.2.10b) is computed, so that a single isostatic compressive member or *simultaneously* several such members are driven to either their critical (Euler) or limiting loads, while all other axial loads satisfy the constraints,

$$-\mathbf{P}_y < \mathbf{P}_1 < \mathbf{P}^0 \tag{3.2.11}$$

as strict inequalities, $(\mathbf{P}_1)$ representing the instantaneous load vector in the loading step (Δw) and $(\mathbf{P}_y)$ and $(\mathbf{P}^0)$ the limiting vector forces. In the physical sense, an isostatic member under its limiting load is referred to as a *critical isostatic* member. When the computation in the loading step is completed, the vector $(\mathbf{P}_1)$ and the load-parameter (w) are updated to reflect the load parameter increase (Δw).

As (w) cannot be increased beyond the first intermediate critical state, without causing one or simultaneously several critical isostatic members to violate the limiting load values, it must be checked if a new isostatic subsystem may be formed, which allows (w) to be increased further, while the existing critical member forces are assumed to be acting on this new subsystem with constant magnitudes. The second isostatic subsystem is selected by reconnecting for each critical isostatic member another unstressed (temporarily disconnected) redundant member in its proper location, as explained earlier, so that the new system of isostatic members is statically consistent and just-stiff. In this step, any member with a zero axial load, or any member which unloads from its limiting value, as (w) is increased from (w_1^0), may be used as a replacement member.

The selection of a new isostatic subsystem, resulting from the replacement of critical isostatic members from the previous loading step by new redundant members, which are activated or reconnected in their respective locations, is equivalent to the *repartitioning* of the $(N \times M)$ statical $(\mathbf{A})$ matrix and the $(N \times 1)$ axial load vector $(\mathbf{P})$, now embodying the bar forces of the new isostatic subsystem. In practice, the equilibrium equations in the form in Eq. (3.2.10a) are obtained by performing a Gauss-Jordan reduction on the previous set of equilibrium equations, also given in the form of Eq. (3.2.10), where now some of the critical axial loads from the previous loading step appear as constant loads in the vector $(\mathbf{P}_2)$, while the axial loads in this vector, associated with the redundant members, which remain disconnected in the current loading step are zero. Consequently, Eq. (3.2.10a) may be used to compute the increases in the vector $(\mathbf{P}_1)$ in the current loading step, as well as the magnitude (Δw) at which the next intermediate critical state is reached.

In general, the following directive should be used: on completing the first loading step the remaining computational procedure to determine the *ultimate* critical state (when no further increase of (w) is possible and all the redundant members have been exhausted as physically possible replacement members) consists essentially of two alternating steps, as follows

step 1: (w) is increased by the amounts (Δw) until one or several isostatic members simultaneously become critically stressed, while all other isostatic members remain subcritical.

step 2: The critical isostatic members determined in step 1 are replaced by acceptable redundant members, which are reconnected to form a new isostatic subsystem, consisting from these and from the subcritical members in step 1, such that (w) can be increased further without violating the axial force constraints defined in Eq. (3.2.7).

The details in carrying out those two steps in a linear programming problem are discussed by **Livesley** in Ref. 37. The ultimate critical value ($w = w^0$) is reached when step 2 cannot be carried out.

At the *ultimate* critical state the equilibrium equations (3.2.10a) may be written in the form

$$\mathbf{P}_1^0 + \mathbf{A}_1^{-1}\mathbf{A}_2\mathbf{P}_2^0 = \mathbf{A}_1^{-1}\boldsymbol{\beta} w^0 \tag{3.2.12}$$

where the matrices $\mathbf{A}_1$ and $\mathbf{A}_2$ are the repartitioned submatrices of ($\mathbf{A}$) based on the *last* isostatic subsystem, before the ultimate critical state is reached. The superscript (0) denotes the values at this state. At the ultimate critical state the equilibrium equations are satisfied on the flexurally undistored geometry of the system in which all the members can be considered as *inextensible*. Axial elastic extensibility of the members is therefore *not* required, to determine the ultimate critical state of a hyperstatic lattice. This is usually *not* the case in conventional elasticity problems. The vectors ($\mathbf{P}_1^0$) and ($\mathbf{P}_2^0$) now contain the axial force values of the members at the ultimate critical state (as opposed to an intermediate critical state). The compressive members in these two vectors, which may buckle, are represented by their Euler loads (alternatively, experimental critical values may be used). Tensile or the other members are represented by the limiting or sub-limiting viz. subcritical load values evaluated at this state. Although the set of compressive critical members, stressed to the Euler loads, is unique in the whole hyperstatic system, the set of critical isostatic members in the last isostatic subsystem, before the ultimate critical state is reached, is *not* unique, as the sequence of the subsequent isostatic subsystems is arbitrary and immaterial for the computation of (w^0).

It remains to establish the criterion which is necessary to show that (w) cannot be increased further, because of the *existence of a kinematically admissible mechanism* at the particular critical state. To this end we adopt an approach, which is similar to that due to **Davister** (Ref. 27). We saw that the criterion, which must be satisfied for a partial or a global buckling mechanism to form, is essentially given by the kinematic conditions (3.2.8). If the relations (3.2.9a) and (3.2.9b) are also employed, we may rewrite the kinematic conditions as follows

$$\mathbf{e}_1 = \mathbf{C}_1\mathbf{u} = \mathbf{A}_1^T\mathbf{u} > \mathbf{0} \tag{3.2.13a}$$

and

$$\mathbf{e}_2 = \mathbf{C}_2\mathbf{C}_1^{-1}\mathbf{e}_1 = \mathbf{A}_2^T(\mathbf{A}_1^T)^{-1}\mathbf{e}_1 = \mathbf{A}_2^T(\mathbf{A}_1^{-1})^T\mathbf{e}_1$$

or

$$\mathbf{e}_2 = (\mathbf{A}_1^{-1}\mathbf{A}_2)^T\mathbf{e}_1 = \mathbf{C}_2\mathbf{C}_1^{-1}\mathbf{e}_1 \geq \mathbf{0} \tag{3.2.13b}$$

If, after the multiplication of $(\mathbf{A}_I^{-1})$ and $(\mathbf{A}_2)$ the transposed product matrix is denoted by $(\mathbf{c})$, i.e.

$$(\mathbf{A}_I^{-1}\mathbf{A}_2)^T = \mathbf{c} \tag{3.2.14}$$

then according to (3.2.13b),

$$\mathbf{e}_2 = (\mathbf{A}_I^{-1}\mathbf{A}_2)^T\mathbf{e}_1 = \left[\begin{array}{ccc|c|cc} c_{11} & c_{12} & \cdot & c_{1\alpha} & \cdot & c_{1N} \\ c_{21} & c_{22} & \cdot & c_{2\alpha} & \cdot & c_{2N} \\ \cdot & \cdot & \cdot & \cdot & \cdot & \cdot \\ \cdot & \cdot & \cdot & \cdot & \cdot & \cdot \\ c_{R1} & c_{R2} & \cdot & c_{R\alpha} & \cdot & c_{RN} \end{array}\right] \begin{bmatrix} e_1 \\ e_2 \\ \cdot \\ -- \\ e_\alpha \\ -- \\ \cdot \\ e_N \end{bmatrix} > \mathbf{0}, \quad \text{(column } \alpha = \mathbf{c}_\alpha) \tag{3.2.14a}$$

where $(R = M - N)$ and a dot represents a zero or nonzero element. Usually, some elements of the vector $(\mathbf{e}_1 = (e_1 e_2 \cdot e_\alpha \cdot e_N)^T)$ are zero, as not all the bars in the last isostatic subsystem are compressive, nor are they all critically stressed. Suppose, that only (c) of the (N) bars are critical, that may buckle (shorten flexurally). Then, in Eq. (3.2.14a), $(N = c)$, and the corresponding shortenings (e_1) to (e_c) are to be considered as the independent quantities (variables) related to the last isostatic subsystem. If only *one* isostatic member (α) buckles, then, because of the condition, $(e_i = 0;\ i \neq \alpha)$, and

$$(\mathbf{A}_I^{-1}\mathbf{A}_2)^T\mathbf{e}_1 = \mathbf{e}_2 = \begin{bmatrix} c_{1\alpha} \\ c_{2\alpha} \\ \cdot \\ \cdot \\ c_{R\alpha} \end{bmatrix} e_\alpha = \mathbf{c}_\alpha e_\alpha > \mathbf{0} \tag{3.2.15}$$

where $(\mathbf{c}_\alpha)$ represents the column vector followed by (e_α), all the coefficients $(c_{i\alpha};\ i = 1, 2, \ldots, R)$, multiplying the flexural contraction of the member (α), must be *positive* if a buckling mechanism exists.

Now, if we consider the axial force in the same member (α) and a single equilibrium equation in the set given in (3.2.12), associated with this critical isostatic member, then, because of the relationship (3.2.14) above,

$$(\mathbf{A}_I^{-1}\mathbf{A}_2)\mathbf{P}_2^0 = \mathbf{c}^T\mathbf{P}_2^0 \tag{3.2.16}$$

and in the critical state considered, the equilibrium equation (3.2.12) associated with the axial load (P_α) of the buckling isostatic member (α) has the form

$$P_\alpha^0 + \mathbf{c}_\alpha^T\mathbf{P}_2^0 = (\mathbf{A}_I^{-1}\boldsymbol{\beta})_\alpha w^0 \tag{3.2.17}$$

where now vector $(\mathbf{c}_\alpha^T)$ represents the (α^{th}) row of the matrix $(\mathbf{A}_I^{-1}\mathbf{A}_2)$, i.e.

$$P_\alpha^0 + c_{1\alpha}P_1^0 + c_{2\alpha}P_2^0 + \cdots + c_{R\alpha}P_R^0 = (\mathbf{A}^{-1}\boldsymbol{\beta})_\alpha w^0 \tag{3.2.17a}$$

and where the forces (P_1^0) to (P_R^0) in the vector $(\mathbf{P}_2^0)$ generally denote the positive critical axial loads in the redundant members, evaluated at the last intermediate critical state, before the current critical state. Clearly, the coefficients $(c_{1\alpha}), \ldots$

etc. must be *positive* because of the kinematic condition (3.2.15), if this current state is also the *ultimate critical state* and a buckling mechanism is kinematically admissible. In the loading step, immediately preceding this state, the *same* equation applies, only without the zero superscript on the axial load in the member (α), which may still be subcritically loaded. The coefficient $(\mathbf{A}_{\mathrm{I}}^{-1}\boldsymbol{\beta})_{\alpha}$ in Eq. (3.2.12) viz. (3.2.17a) represents the corresponding element (α) of the product vector $(\mathbf{A}_{\mathrm{I}}^{-1}\boldsymbol{\beta})$, which must be positive in order that (P_{α}) may increase as (w) increases, indicating that the member associated with the axial load (P_{α}) would ultimately *buckle*, if none of the redundant limiting forces (P_{I}^{0} to P_{R}^{0}) unloaded. This situation exists when the coefficients ($c_{i\alpha}$; $i = 1, 2, \ldots, R$) are positive or zero. If (P_{α}^{0}) is a tensile limiting load, where (P_{α}^{0}) is now negative, then, according to the kinematic relation (3.2.15), the matching coefficients $(c_{i\alpha})$ must be negative, if under a plastic extension of the bar (α) a kinematically admissible mechanism is to form. In that case, the coefficient $(A_{\mathrm{I}}^{-1}\beta)_{\alpha}$ must also be negative, for (P_{α}) to attain a negative limiting yield value under an increasing load parameter (w). Thus, in order that $(\mathbf{e}_2)$ may represent a kinematically admisible field of the nodes of the hyperstatic lattice, all the shortenings of the redundant members at the compressive limit of positive limiting loads must be positive or zero, whereas the shortenings at the tensile limit of negative limiting loads must be negative or zero.

Then, as (P_{α}) reaches its critical value (P_{α}^{0}), at least one *kinematically admissible mechanism* may develop with one member buckling, as required by the kinematic admissibility condition (3.2.15). This also means that the set of physically possible replacement members has been *exhausted*, when the signs of the coefficients $(c_{i\alpha})$ and $(\mathbf{A}_{\mathrm{I}}^{-1}\boldsymbol{\beta})_{\alpha}$ in the loading step (Δw) match the signs of the axial forces (P_{α}^{0}) in $(\mathbf{P}_{\mathrm{I}}^{0})$, i.e. the problem variables in Eq. (3.2.12).

In other words, the sufficient condition that (w) cannot be increased further without violating the limiting force constraints and that a kinematically *admissible* buckling mechanism may form is that all the non-zero coefficients $(c_{i\alpha})$ in the matrix

$$\mathbf{c} = (\mathbf{A}_{\mathrm{I}}^{-1}\mathbf{A}_2)^T = \mathbf{C}_2\mathbf{C}_{\mathrm{I}}^{-1}$$

in Eq. (3.2.13b), and, therefore, in the matrix

$$\mathbf{c}^T = (\mathbf{A}_{\mathrm{I}}^{-1}\mathbf{A}_2) = (\mathbf{C}_2\mathbf{C}_{\mathrm{I}}^{-1})^T$$

associated with compressive forces (P_{α}^{0}) in the equilibrium Eqs. (3.2.12), related to the ultimate critical state, must be positive or of the *same* sign as the corresponding axial loads (P_{α}^{0}) in the vector $(\mathbf{P}_{\mathrm{I}}^{0})$.

This criterion is also sufficient to prove that the critical state, determined by the solution of the linear programming problem defined by the Eqs. (3.2.1) and the inequality conditions (3.2.11), represents the ultimate critical state, beyond which (w) cannot be increased further and at which *at least one* kinematically admissible post-buckling mechanism with at least one isostatic member buckling, may develop. Although the set of members, which are stressed to their critical (Euler) loads in the hyperstatic structure is unique, the set of critical isostatic members in the last isostatic subsystem is *not* unique, as the sequence in choosing the isostatic subsystems is immaterial for the computation of (w^0).

This buckling process usually takes place in an unstable motion, if the applied loads are not adjusted in the post-buckling process to maintain the system in statical equilibrium. If this adjustment can be made, the equilibrium in post-buckling is usually statically unstable, as in the case of statically determinate lattices. However, unstable equilibrium may be maintained only under controlled laboratory conditions, as explained in Chapters 3 to 7 in Ref. 15. For a further discussion of this point the reader is referred to Ref. 15, p. 74.

3.3 Simplified Kinematic Admissibility Conditions for the Buckled Hyperstatic Lattice

We have already discussed the kinematic admissibility conditions at the ultimate critical state in terms of the flexural shortenings of the compressive members contained within the last isostatic subsystem, just before the ultimate critical state is reached. We showed that the vector $(\mathbf{e})$ of flexural contractions can be partitionated into $(\mathbf{e}_1)$, in which the contractions of the (N) isostatic buckling members within the last isostatic subsystem are contained, and the subvector $(\mathbf{e}_2)$ in which the remaining (R) contractions of the redundant buckling members appear. The kinematic $(M \times N)$ matrix $(\mathbf{C})$, where

$$M = N + R \tag{3.3.1}$$

which relates the vector $(\mathbf{e})$ to the nodal displacement vector $(\mathbf{u})$ in the basic kinematic relation (2.3.6) was then partitioned accordingly to give

$$\mathbf{e} = \mathbf{C}\mathbf{u} = \begin{pmatrix} \mathbf{e}_1 \\ \hline \mathbf{e}_2 \end{pmatrix} = \begin{pmatrix} \mathbf{C}_1 \\ \hline \mathbf{C}_2 \end{pmatrix} \mathbf{u} \qquad (3.2.3)\text{bis}$$

and, hence,

$$\mathbf{e}_1 = \mathbf{C}_1 \mathbf{u} \qquad (3.2.4)\text{bis}$$

or, after the inversion

$$\mathbf{u} = \mathbf{C}_1^{-1} \mathbf{e}_1 \qquad (3.2.6)\text{bis}$$

and, further, according to (3.2.3),

$$\mathbf{e}_2 = \mathbf{C}_2 \mathbf{u} = \mathbf{C}_2 \mathbf{C}_1^{-1} \mathbf{e}_1 \qquad (3.2.7)\text{bis}$$

It is possible to simplify these expressions further, if new subdivisions or partitions of the vectors $(\mathbf{e}_1)$ and $(\mathbf{e}_2)$ are introduced. Let, the flexural shortenings of the (c) compressive isostatic members (referred to the last isostatic subsystem before the ultimate critical state) be denoted by the subvector $(\mathbf{e}_c)$ and those of the (s) tensile or other members, which do not shorten flexurally, by $(\mathbf{e}_s)$, then $(N = c + s)$. Let us put the subvector $(\mathbf{e}_c)$ into the leading first (c) elements of the vector $(\mathbf{e}_1)$. Then the inverse kinematic relation (3.2.6) requires that also the kinematic submatrix $(\mathbf{C}_1^{-1})$ be regrouped and partitioned accordingly. In this way we obtain

$$\mathbf{u} = \mathbf{C}_1^{-1}\mathbf{e}_1 = (\mathbf{C}'_{1c} \mid \mathbf{C}'_{1s})\begin{pmatrix} \mathbf{e}_c \\ \hline \mathbf{e}_s \end{pmatrix} = \mathbf{C}'_{1c}\mathbf{e}_c \tag{3.3.2}$$

where $(\mathbf{e}_s = \mathbf{0})$, as defined. The prime on $(\mathbf{C}'_{1c})$ is introduced to remind the reader that the submatrix is a part of the *inverse* kinematic matrix $(\mathbf{C}_1^{-1})$ of *one* of the possible choices of isostatic subsystems. The advantage of this reduction is in that the number (c) of the buckling isostatic members is much smaller than the number $(n = N)$ of the freedoms or the nodal displacement components $(\mathbf{u})$ of the system, which are considered to depend on $(\mathbf{e}_c)$.

The kinematic relations may be simplified further, if the vector $(\mathbf{e}_2)$ is partitioned similarly. Let, the (γ) flexural contractions of the buckling redundant compressive members be denoted by $(\mathbf{e}_\gamma)$ and placed into the leading position of the vector $(\mathbf{e}_2)$. The (σ) shortenings of those redundant bars, which do not buckle are denoted by $(\mathbf{e}_\sigma)$. Then

$$\gamma + \sigma + u = R \tag{3.3.3}$$

where (R) represents the number of redundant members in the lattice.

Combining the Eqs. (3.3.1), (3.3.2), (3.3.3) we conclude that

$$c + s + \gamma + \sigma + u = M \tag{3.3.4}$$

This division results in *five* characteristic sets of members in a hyperstatic lattice in its post-buckling configuration.

In a buckling mechanism $(b + \gamma)$ members, where $(b \leq c)$ may buckle and shorten flexurally, whereas $(s' + \sigma)$ members do not contract *flexurally*. The (u) unstressed or unused members move and rotate as rigid bodies in a *partial* buckling mechanism. Here, as before $(b + s' = c + s)$, and (s') is the total number of non-buckling members within the *last* isostatic subsystem.

To simplify the kinematic conditions further, it is necessary to partition the vector $(\mathbf{e}_2)$ and kinematic submatrix $(\mathbf{C}_2)$, as required by the relation (3.2.7) and (3.3.4) as follows

$$\mathbf{e}_2 = \begin{pmatrix} \mathbf{e}_\gamma \\ -- \\ \mathbf{e}_\sigma \\ -- \\ \mathbf{e}_u \end{pmatrix} = \mathbf{C}_2\mathbf{u} = \begin{pmatrix} \mathbf{C}_\gamma \\ -- \\ \mathbf{C}_\sigma \\ -- \\ \mathbf{C}_u \end{pmatrix}\mathbf{u} = \begin{pmatrix} \mathbf{C}_\gamma \\ -- \\ \mathbf{C}_\sigma \\ -- \\ \mathbf{C}_u \end{pmatrix}\mathbf{C}'_{1c}\mathbf{e}_c \tag{3.3.5}$$

Finally, it follows from the last expression, that the following inequality and equality constraints must be satisfied, in order that a deformation state of the hyperstatic lattice be kinematically admissible.

$$\begin{array}{ll} \mathbf{e}_c \geq \mathbf{0} & \text{(a)} \\ \hline \mathbf{e}_\gamma = \mathbf{C}_\gamma \mathbf{C}'_{1c} \mathbf{e}_c \geq \mathbf{0} & \text{(b)} \\ \hline \mathbf{e}_\sigma = \mathbf{C}_\sigma \mathbf{C}'_{1c} \mathbf{e}_c \geq \mathbf{0} & \text{(c)} \\ \mathbf{e}_u = \mathbf{C}_u \mathbf{C}'_{1c} \mathbf{e}_c \geq \mathbf{0} & \text{(d)} \end{array} \qquad (3.3.6)$$

where ($\mathbf{e}_\sigma > \mathbf{0}$) implies that a nonbuckling member may extend or shorten *axially* under a limiting load. Clearly, the unstressed members suffer no change of length and, therefore, ($\mathbf{e}_u = \mathbf{0}$). *Usually*, the redundant nonbuckling members do not shorten so that also ($\mathbf{e}_\sigma = \mathbf{0}$). The flexural contractions of the redundant members are generally nonzero, as they all buckle and shorten flexurally. Therefore, the most important equation above is the Eq. (3.3.6b), which makes it possible to calculate the flexural contractions ($\mathbf{e}_\gamma$) of the redundant buckling members in the system in any kinematically, or a kinematically and statically admissible post-buckling mode.

3.4 Matrix Formulation of the General Law for Prismatic Pin-Jointed Members in a Hyperstatic Lattice

The general law introduced in Section 2.1, Eq. (2.1.1), in Chapter 2 may be generalized to a hyperstatic lattice similarly as this was done for isostatic lattices in Section 2.9 in the last chapter. Writing this law for a typical member (α) between the joints (i) and (j) we have

$$\delta P_\alpha = K_\alpha P_\alpha L_\alpha^{-1} e_\alpha; \qquad \alpha = 1, 2, \ldots, n \qquad (3.4.1)$$

Similarly as before, Section 2.9, we define the ($n \times n$) diagonal matrix

$$\mathbf{T}(\text{diag.}) = \mathbf{K}\mathbf{P}^0\mathbf{L}^{-1} \qquad (3.4.2),\ (2.9.6)\text{bis}$$

where ($\mathbf{K}$), ($\mathbf{P}^0$), ($\mathbf{L}^{-1}$) and ($\mathbf{T}$) are then given by the diagonal matrices in Eqs. (2.9.3), (2.9.4), (2.9.5) and (2.9.6) in Section 2.9 in Chapter 2, except that now (n) is not necessarily equal to either (N) or (M). Again, we define the diagonal matrix ($\mathbf{T}'$), where

$$\mathbf{T}' = \mathbf{P}^0\mathbf{L}^{-1} \qquad (3.4.3),\ (2.9.6\text{a})\text{bis}$$

so that

$$\mathbf{T} = \mathbf{K}\mathbf{T}' \qquad (3.4.4),\ (2.9.6\text{b})\text{bis}$$

as before. Then, the general law may be written for the entire system in the matrix form

$$\delta\mathbf{P} = \mathbf{T}\mathbf{e} \qquad (3.4.5)$$

where ($\delta\mathbf{P}$) represents the incremental axial loads beyond the ultimate critical state in the relevant set of members of the hyperstatic system, the tensile and the nonbuckling members included. The elements of $(\mathbf{e})$ are the flexural contractions of all the members, which are of the same order of magnitude as the corresponding elements of the vector $(\delta\mathbf{P})$. If a member buckles and it is prismatic the buckling contant $(K_\alpha = 1/2)$, whereas otherwise the (K_α) takes the values prescribed by this law in Section 2.1.

We next divide the members into the five categories defined in the last section. In a post-buckling mode, the maximum number of members which may buckle is then equal to $(c + \gamma)$. But, generally not all of the (c) critical isostatic members (referred to the last isosatic subsystem) shorten flexurally in a post-buckling mode, as some may unload and remain straight. Those subcritical members which remain straight may be considered as belonging to the set of (s') non-buckling members. If $(b \leq c)$ refers only to those critically stressed members which actually buckle and shorten flexurally, the general law in matrix form may be applied only to these (b) members to give

$$\delta\mathbf{P}_b = \mathbf{T}_b\mathbf{e}_b; \qquad b \leq c \tag{3.4.6}$$

where (c) is the number of the critical *isostatic* members which are on the point of buckling in the ultimate critical state. $(\mathbf{T}_b)$ in this case is a $(b \times b)$ nonsingular diagonal matrix, so that the incremental forces $(\delta\mathbf{P}_b)$ are considered as the independent variables, and the flexural shortenings $(\mathbf{e}_b)$ may be calculated from the last relation. Similarly, the general law applied to the redundant buckling members, in which the incremental axial loads are given by

$$\delta\mathbf{P}_\gamma = \mathbf{T}_\gamma\mathbf{e}_\gamma \tag{3.4.7}$$

where, again $(\mathbf{T}_\gamma)$ is a nonsingular $(\gamma \times \gamma)$ diagonal matrix which may be inverted if the flexural shortenings of the redundant members need be calculated. From the last two equations, it follows that, after the inversion,

$$\mathbf{e}_b = \mathbf{T}_b^{-1}\,\delta\mathbf{P}_b \tag{3.4.6a}$$

and

$$\mathbf{e}_\gamma = \mathbf{T}_\gamma^{-1}\,\delta\mathbf{P}_\gamma \tag{3.4.7a}$$

where the incremental axial loads are considered as the independent variables. We shall see that this approach leads again to a significant reduction of the governing equilibrium equations in a post-buckling mode to be solved numerically, similarly as this was the case for the isostatic lattices treated in Chapter 2.

3.5 Matrix Formulation of the General Post-Buckling Equilibrium Equations for Hyperstatic Pin-Jointed Lattices

The development of the equilibrium equations of the buckled hyperstatic lattice proceeds similarly as in Section 2.10 in the case of isostatic lattices. The only difference is in the total number (M) of members which is greater than the number of freedoms of the nodes. The number of freedoms and the number of the independent nodal displacement components in the vector ($\mathbf{u}$) is now, as before, equal to ($n = N$). Therefore, the Eq. (2.10.1), which refers to a single member is valid now as before, without any modifications. But, the augmentation of the (3×1) direction cosine vector ($\mathbf{C}_{ij}$) in the first term of the Eq. (2.10.1), must now be carried out by adding $(M - 1)$ null (3×1) column vectors to account for the alien members of the lattice, which are now $(M - 1)$ in number. Also, the $(3 \times M)$ ($\mathbf{C}^a_{ij}$) submatrix in Eq. (2.10.5) is augmented by $(N - 6)$ null rows and a ($-\mathbf{C}^a_{ij}$) submatrix to yield a $(N \times M)$ ($\mathbf{C}^{aa}_\alpha$) matrix, as explained in Eq. (2.10.12). Accordingly, the vector ($\delta\mathbf{P}$) is augmented by $(M - 1)$ zero elements. The superscripts (a) and (aa) indicate that the matrix is once or twice augmented as defined in section (2.10). The (3×6) matrices $(\mathbf{C}_{ij}\mathbf{C}^T_{ij} \mid -\mathbf{C}_{ij}\mathbf{C}^T_{ij})$ are again augmented by only $(N - 6)$ null (3×1) column vectors to account for the displacements in the vector ($\mathbf{u}$) which are alien to the member (α). Similarly, the (3×6) submatrix $(\mathbf{I}_{ij} \mid -\mathbf{I}_{ij})$ must be augmented by only $(N - 6)$ null (3×1) columns, so that submatrices ($\mathbf{C}^A_{ij}$) and ($\mathbf{I}^A$) are again of the order $(3 \times N)$ as defined in Eq. (2.10.3). Therefore, in a hyperstatic lattice of (M) members and ($n = N$) degrees of freedom the basic equations (2.10.8) and (2.10.9) embodying the incremental component forces at the node (i) derived from the adjoining member (α) between the joints (i) and (j) are valid as they stand, i.e.

$$\delta\mathbf{F}_{ij} = \delta\mathbf{F}_{\alpha i} = \mathbf{C}^a_{\alpha i}\,\delta\mathbf{P} + \frac{P^0_\alpha}{L_\alpha}\mathbf{C}^A_{\alpha i}\,\mathbf{u} - \frac{P^0_\alpha}{L_\alpha}\mathbf{I}^A_{\alpha i}\,\mathbf{u} \qquad (2.10.8)\text{bis}, (3.5.1)$$

and

$$\delta\mathbf{F}_{ji} = \delta\mathbf{F}_{\alpha j} = -\mathbf{C}^a_{\alpha i}\,\delta\mathbf{P} - \frac{P^0_\alpha}{L_\alpha}\mathbf{C}^A_{\alpha i}\,\mathbf{u} + \frac{P^0_\alpha}{L_\alpha}\mathbf{I}^A_{\alpha i}\,\mathbf{u} \qquad (2.10.9)\text{bis}, (3.5.2)$$

except that the submatrices ($\mathbf{C}^a_{\alpha i}$) and the vector ($\delta\mathbf{P}$) have the new dimensions $(3 \times M)$ and $(M \times 1)$ respectively.

The remaining development of these equations, as shown in Section 2.10, is the same as before. The totality of Cartesian components of the internal axial incremental loads at different joints is now equal to ($3M$), the direction-cosines of the additional components at a typical joint (i) now appearing in the appropriate $(M - N)$ additional columns of the matrix ($\mathbf{C}^a_{\alpha i}$). The number ($n = N$) of free directions of load application at the joints is the same as before. Therefore, the Eq. (2.10.12) holds unchanged except that the matrix ($\mathbf{C}^{aa}_\alpha$) is now a $(N \times M)$ matrix and, the vector ($\delta\mathbf{P}$) a $(M \times 1)$ column vector and ($\delta\mathbf{F}_\alpha$) a $(N \times 1)$ vector, as shown before in Section (2.10), i.e.

$$\delta\mathbf{F}_\alpha = \mathbf{C}^{aa}_\alpha\,\delta\mathbf{P} + \frac{P^0_\alpha}{L_\alpha}\mathbf{C}^{AA}_\alpha\,\mathbf{u} - \frac{P^0_\alpha}{L_\alpha}\mathbf{I}^{AA}_\alpha\,\mathbf{u} \qquad (2.10.13)\text{bis}, (3.5.3)$$

The matrices ($\mathbf{C}^{AA}_\alpha$) and ($\mathbf{I}^{AA}_\alpha$) are of dimension $(N \times N)$. It can be verified similarly as before that the relation (2.10.14) holds, i.e. that

$$\mathbf{C}^{aa}_\alpha\,\mathbf{C}^{aaT} = \mathbf{C}^{AA}_\alpha \qquad (2.16.14)\text{bis}, (3.5.4)$$

where $(\mathbf{C}_\alpha^{AA})$ is a *symmetrical* matrix and therefore

$$\frac{P_\alpha^0}{L_\alpha}\mathbf{C}_\alpha^{AA} = \frac{P_\alpha^0}{L_\alpha}\mathbf{C}_\alpha^{aa}\,\mathbf{C}_\alpha^{aaT} = \mathbf{C}_\alpha^{aa}\,\mathbf{T}'\mathbf{C}_\alpha^{aaT} \qquad (2.10.18)\text{bis}, (3.5.5)$$

where the $(\mathbf{T}')$ matrix is defined by the Eq. (3.4.3). With this simplification, the expression (3.5.3) becomes

$$\delta\mathbf{F}_\alpha = \mathbf{C}_\alpha^{aa}\,\delta\mathbf{P} + \mathbf{C}_\alpha^{aa}\mathbf{T}'\mathbf{C}_\alpha^{aaT}\,\mathbf{u} - \frac{P_\alpha^0}{L_\alpha}\mathbf{I}_\alpha^{AA}\,\mathbf{u} \qquad (3.5.6)$$

which is then identical to (2.10.19), except that now the matrix $(\mathbf{C}_\alpha^{aa})$ is of dimension $(N \times M)$.

The statical equilibrium equations for the distorted and buckled lattice are now written on the displaced configuration of the hyperstatic lattice similarly as this was done for an isostatic lattice in Section 2.11. $(\boldsymbol{\beta}')$ denotes again the load proportionality vector in post-buckling, as opposed to $(\boldsymbol{\beta})$, which was used in the subcritical range, (in most cases, however, $(\boldsymbol{\beta}' = \boldsymbol{\beta})$). By equating the incremental external load vector $(\delta\mathbf{W} = \boldsymbol{\beta}'\,\delta w)$ to the vector $(\delta\mathbf{F})$ of internal incremental loads we get the equilibrium conditions at all nodes, similarly as in the case of isostatic lattices, i.e.

$$\delta\mathbf{W} = \boldsymbol{\beta}'\,\delta w = \delta\mathbf{F} = \sum_{\alpha=1}^{M}\delta\mathbf{F}_\alpha; \qquad \alpha = 1, 2, \ldots, M \qquad (3.5.7)$$

where (M) is the total number of members. Evaluating $(\delta\mathbf{F})$, we find

$$\delta\mathbf{F} = \sum_{\alpha=1}^{M}\delta\mathbf{F}_\alpha = \sum_{\alpha=1}^{M}\mathbf{C}_\alpha^{aa}\,\delta\mathbf{P} + \sum_{\alpha=1}^{M}(\mathbf{C}_\alpha^{aa}\,\mathbf{T}'\mathbf{C}_\alpha^{aaT})\mathbf{u} - \sum_{\alpha=1}^{M}\left(\frac{P_\alpha^0}{L_\alpha}\mathbf{I}_\alpha^{AA}\right)\mathbf{u} \qquad (3.5.8)$$

The last equation is then identical to the Eq. (2.11.3), except that now the summation is carried out over (M) instead of (N) members. It is also clear that

$$\sum_{\alpha=1}^{M}\mathbf{C}_\alpha^{aa} = \mathbf{A} = \mathbf{C}^T \qquad (3.5.9)$$

now represents the statical matrix which is of the order $(N \times M)$ and which is identical to the kinematic matrix $(\mathbf{C})$ of order $(M \times N)$ transposed, as stipulated by Eq. (3.2.9). Then also, similarly as in Eqs. (2.11.8) and (2.11.9) in Chapter 2,

$$\sum_{\alpha=1}^{M}\mathbf{C}_\alpha^{aa}\,\mathbf{T}'\mathbf{C}_\alpha^{aaT} = \mathbf{C}^T\mathbf{T}'\mathbf{C} \qquad (3.5.10)$$

where $(\mathbf{C})$ represents the $(M \times N)$ kinematic matrix of the hyperstatic lattice. Because of the relation (3.2.9) the last result becomes

$$\sum_{\alpha=1}^{M}\mathbf{C}_\alpha^{aa}\mathbf{T}'\mathbf{C}_\alpha^{aaT} = \mathbf{A}\mathbf{T}'\mathbf{A}^T \qquad (3.5.11)$$

where $(\mathbf{A})$ is the $(N \times M)$ statical matrix of the hyperstatic lattice introduced in Eq. (3.2.1). The last term in Eq. (3.5.8) represents the $(\mathbf{P}_L^0)$ matrix, first introduced in Eq. (2.11.11), where now the summation embodies all the members, i.e.

$$\mathbf{P}_L^0 = \sum_{\alpha=1}^{M} \frac{P_\alpha^0}{L_\alpha} \mathbf{I}_\alpha^{AA} \tag{3.5.12}$$

Note that $(\mathbf{P}_L^0)$ is a *symmetrical* matrix about the main diagonal, because it is a sum of symmetrical matrices. Thus the result of Eq. (3.5.8) may be written simply

$$\delta \mathbf{F} = \mathbf{A}\, \delta \mathbf{P} + (\mathbf{A}\mathbf{T}'\mathbf{A}^T - \mathbf{P}_L^0)\mathbf{u} \tag{3.5.13}$$

The last result is formally identical to the result of Eq. 2.11.12 for isostatic systems, except that now in hyperstatic lattices the statical matrix $(\mathbf{A})$ has the dimension $(N \times M)$, the diagonal matrix $(\mathbf{T}')$ the dimension $(M \times M)$ and the matrix $(\mathbf{P}_L^0)$, embodying the contributions of the isostatic and the redundant members, again the dimension $(N \times N)$. The *equilibrium condition* of the buckled *hyperstatic* lattice under the applied loads is then obtained by equating

$$\delta \mathbf{F} = \sum_{\alpha=1}^{M} \delta \mathbf{F}_\alpha = \mathbf{A}\, \delta \mathbf{P} + (\mathbf{A}\mathbf{T}'\mathbf{A}^T - \mathbf{P}_L^0)\mathbf{u} = \boldsymbol{\beta}'\, \delta w \tag{3.5.14}$$

If also $(\mathbf{A})$ is replaced by $(\mathbf{A} = \mathbf{C}^T)$ we get the general form of the equilibrium equations of a hyperstatic buckled lattice in the form

$$\boxed{\mathbf{C}^T \delta \mathbf{P} + (\mathbf{C}^T\mathbf{T}'\mathbf{C} - \mathbf{P}_L^0)\mathbf{u} = \boldsymbol{\beta}'\, \delta w} \tag{3.5.15}$$

This form is identical to that in Eq. (2.11.15), except for the dimension of the matrix $(\mathbf{C})$, which is $(M \times N)$, and of the vector $(\delta \mathbf{P})$ which is $(M \times 1)$. We observe that the matrix $(\mathbf{C}^T\mathbf{T}'\mathbf{C} - \mathbf{P}_L^0)$ is symmetrical about its main diagonal. This follows from Eq. (3.5.12) and the result that the symmetrical matrix $(\mathbf{T}')$ (diag.) premultiplied by the transpose of the matrix by which it is post-multiplied, is also a symmetrical $(N \times N)$ matrix.

3.6 Reduction of the General Equilibrium Equations of the Hyperstatic Lattice and Their Solution

The reduction of the equilibrium equations given by (3.5.15) proceeds similarly as in the case of isostatic lattices, except that now there are five instead of only two groups of members according to the divisions (c), (s), (γ), (σ) and (u). The groups $(c + s = N)$ constitute the isostatic subsystem before the ultimate critical state is reached.

We partition the matrices $(\mathbf{C})$, $(\mathbf{T}')$ and $(\delta \mathbf{P})$ as follows

$$\mathbf{C} = \begin{bmatrix} \mathbf{C}_1 \\ \hline \mathbf{C}_\gamma \\ \hline \mathbf{C}_\sigma \\ \hline \mathbf{C}_u \end{bmatrix}; \quad \mathbf{T}' = \left[\begin{array}{c|c|c|c} \mathbf{T}'_N & \mathbf{0} & \mathbf{0} & \mathbf{0} \\ \hline \mathbf{0} & \mathbf{T}'_\gamma & \mathbf{0} & \mathbf{0} \\ \hline \mathbf{0} & \mathbf{0} & \mathbf{T}'_\sigma & \mathbf{0} \\ \hline \mathbf{0} & \mathbf{0} & \mathbf{0} & \mathbf{T}'_u \end{array}\right]; \quad \delta \mathbf{P} = \begin{bmatrix} \delta \mathbf{P}_N \\ \hline \delta \mathbf{P}_\gamma \\ \hline \delta \mathbf{P}_\sigma \\ \hline \delta \mathbf{P}_u \end{bmatrix}. \tag{3.6.1}$$

where $(\mathbf{C}_1)$ represents the already familiar $(N \times N)$ submatrix referred to the last isostatic subsystem and the submatrices $(\mathbf{C}_\gamma)$, $(\mathbf{C}_\sigma)$, $(\mathbf{C}_u)$, respectively, constitute the kinematic submatrix $(\mathbf{C}_2)$ of order $(M-N) \times N$ in Eq. (3.2.3). The indices (N), (γ), (σ) and (u) refer to the last set of the isostatic members, the redundant compressive, the tensile and the non-buckling compressive straight redundant and the unstressed redundant members, respectively. Accordingly,

$$\mathbf{C}_1 = \begin{pmatrix} \mathbf{C}_c \\ \hline \mathbf{C}_s \end{pmatrix} \quad \text{and} \quad \mathbf{C}_2 = \begin{bmatrix} \mathbf{C}_\gamma \\ \hline \mathbf{C}_\sigma \\ \hline \mathbf{C}_u \end{bmatrix} \tag{3.6.2}$$

where $(\mathbf{C}_1)$ is of order $(N \times N)$, $(\mathbf{C}_c)$ of order $(c \times N)$ and $(\mathbf{C}_s)$ of order $(s \times N)$. On substituting in (3.5.15) and carrying out the matrix multiplications, we get

$$\begin{aligned} &\mathbf{C}_1^T\, \delta\mathbf{P}_N + \mathbf{C}_\gamma^T\, \delta\mathbf{P}_\gamma + \mathbf{C}_\sigma^T\, \delta\mathbf{P}_\sigma + \mathbf{C}_u^T\, \delta\mathbf{P}_u + (\mathbf{C}_1^T\mathbf{T}'_N\mathbf{C}_1 + \mathbf{C}_\gamma^T\mathbf{T}'_\gamma\mathbf{C}_\gamma \\ &\quad + \mathbf{C}_\sigma^T\mathbf{T}'_\sigma\mathbf{C}_\sigma + \mathbf{C}_u^T\mathbf{T}'_u\mathbf{C}_u - \mathbf{P}_L^0)\mathbf{u} = \boldsymbol{\beta}'\, \delta w \end{aligned} \tag{3.6.3}$$

The diagonal matrix $(\mathbf{T}'_u = \mathbf{0})$, since in the unused redundant members, in the case of a partial mechanism, $(\mathbf{P}_u^0 = \mathbf{0})$. Thus, the term, involving $(\mathbf{T}'_u)$ in Eq. (3.6.3) vanishes. On premultiplying the last equation by the inverse matrix $(\mathbf{C}_1^T)^{-1} = (\mathbf{C}_1^{-1})^T$, one finds

$$\begin{aligned} &\mathbf{I}\, \delta\mathbf{P}_N + (\mathbf{C}_1^{-1})^T\mathbf{C}_\gamma^T\, \delta\mathbf{P}_\gamma + (\mathbf{C}_1^{-1})^T\mathbf{C}_u^T\, \delta\mathbf{P}_u + (\mathbf{C}_1^{-1})^T\mathbf{C}_\sigma^T\, \delta\mathbf{P}_\sigma \\ &\quad + [\mathbf{T}'_N\mathbf{C}_1 + (\mathbf{C}_1^{-1})^T\mathbf{C}_\gamma^T\mathbf{T}'_\gamma\mathbf{C}_\gamma + (\mathbf{C}_1^{-1})^T\mathbf{C}_\sigma^T\mathbf{T}'_\sigma\mathbf{C}_\sigma - (\mathbf{C}_1^{-1})^T\mathbf{P}_L^0]\mathbf{u} \\ &= (\mathbf{C}_1^{-1})^T\boldsymbol{\beta}'\, \delta w \end{aligned} \tag{3.6.4}$$

Partitioning now the $(\mathbf{C}_1^{-1})$ matrix to match the partitioning of the vector $(\mathbf{e}_1 = \mathbf{e}_N)$ into the subvectors $(\mathbf{e}_c)$ and $(\mathbf{e}_s = \mathbf{0})$, according to Eq. (3.3.2), and similarly the diagonal matrix $(\mathbf{T}'_N)$ and the vector $(\delta\mathbf{P}_N)$, i.e.

$$\mathbf{C}_1^{-1} = (\mathbf{C}'_{1c} \mid \mathbf{C}'_{1s}) \qquad (3.6.5),\ (3.4.3)\text{bis}$$

and

$$\mathbf{T}'_N = \begin{pmatrix} \mathbf{T}'_c & \mathbf{0} \\ \mathbf{0} & \mathbf{T}'_s \end{pmatrix} \tag{3.6.6}$$

$$\delta\mathbf{P}_N = \begin{pmatrix} \delta\mathbf{P}_c \\ \hline \delta\mathbf{P}_s \end{pmatrix} \tag{3.6.7}$$

we get

$$\begin{aligned} &\begin{pmatrix} \delta\mathbf{P}_c \\ \hline \delta\mathbf{P}_s \end{pmatrix} + \begin{pmatrix} \mathbf{C}'^T_{1c} \\ \hline \mathbf{C}'^T_{1s} \end{pmatrix}\mathbf{C}_\gamma^T\, \delta\mathbf{P}_\gamma + \begin{pmatrix} \mathbf{C}'^T_{1c} \\ \hline \mathbf{C}'^T_{1s} \end{pmatrix}\mathbf{C}_\sigma^T\, \delta\mathbf{P}_\sigma + \begin{pmatrix} \mathbf{C}'^T_{1c} \\ \hline \mathbf{C}'^T_{1s} \end{pmatrix}\mathbf{C}_u^T\, \delta\mathbf{P}_u \\ &\quad + \begin{pmatrix} \mathbf{T}'_c & \mathbf{0} \\ \mathbf{0} & \mathbf{T}'_s \end{pmatrix}\begin{pmatrix} \mathbf{C}_c \\ \hline \mathbf{C}_s \end{pmatrix}\mathbf{u} + \begin{pmatrix} \mathbf{C}'^T_{1c} \\ \hline \mathbf{C}'^T_{1s} \end{pmatrix}\mathbf{C}_\gamma^T\mathbf{T}'_\gamma\mathbf{C}_\gamma\mathbf{u} \\ &\quad + \begin{pmatrix} \mathbf{C}'^T_{1c} \\ \hline \mathbf{C}'^T_{1s} \end{pmatrix}\mathbf{C}_\sigma^T\mathbf{T}'_\sigma\mathbf{C}_\sigma\mathbf{u} - \begin{pmatrix} \mathbf{C}'^T_{1c} \\ \hline \mathbf{C}'^T_{1s} \end{pmatrix}\mathbf{P}_L^0\mathbf{u} = \begin{pmatrix} \mathbf{C}'^T_{1c} \\ \hline \mathbf{C}'^T_{1s} \end{pmatrix}\boldsymbol{\beta}'\, \delta w \end{aligned} \tag{3.6.8}$$

Performing now the matrix multiplication and separating the upper from the lower submatrices we obtain two sets of *uncoupled* equilibrium equations in the form

$$\begin{aligned}&\delta\mathbf{P}_c + \mathbf{C}'^T_{1c}\mathbf{C}^T_\gamma\,\delta\mathbf{P}_\gamma + \mathbf{C}'^T_{1c}\mathbf{C}^T_\sigma\,\delta\mathbf{P}_\sigma + \mathbf{C}'^T_{1c}\mathbf{C}^T_u\,\delta\mathbf{P}_u + \mathbf{T}'_c\mathbf{C}_c\mathbf{u}\\&\quad + \mathbf{C}'^T_{1c}\mathbf{C}^T_\gamma\mathbf{T}'_\gamma\mathbf{C}_\gamma\mathbf{u} + \mathbf{C}'^T_{1c}\mathbf{C}^T_\sigma\mathbf{T}'_\sigma\mathbf{C}_\sigma\mathbf{u} - \mathbf{C}'^T_{1c}\mathbf{P}^0_L\mathbf{u} = \mathbf{C}'^T_{1c}\boldsymbol{\beta}'\,\delta w \qquad (3.6.9)\end{aligned}$$

and

$$\begin{aligned}&\delta\mathbf{P}_s + \mathbf{C}'^T_{1s}\mathbf{C}^T_\gamma\,\delta\mathbf{P}_\gamma + \mathbf{C}'^T_{1s}\mathbf{C}^T_\sigma\,\delta\mathbf{P}_\sigma + \mathbf{C}'^T_{1s}\mathbf{C}^T_u\,\delta\mathbf{P}_u + \mathbf{T}'_s\mathbf{C}_s\mathbf{u}\\&\quad + \mathbf{C}'^T_{1s}\mathbf{C}^T_\gamma\mathbf{T}'_\gamma\mathbf{C}_\gamma\mathbf{u} + \mathbf{C}'^T_{1s}\mathbf{C}^T_\sigma\mathbf{T}'_\sigma\mathbf{C}_\sigma\mathbf{u} - \mathbf{C}'^T_{1s}\mathbf{P}^0_L\mathbf{u} = \mathbf{C}'^T_{1s}\boldsymbol{\beta}'\,\delta w \qquad (3.6.10)\end{aligned}$$

If only ($b \leq c$) out of the (c) critical compressive members in the last isostatic subsystem buckle when the ultimate critical state is reached, the first set (3.6.9) can be reduced further, if the general law for the critical compressive members is invoked at this stage. Then, writing, according to (3.4.6), (3.3.2), (3.4.7) and (3.3.6(b)),

$$\delta\mathbf{P}_c = \mathbf{T}_c\mathbf{e}_c \qquad (3.6.11)$$

$$\mathbf{u} = \mathbf{C}'_{1c}\mathbf{e}_c \qquad (3.3.2)\text{bis}$$

and

$$\delta\mathbf{P}_\gamma = \mathbf{T}_\gamma\mathbf{e}_\gamma = \mathbf{T}_\gamma\mathbf{C}_\gamma\mathbf{C}'_{1c}\mathbf{e}_c \qquad (3.6.12)$$

and substituting these expressions in (3.6.9)

$$\begin{aligned}&\mathbf{T}_c\mathbf{e}_c + \mathbf{C}'_{1c}\mathbf{C}^T_\gamma\mathbf{T}_\gamma\mathbf{C}_\gamma\mathbf{C}'_{1c}\mathbf{e}_c + \mathbf{C}'^T_{1c}\mathbf{C}^T_\sigma\,\delta\mathbf{P}_\sigma + \mathbf{C}'^T_{1c}\mathbf{C}^T_u\,\delta\mathbf{P}_u + \mathbf{T}'_c\mathbf{C}_c\mathbf{C}'_{1c}\mathbf{e}_c\\&\quad + \mathbf{C}'^T_{1c}\mathbf{C}^T_\gamma\mathbf{T}'_\gamma\mathbf{C}_\gamma\mathbf{C}'_{1c}\mathbf{e}_c + \mathbf{C}'^T_{1c}\mathbf{C}^T_\sigma\mathbf{T}'_\sigma\mathbf{C}_\sigma\mathbf{C}'_{1c}\mathbf{e}_c - \mathbf{C}'^T_{1c}\mathbf{P}^0_L\mathbf{C}'_{1c}\mathbf{e}_c\\&\quad = \mathbf{C}'^T_{1c}\boldsymbol{\beta}'\,\delta w \qquad (3.6.13)\end{aligned}$$

now,

$$\mathbf{C}_c\mathbf{C}'_{1c} = \mathbf{I} \qquad (3.6.14)$$

because ($\mathbf{C}'_{1c}$) is the inverse of ($\mathbf{C}_c$) and if the substitution according to the Eq. (3.4.4) is made

$$\mathbf{T}' = \mathbf{K}^{-1}\mathbf{T} \qquad (3.6.15)$$

where, for the buckling of prismatic members

$$\mathbf{K}^{-1} = \frac{1}{K}\mathbf{I} \qquad (3.6.15\text{a})$$

where ($K = 1/2$) is constant, then, Eq. (3.6.13) simplifies further to,

$$\begin{aligned}&\mathbf{T}_c\mathbf{e}_c + \mathbf{C}'^T_{1c}\mathbf{C}^T_\gamma\mathbf{T}_\gamma\mathbf{C}_\gamma\mathbf{C}'_{1c}\mathbf{e}_c + \mathbf{C}'^T_{1c}\mathbf{C}^T_\sigma\,\delta\mathbf{P}_\sigma + \mathbf{C}'^T_{1c}\mathbf{C}^T_u\,\delta\mathbf{P}_u + \mathbf{K}^{-1}_c\mathbf{T}_c\mathbf{e}_c\\&\quad + \mathbf{C}'^T_{1c}\mathbf{C}^T_\gamma\mathbf{K}^{-1}_\gamma\mathbf{T}_\gamma\mathbf{C}_\gamma\mathbf{C}'_{1c}\mathbf{e}_c + \mathbf{C}'^T_{1c}\mathbf{C}^T_\sigma\mathbf{T}'_\sigma\mathbf{C}_\sigma\mathbf{C}'_{1c}\mathbf{e}_c - \mathbf{C}'^T_{1c}\mathbf{P}^0_L\mathbf{C}'_{1c}\mathbf{e}_c\\&\quad = \mathbf{C}'^T_{1c}\boldsymbol{\beta}'\,\delta w \qquad (3.6.16)\end{aligned}$$

If all the members which are on the point of buckling are prismatic, relation (3.6.15a) holds and the post-buckling equilibrium equations simplify further to

$$\mathbf{E}_c\mathbf{e}_c + \mathbf{C}'^T_{1c}\mathbf{C}^T_\sigma\,\delta\mathbf{P}_\sigma + \mathbf{C}'^T_{1c}\mathbf{C}^T_u\,\delta\mathbf{P}_u = \mathbf{C}'^T_{1c}\boldsymbol{\beta}'\,\delta w \qquad (3.6.17)$$

where

$$\mathbf{E}_c = (\mathbf{I}+\mathbf{K}_c^{-1})\mathbf{T}_c + \mathbf{C}_{1c}'^T\mathbf{C}_\gamma^T(\mathbf{I}+\mathbf{K}_\gamma^{-1})\mathbf{T}_\gamma\mathbf{C}_\gamma\mathbf{C}_{1c}' + \mathbf{C}_{1c}'^T\mathbf{C}_\sigma^T\mathbf{T}_\sigma'\mathbf{C}_\sigma\mathbf{C}_{1c}' - \mathbf{C}_{1c}'^T\mathbf{P}_L^0\mathbf{C}_{1c}' \tag{3.6.18}$$

In particular, for prismatic buckling members only,

$$\mathbf{E}_c = (1+1/K)(\mathbf{T}_c + \mathbf{C}_{1c}'^T\mathbf{C}_\gamma^T\mathbf{T}_\gamma\mathbf{C}_\gamma\mathbf{C}_{1c}') + \mathbf{C}_{1c}'^T\mathbf{C}_\sigma^T\mathbf{T}_\sigma'\mathbf{C}_\sigma\mathbf{C}_{1c}' - \mathbf{C}_{1c}'^T\mathbf{P}_L^0\mathbf{C}_{1c}' \tag{3.6.19}$$

At this stage we note another important property of the matrix ($\mathbf{E}_c$). The matrix ($\mathbf{T}_\gamma$) is a diagonal matrix and, therefore, pre- and post-multiplied twice by transposed matrices yields a symmetrical matrix. Similarly, ($\mathbf{T}_\sigma'$). Then, ($\mathbf{T}_c$) is diagonal and therefore also symmetrical. We saw before that ($\mathbf{P}_L^0$) is symmetrical and, therefore, after post- and pre-multiplication by ($\mathbf{C}_{1c}'$) and its transpose yields a symmetrical matrix. Thus, also the sum and the matrix ($\mathbf{E}_c$) itself must be *symmetrical* about its main diagonal. This result will be of importance later when we discuss the stability of the system in the critical state.

If the axial plastic deformation of the (σ) redundant members takes place under constant axial limiting loads, then ($\delta\mathbf{P}_\sigma = \mathbf{0}$). If this is not the case ($\delta\mathbf{P}_\sigma$) must be known from the law of plastic deformation. Similarly, the unstressed members usually remain unstressed in the post-buckling range in the case of a partial mechanism. In that case also ($\delta\mathbf{P}_u = \mathbf{0}$). With these conditions the post-buckling equilibrium equations (3.6.17) simplify further to

$$\boxed{\mathbf{E}_c\mathbf{e}_c = \mathbf{C}_{1c}'^T\boldsymbol{\beta}'\,\delta w} \tag{3.6.20}$$

where ($\mathbf{E}_c$) is given by the Eq. (3.6.18) or (3.6.19). ($\mathbf{E}_c$) is called the post-buckling *stiffness or stability matrix.*

These equilibrium equations apply also to isostatic lattices made of slender members. In that case there are *no* redundant members and consequently ($\mathbf{T}_\gamma = \mathbf{0}$) and ($\mathbf{T}_\sigma = \mathbf{0}$). In that case the Eqs. (3.6.20) and (3.6.18) reduce to

$$\mathbf{E}_c\mathbf{e}_c = \mathbf{C}_{1c}^T\boldsymbol{\beta}'\,\delta w \tag{3.6.21}$$

where now, according to (2.8.3), ($\mathbf{u} = \mathbf{C}_{1c}\mathbf{e}_c$), and by comparing with (3.3.2), it follows that,

$$\mathbf{C}_{1c}' = \mathbf{C}_{1c} \tag{3.6.22}$$

and

$$\begin{aligned}\mathbf{E}_c &= (\mathbf{I}+\mathbf{K}_c^{-1})\mathbf{T}_c - \mathbf{C}_{1c}'^T\mathbf{P}_L^0\mathbf{C}_{1c}' \\ &= (\mathbf{I}+\mathbf{K}_c^{-1})\mathbf{T}_c - \mathbf{C}_{1c}^T\mathbf{P}_L^0\mathbf{C}_{1c}\end{aligned} \tag{3.6.23}$$

But, the last result is identical to the Eq. (2.12.13a) derived earlier for isostatic reticulated systems. The last result also serves as a check on the development in this section.

The Eqs. (3.6.20) must be next solved. This solution is now easier since the number of unknowns is at most equal to (c) which is usually less than (M) or (N). Therefore, this also results in a saving of the computer storage and time. Assuming that (b) out of (c) critical compressive members actually buckle in an equilibrium mode and shorten flexurally, the remaining ($c-b$) members are then treated as straight and those are included into the group of (s') non-buckling bars. Eqs. (3.6.20) then reduce to the final form

$$\boxed{\mathbf{E}_b \mathbf{e}_b = \mathbf{C}'^T_{1b} \boldsymbol{\beta}' \, \delta w \, ; \qquad b \le c} \tag{3.6.24}$$

where

$$\mathbf{E}_b = (\mathbf{I} + \mathbf{K}_b^{-1})\mathbf{T}_b + \mathbf{C}'^T_{1b} \mathbf{C}^T_\gamma (\mathbf{I} + \mathbf{K}_\gamma^{-1}) \mathbf{T}_\gamma \mathbf{C}_\gamma \mathbf{C}'_{1b} + \mathbf{C}'^T_{1b} \mathbf{C}^T_\sigma \mathbf{T}'_\sigma \mathbf{C}_\sigma \mathbf{C}'_{1b} \\ - \mathbf{C}'^T_{1b} \mathbf{P}^0_L \mathbf{C}'_{1b} \tag{3.6.25a}$$

or, in the case when the members are prismatic

$$\mathbf{E}_b = (1 + 1/K)(\mathbf{T}_b + \mathbf{C}'^T_{1b} \mathbf{C}^T_\gamma \mathbf{T}_\gamma \mathbf{C}_\gamma \mathbf{C}'_{1b}) + \mathbf{C}'^T_{1b} \mathbf{C}^T_\sigma \mathbf{T}'_\sigma \mathbf{C}_\sigma \mathbf{C}'_{1b} \\ - \mathbf{C}'^T_{1b} \mathbf{P}^0_L \mathbf{C}'_{1b} \tag{3.6.25b}$$

The matrix ($\mathbf{E}_c$) is of order ($c \times c$) and ($\mathbf{E}_b$) of order ($b \times b$). The solution of these equations follows on inverting the ($b \times b$) matrix ($\mathbf{E}_b$). Then, from (3.6.24)

$$\mathbf{e}_b = \mathbf{E}_b^{-1} \mathbf{C}'^T_{1b} \boldsymbol{\beta}' \, \delta w \ge \mathbf{0} \tag{3.6.26}$$

Here (δw) must be of *one* sign and ($\mathbf{e}_b$) must satisfy the above inequality. Otherwise the buckling mode is *statically inadmissible* under the prescribed loads. Once the flexural contractions ($\mathbf{e}_b$), in an equilibrium post-buckling mode of the system are known in terms of the load-parameter change (δw), the remaining variables in the problem can be found in terms of (δw) from the relations derived previously. The partitioning of the (N) isostatic members is now carried out according to the (b) buckling isostatic compressive members, referred to the last isostatic subsystem before the ultimate critical state is reached, and the (s') straight members of this subsystem, which include those critical compressive bars which unload in post-buckling.

The incremental axial loads in these bars are next calculated from the second uncoupled set of equilibrium Eqs. (3.6.10), by a direct elimination of the vector ($\delta \mathbf{P}_s$) which now becomes ($\delta \mathbf{P}_{s'}$). The flexural contractions ($\mathbf{e}_\gamma$) of the redundant members are found from Eq. (3.3.6b) and the incremental axial loads in these members from (3.4.7). Then

$$\mathbf{e}_\gamma = \mathbf{C}_\gamma \mathbf{C}'_{1b} \mathbf{e}_b \ge 0 \tag{3.4.7)bis, (3.6.27}$$

If ($\mathbf{e}_b$) does not satisfy the above inequality, the buckling mode is *kinematically inadmissible*. Similarly,

$$\delta \mathbf{P}_\gamma = \mathbf{T}_\gamma \mathbf{e}_\gamma = \mathbf{T}_\gamma \mathbf{C}_\gamma \mathbf{C}'_{1b} \mathbf{e}_b \tag{3.4.7)bis, (3.6.28}$$

and

$$\mathbf{u} = \mathbf{C}'_{1b} \mathbf{e}_b \, ; \qquad b \le c \tag{3.3.2)bis, (3.6.29}$$

Then, in this case, from Eq. (3.6.10),

$$\delta \mathbf{P}_{s'} = -\mathbf{C}'^T_{1s'}\mathbf{C}^T_\gamma \mathbf{T}_\gamma \mathbf{C}_\gamma \mathbf{C}'_{1b}\mathbf{e}_b - \mathbf{T}'_s \mathbf{C}_{s'}\mathbf{C}'_{1b}\mathbf{e}_b - \mathbf{C}'^T_{1s'}\mathbf{C}^T_\gamma \mathbf{T}'_\gamma \mathbf{C}_\gamma \mathbf{C}'_{1b}\mathbf{e}_b$$
$$- \mathbf{C}'^T_{1s'}\mathbf{C}^T_\sigma \mathbf{T}'_\sigma \mathbf{C}_\sigma \mathbf{C}'_{1b}\mathbf{e}_b + \mathbf{C}'^T_{1s'}\mathbf{P}^L_0 \mathbf{C}'_{1b}\mathbf{e}_b + \mathbf{C}'^T_{1s'}\boldsymbol{\beta}'\,\delta w$$

or by Eq. (3.6.26)

$$\delta \mathbf{P}_{s'} = \boldsymbol{\gamma}_{s'}\,\delta w \tag{3.6.30}$$

where the $(s' \times 1)$ vector $(\boldsymbol{\gamma}_{s'})$ is obtained from the expression

$$\boldsymbol{\gamma}_{s'} = [\mathbf{C}'^T_{1s'} - (\mathbf{C}'^T_{1s'}\mathbf{C}^T_\gamma \mathbf{T}_\gamma \mathbf{C}_\gamma \mathbf{C}'_{1b} + \mathbf{T}'_{s'}\mathbf{C}_{s'}\mathbf{C}'_{1b} + \mathbf{C}'^T_{1s'}\mathbf{C}^T_\gamma \mathbf{T}'_\gamma \mathbf{C}_\gamma \mathbf{C}'_{1b}$$
$$+ \mathbf{C}'^T_{1s'}\mathbf{C}^T_\sigma \mathbf{T}'_\sigma \mathbf{C}_\sigma \mathbf{C}'_{1b} - \mathbf{C}'^T_{1s'}\mathbf{P}^0_L \mathbf{C}'_{1b})\mathbf{E}^{-1}_b \mathbf{C}'^T_{1b}]\boldsymbol{\beta}' \tag{3.6.31}$$

The last result may be re-written by Eq. (3.4.4) as,

$$\boldsymbol{\gamma}_{s'} = \{\mathbf{C}'^T_{1s'} - [\mathbf{C}'^T_{1s'}\mathbf{C}^T_\gamma(\mathbf{I}_\gamma + \mathbf{K}_\gamma)\mathbf{T}'_\gamma \mathbf{C}_\gamma \mathbf{C}'_{1b} + \mathbf{T}'_{s'}\mathbf{C}_{s'}\mathbf{C}'_{1b}$$
$$+ \mathbf{C}'^T_{1s'}\mathbf{C}^T_\sigma \mathbf{T}'_\sigma \mathbf{C}_\sigma \mathbf{C}'_{1b} - \mathbf{C}'^T_{1s'}\mathbf{P}^0_L \mathbf{C}'_{1b}]\mathbf{E}^{-1}_b \mathbf{C}'^T_{1b}\}\boldsymbol{\beta}'$$

If all the buckling redundant members are *prismatic*, the above expression simplifies, since $(\mathbf{K}_\gamma = 1/2\mathbf{I}_\gamma)$ and $(\mathbf{I}_\gamma + \mathbf{K}_\gamma = 3/2\mathbf{I}_\gamma)$, where $(\mathbf{I}_\gamma)$ is a $(\gamma \times \gamma)$ unit matrix.

Those critical members which unload and remain straight are now reckoned to belong to the group (s'). The $(c - b)$ incremental axial loads $[\delta P_i;\ i = 1, \ldots, (c - b)]$ in those members must be negative for the mode to be statically admissible.

With this evaluation of the problem variables, the solution of the post-buckling equations (3.5.15) or the reduced equations (3.6.24) and (3.6.10) is complete. However, the equations must be solved for *each* physically possible buckling mode $(1, 2, \ldots, b \le c)$. The number of combinations that (b) can assume is considerable as, in principle, all the combinations, which may be formed out of (c) critical compressive members in sets of (b) members, must be tested. If in a mode, (i) isostatic critical compressive members buckle, the question arises how many combinations of (i) members out of (c) there are. The answer is

$$C^c_i = \frac{c!}{(c - i)!\,i!} \tag{3.6.32}$$

where (i) may take the values from 1 to c and (C^c_i) stands for the number of combinations. The total number (m_c) of modes is then equal to

$$m_c = \sum_{i=1}^{c} C^c_i = \sum_{i=1}^{c} \frac{c!}{(c - i)!\,i!} \tag{3.6.33}$$

For example, if $c = 5$

$$m_c = \sum_{i=1}^{5} C^5_i = 5 + 10 + 10 + 5 + 1 = 31$$

Not all of those modes are necessarily statically and kinematically admissible. Only those, which satisfy the inequality conditions (3.6.26) and (3.3.6b) may actually develop in statical equilibrium.

The question now arises which of the possible modes actually develops in practice under the influence of the intrinsic geometrical and other imperfections inherent in the system. To answer this question one must resort to some empirical verification. Experiments show that, usually, that mode develops, *which involves the greatest conversion of the total potential into the kinetic energy*. Such modes usually take place in *unstable motion* under constant external loads which act on the structure in the ultimate critical state. However, some of these modes may be realized under the equilibrium conditions, if the external loads are appropriately adjusted by the vector ($\boldsymbol{\beta}' \, \delta w$), as explained earlier.

Therefore, it is necessary to distinguish between the most degrading *equilibrium* modes and the most degrading *dynamic collapse* modes, taking place from the ultimate critical *equilibrium* state. In the equilibrium modes, the conversion of the total potential energy is measured according to the amount that would be converted into the kinetic energy, if in the same mode, *under the same deformation of the system*, the system were to collapse by motion under the original *constant* critical loads, acting on the system in its ultimate critical state just before the onset of flexural buckling. It must be emphasized that these two types of most degrading modes can be geometrically or kinematically *very different*. How this converted energy can be calculated from an equilibrium test, in the case of the most degrading equilibrium mode, is shown briefly in Section 1.2, p. 11 and in Ref. 15, Chapter 1-7., *Theorem 5*, p. 74 and the reader is advised to refer to this source for a better understanding of the physical process under discussion, which is explained there on several examples in Chapters 3, 6 and 7.

Our next task is to explore how such a *most degrading* mode may be *evaluated directly*, without resorting to the comparison of the modes obtained by the solution of the Eqs. (3.6.24) and the use of Theorem 5 in Ref. 15, p. 74 or the Theorem 1 in Chapter 1, p. 13.

Example:
Model Reticulated Shell

Before discussing this direct evaluation of the most degrading mode, the theory developed so far is illustrated on a model reticulated shell, Figs. 3.6.1a to 3.6.1d, intended for subsea applications, Ref. 20, which, in its given form, is referred to as the unoptimized model or Model No. 1. In practice, the shell lattice would be enveloped by a continuous cover-shell supported at the peripheral outer nodes. These supports permit a transmission of the hydrostatic pressure onto the lattice at these nodes. However, the structural effects, such as a possible stiffening of the lattice by this cover-shell, are not considered in the present analysis.

The lattice of the shell is made up of tetrahedron and octahedron space elements, which, in turn, are composed of flexible bars of circular or hollow circular cross-sections made of high-quality tensile steel. These polyhedral space elements fill the space bounded on top and bottom by five equal spherical triangles of an

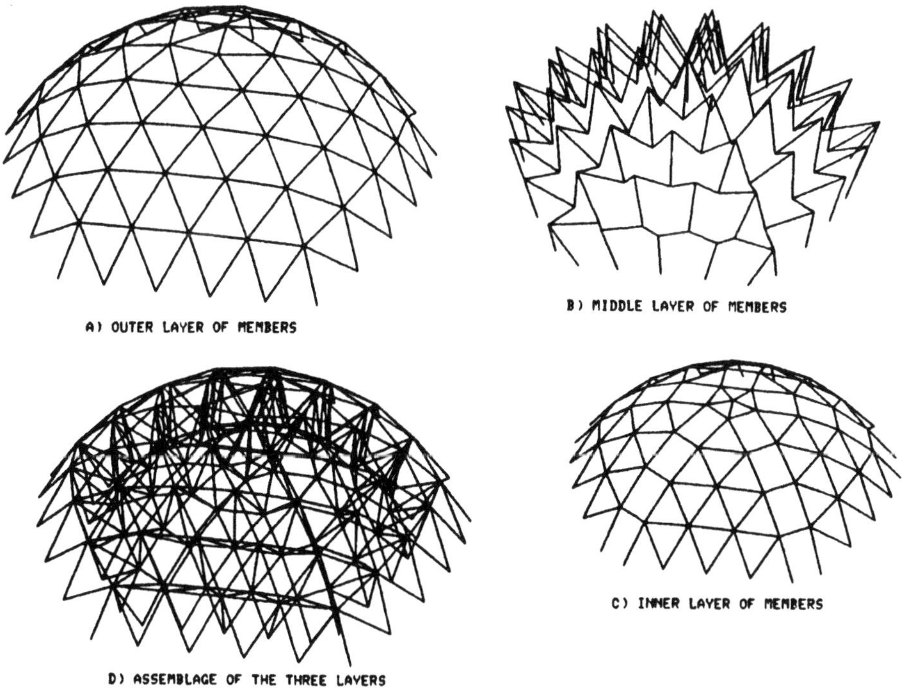

Figure 3.6.1
Geometry of the model reticulated shell and its three divisions (layers)

icosahedron. Twenty such triangles form the icosahedron on a complete sphere. The spaces along the boundaries between the five spherical triangles are filled with pentahedron space elements. Each spherical triangle on the outer sphere is divided into five equal parts. The dividing points are joined to adjacent points by great circles to yield further triangles. The subdivision was performed so that the least number of different member lengths resulted from it. Each subdivision then contains a straight member between its ends. The inner shell was divided similarly, except that triangles were segmented and separated between the adjacent icosahedrons by the bars of the pentahedron space elements. The cover shell on the outer layer is not shown, as it was not considered in the analysis.

The members were conceived as hollow or solid tubular bars made of high tensile steel with a yield point of 50 kips/in.^{2} (3,515.35 kg*[3]/cm^2). The diameter of each member was selected to be 1/20 of the member's length, allowing each compressive member to buckle before yielding. A list of member lengths and diameters is presented in Table 1. On the outer sphere, Fig. 3.6.1a, five basic bar lengths were used. The inner sphere, Fig. 3.6.1c, was subdivided using eight different bar lengths and the middle layer, Fig. 3.6.1b, i.e. the thickness of the shell lattice, which is the space between the two single layer shells (outer and inner), was filled with an array of members resulting in a series of tetrahedral, octahedral and

[3] * implies kg weight.

Table 1
List of member properties for the reticulated shell subjected to hydrostatic pressure. (Unoptimized Model No. 1), (Britvec, S.J. and Davister, M.D., Ref. 23)

Member set (1)	Number of members in set (2)	Member length, in feet (3)	Member outside diameter, in inches (4)
In outer layer			
1	25	4.4196	2.652
2	15	5.1673	3.098
3	60	4.7172	2.830
4	30	5.0843	3.051
5	45	4.9378	2.963
Subtotal	175		
In middle layer			
6	1	4.0000	2.400
7	5	4.6691	2.801
8	20	4.6914	2.815
9	20	4.7092	2.826
10	20	4.7215	2.833
11	30	4.6721	2.803
12	20	4.7278	2.837
13	30	4.7038	2.822
14	15	4.7153	2.829
15	15	4.6654	2.799
Subtotal	176		
In inner layer			
16	5	2.7963	1.678
17	5	2.1542	1.292
18	5	2.5265	1.516
19	40	3.5356	2.121
20	10	2.7280	1.673
21	15	3.8892	2.334
22	45	3.7064	2.224
23	30	3.8233	2.294
Subtotal	155		

pentahedral space elements, erected on the inside of the outer shell in such a way, that each element fitted exactly between the segments of the two divided concentric spheres. To this end ten different sets of members were used as shown in Table 1.

The overall dimensions of the shell model may be characterized by the outside diameter of its base (35.75 ft. = 10.8966 m) and by its height which varied along the boundary of the shell-lattice between (9.55 ft. = 2.911 m) to (11.06 ft. = 3.371 m). The diameter of the outer sphere of the shell was (40 ft. = 12.192 m) and that of the inner (separate) shell (32 ft. = 9.754 m). Thus the thickness of the shell-lattice amounted to (4 ft. or 1.219 m).

The lattice consisted of 506 members ($M = 506$) and 306 degrees of freedom ($n = N = 306$). Therefore, 200 members ($M - N = R = 200$) were redundant. The lattice had 152 joints, 50 of which were fixed to rigid foundation (such as a ring-footing). Their identification is shown in Figs. 3.6.2a to 3.6.2c. The identification of the members in the three layers is shown in Figs. 3.6.3a to 3.6.3c.

Three versions of the reticulated shell were analysed. The first is referred to as the *unoptimized* model, the second, as the *first optimized* and the third, as the *second optimized* model shell. The number of critically stressed members in the last isostatic subsystem of the unoptimized shell used in this computation was 5 ($c = 5$). For the first and the second optimized model reticulated shells this number was 10 ($c = 10$). The unoptimized shell was found capable of carrying some (272 ft. = 82.90 m) of seawater, the first optimized model (280 ft. = 85.34 m) and the second optimized model shell about (417 ft. = 127.10 m). The optimization in this case consisted in stiffening selected members in the lattice, as it is explained later. In the ultimate critical state of the unoptimized model 200 compressive members were found to be on the point of buckling, 5 of these being part of the last isostatic subsystem, just before this state was reached. Only 8 admissible equilibrium modes were found in the unoptimized model shell. Thus, there were 195 redundant buckling members in each mode in the unoptimized model shell and the number of isostatic buckling bars varied from 3 to 5, as indicated in Figs. 3.6.4a to 3.6.4c and in Fig. 3.6.5 and Fig. 3.6.6. Altogether 506

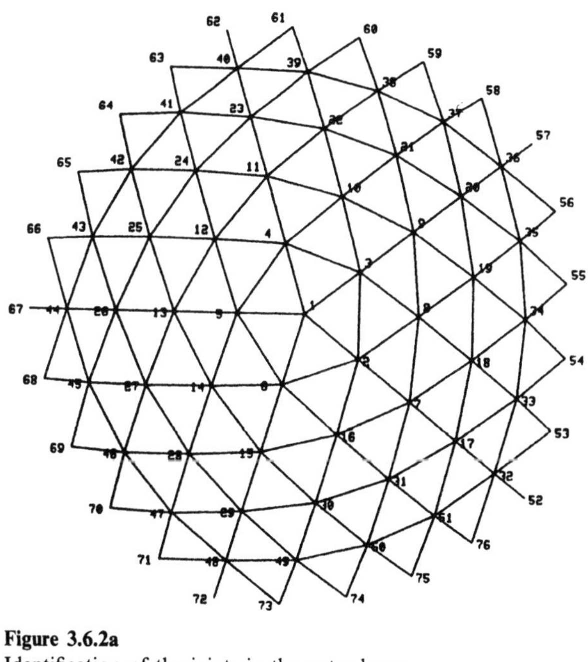

Figure 3.6.2a
Identification of the joints in the outer layer

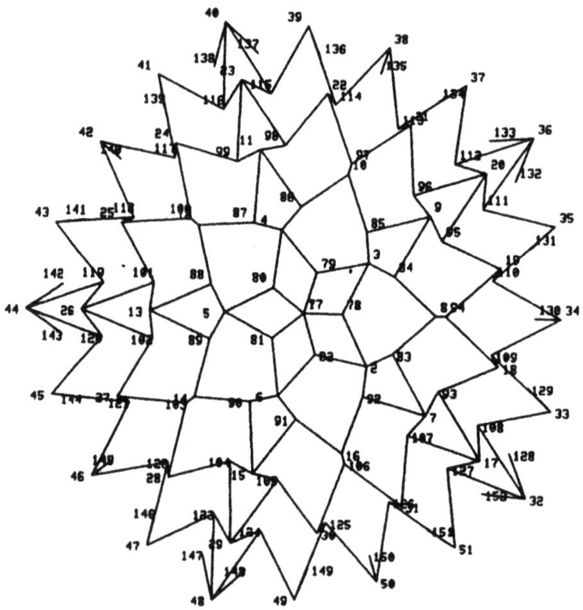

Figure 3.6.2b
Identification of the joints in the middle layer

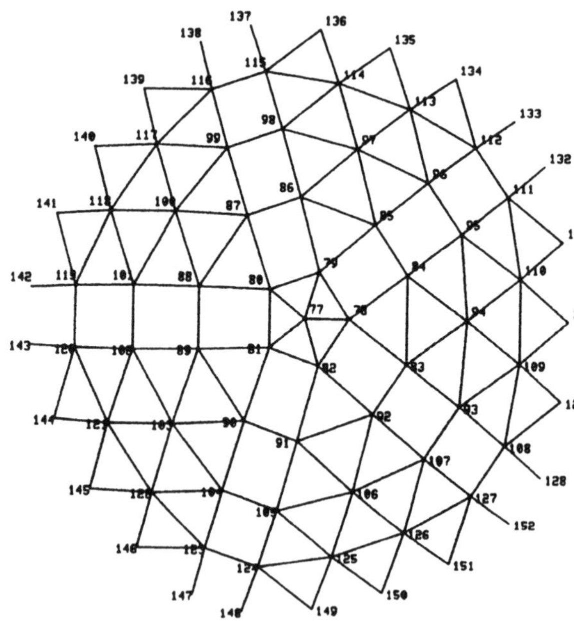

Figure 3.6.2c
Identification of the joints in the inner layer

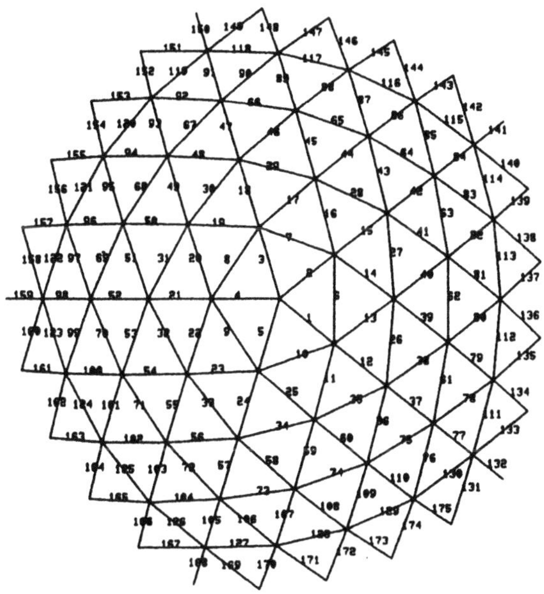

Figure 3.6.3a
Identification of the members in the outer layer

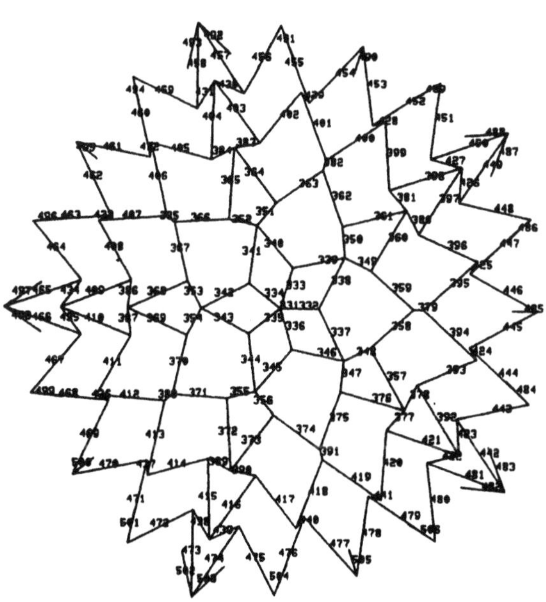

Figure 3.6.3b
Identification of the members in the middle layer

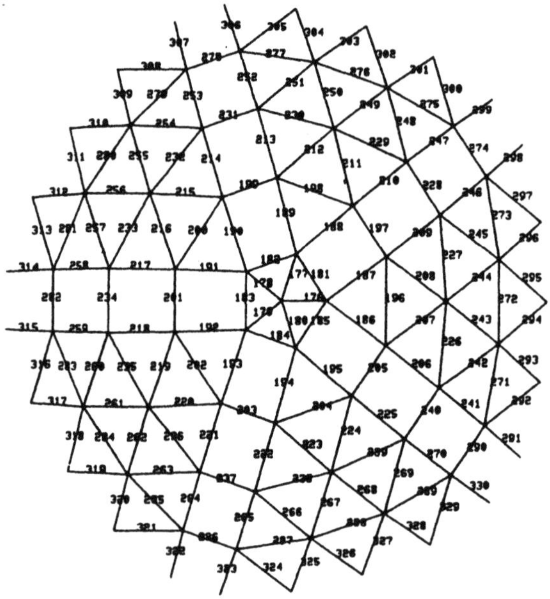

Figure 3.6.3c
Identification of the members in the inner layer

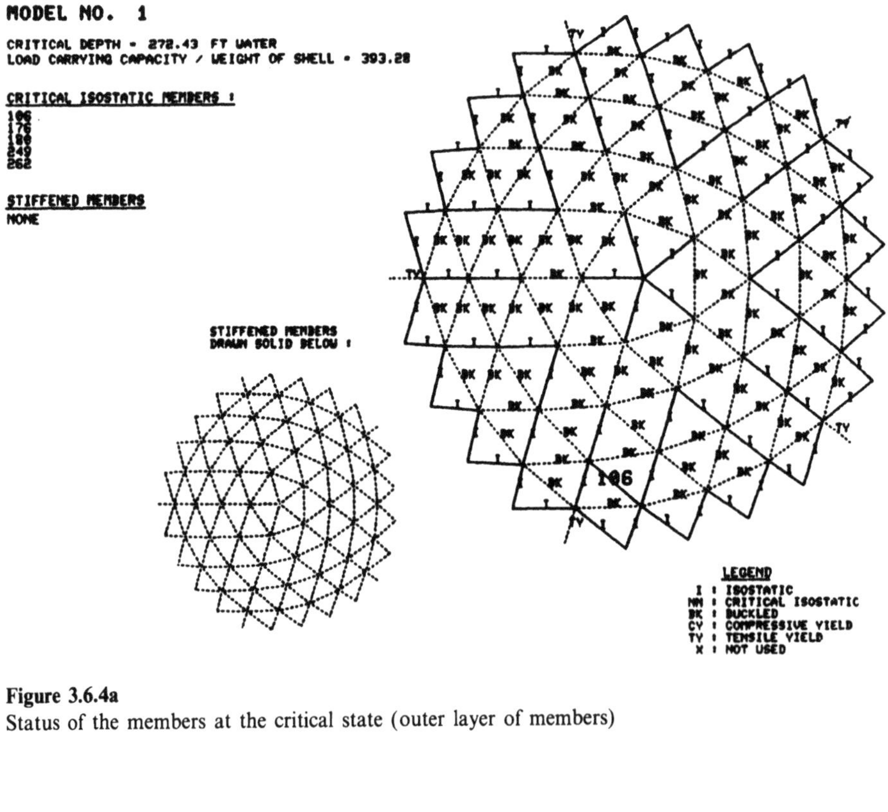

Figure 3.6.4a
Status of the members at the critical state (outer layer of members)

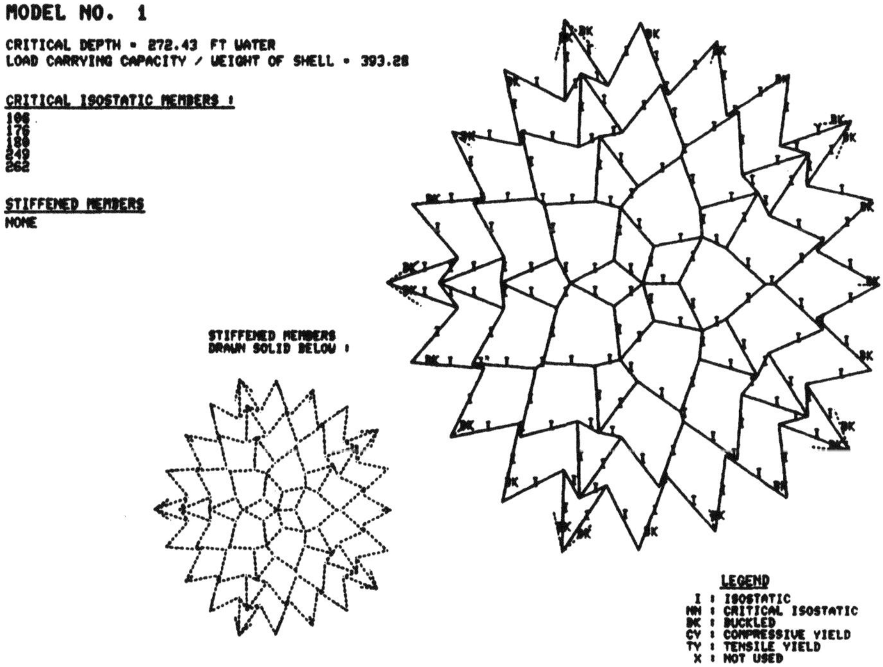

Figure 3.6.4b
Status of the members at the critical state (middle layer of members)

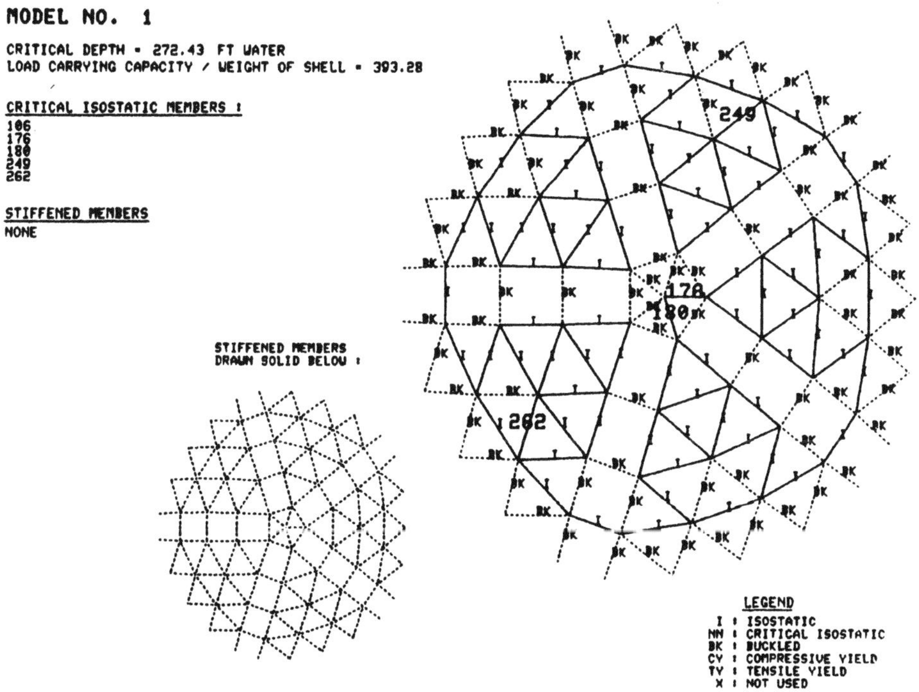

Figure 3.6.4c
Status of the members at the critical state (inner layer of members)

members were involved in each buckling mechanism. All mechanisms were found to be global.

Thus, only 3 to 5 equations of the type (3.6.24) had to be solved, instead of the 306, to find the modal vectors ($\mathbf{e}_b$), which represented all the combinations of the elements of the vector ($\mathbf{e}_c$). In the case of ($c = 5$) there were 31 independent combinations of the vector ($\mathbf{e}_b$) but only 8 were found to be statically and kinematically admissible. These modes were numbered 16, 17, 20, 26, 27, 28, 29, and 31. Only the modes 16 and 31 are presented in Figs. 3.6.4 and 3.6.5. In the case of the second optimized model the number of critical compressive members varied from 5 to 10. In the case of ($c = 10$), for example, 1023 such combinations, defining the independent post-buckling modes in equilibrium were discovered but only 116 of these were found to be kinematically and statically admissible. To evaluate these modes, in the first case, 5 equations were solved 31 times on a digital computer, and in the second case (second optimized model), 10 equations were solved 1023 times. However, only 8 of the 31 combinations yielded statically and kinematically admissible post-buckling modes (these were numbered 16, 17, 20, 26, 27, 28, 29 and 31).

A similar procedure would have to be carried out for any other reference isostatic subsystem (last subsystem), which might have been obtained in a different sequence of intermediate isostatic subsystems in the evaluation of the ultimate critical state, in order that all possible post-buckling modes could be evaluated. The proliferation of possible post-buckling mechanisms in equilibrium becomes enormous, as the number of members increases. Therefore, *the direct evaluation of the most degrading post-buckling mode, in one computer run, gains in its significance*. We shall consider this approach in the next section. The numerical solution of the problem was accomplished by means of a computer program based on the theory expounded in this chapter. The theory was programmed by **M.D. Davister** and the numerical procedures adopted in the solution are described in detail in Refs. 22 and 27. All the matrices in the Eqs. (3.6.21) and (3.6.23) or in Eqs. (3.6.24) and (3.6.25) respectively can be computed once the initial geometry of the lattice is known and the last isostatic subsystem in the sequence just before the ultimate critical state is reached, is evaluated. The choice of the initial isostatic subsystem in the sequence is arbitrary, although some such subsystems are computationally better than others. The gradients of the equilibrium paths, may be readily computed at any joint using the solutions for the vector ($\mathbf{e}_b$) and the kinematic relations (3.3.2).

The status of the members i.e. isostatic, critical isostatic, buckled, compressive yielding, tensile yielding and not used for the unoptimized model is shown in Fig. 3.6.4a to 3.6.4c. Two typical equilibrium post-buckling modes (mode 16) and (mode 31) are reproduced here to illustrate the response of the shell to hydrostatic pressure, Figs. 3.6.5 and 3.6.6.

A summary of the ranges of the post-buckling equilibrium slopes (gradients) at the 51 loaded outer peripheral joints, based on the average pressure area of

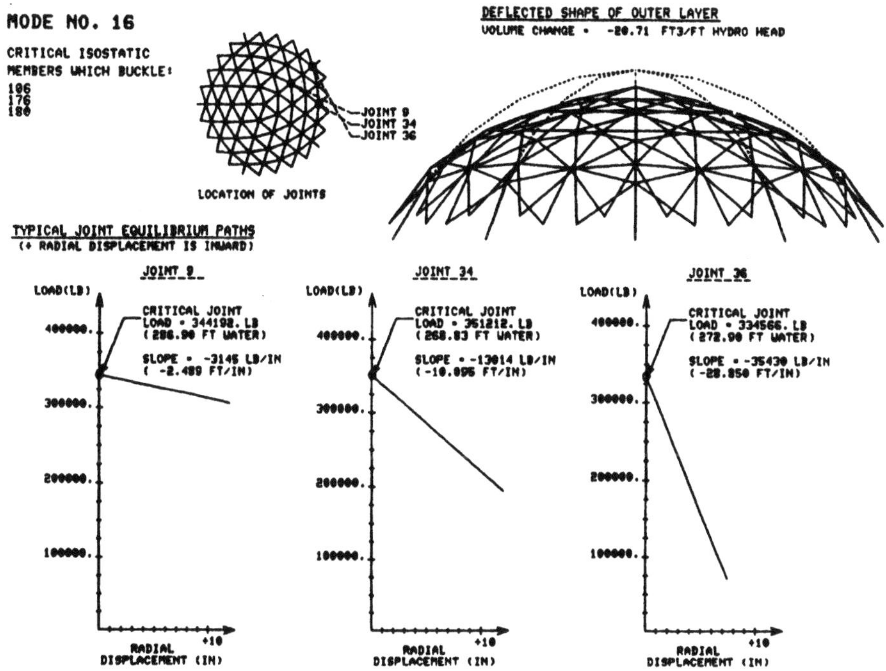

Figure 3.6.5
Plot of the deflected shell and of typical post-buckling equilibrium paths. Mode No. 16

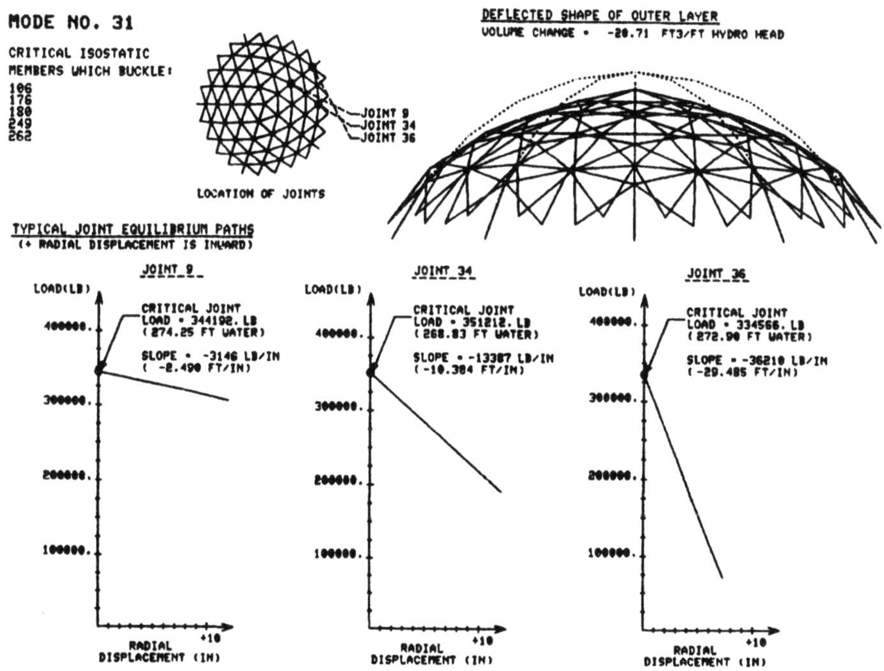

Figure 3.6.6
Plot of the deflected shell and of typical post-buckling equilibrium paths. Mode No. 31

20 sq.ft. = 1.858 m^2, computed for the 8 admissible post-buckling modes of the *unoptimized* model shell and compared to the average critical loads or hydrostatic loads at these joints of the shell-lattice, is shown in Table 2. A computed graphical representation of the superposition of the gradients in all the 8 modes at the 51 loaded outer nodes and of the deflected outer layer is shown in Fig. 3.6.7. The steepest slopes of these equilibrium paths (gradients) were discovered at the nodes near the shell boundaries. These slopes tend to diminish at the joints closer to the top of the shell. Their magnitudes varied considerably and they were all found to be negative (some having fairly large negative values). This indicates that the unoptimized shell would certainly fail by buckling in an unstable collapse motion at this depth. These gradients can be meaningfully interpreted if they are related to the actual total critical loads or hydrostatic heads at the joints. Their steepness is then meaningful only in relation to the total load. For example, in the mode 16 these gradients varied between −31.33 ft. (seawater)/inch deflection = −3.76 m (seawater)/cm (elastic displacement of the node) to −2.04 ft. (seawater)/inch deflection = −0.245 m (seawater)/cm (elastic deflection). This means that, theoretically the shell would have to be raised by so many feet to a lesser depth (and stabilized), if it were to deflect radially by (1 in. = 2.54 cm) at the particular node. A radial displacement of (10 in. = 25.4 cm) was assumed to represent a sufficiently large distortion of the shell lattice, so that the imperfection-sensitivity of the shell could be judged in this range of elastic deformation.

The gradients of the stability boundaries were not shown on these graphs, but, as explained earlier (Section 2.4), they are always equal to *3 times the gradient of the equilibrium path* at the critical state *in the same mode*. Thus, to obtain an idea where the instability may set in under the influence of initial geometrical and other imperfections at the nodal deflection of (1 in. = 2.54 cm) the figures in the last column in Table 2 should be multiplied by a factor of 3. Thus, for example, the instability in mode 16 in practice could occur at the minimum depth of (272.43 ft. = 83.04 m), which means at a depth of some (272.43 − 93.99 =

Table 2
Summary of average critical loads, hydrostatic heads and post-buckling slopes

Mode Number	Average critical load at outer joints (lb.)	Post-buckling slope (lb.)/(in.)		Average critical Hydrostatic Head (ft.)	Post-buckling slope ft.(water)/in.	
		Minimum	Maximum		Minimum	Maximum
16	340000	−39096	−2551	272.43	−31.33	−2.04
17	340000	−39782	−2546	272.43	−31.88	−2.04
20	340000	−39782	−2546	272.43	−31.88	−2.04
26	340000	−39383	−2536	272.43	−31.56	−2.03
27	340000	−39383	−2536	272.43	−31.56	−2.03
28	340000	−39597	−2546	272.43	−31.73	−2.04
29	340000	−39597	−2546	272.43	−31.73	−2.04
31	340000	−36798	−2558	272.43	−29.49	−2.05

ALL ADMISSIBLE MODES OF MODEL NO. 1

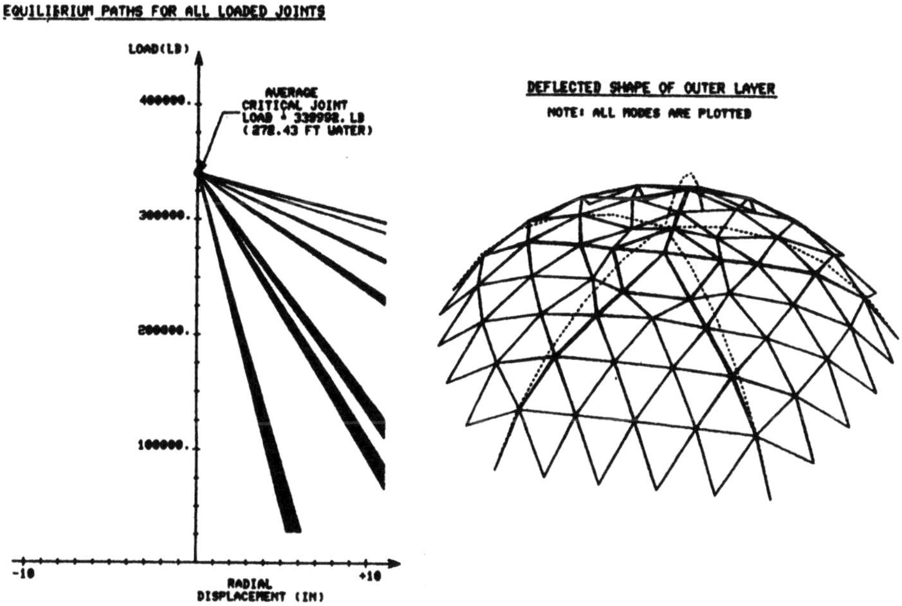

Figure 3.6.7
Superposition of all modes displaying all post-buckling slopes and the deflected outer layer of model No. 1.

178.44 ft. = 54.38 m); which is considerably less than the theoretical critical depth. These slopes indicate also that different degrees of imperfection-sensitivity may be expected in different locations of the shell-lattice. But, this sensitivity seemed to be stronger towards the boundaries of the shell.

The eight different modes which were investigated did not reveal very much difference in these slopes at different nodes, as the figures in Table 2 indicate. The nodes at which these gradients were evaluated are indicated in the left top corner in the diagrams. The deflected shape of the outer layer of Model No. 1 only is indicated in the right top corners in Figs. 3.6.5 and 3.6.6. The actually buckling isostatic members are also shown in the figures. These are referred to the numbering sequence in the identification chart of the members, Figs. 3.6.3a to c. This chart may be found in Ref. 20 and Ref. 27. For further details, the reader is referred to these sources.

The load-carrying capacity of the shell based on its theoretical critical load to self-weight (LCC/weight) ratio for this unoptimized lattice was found to be 393.28. Under its imperfection-sensitivity, as explained above, this ratio may be only of the order of $(178.44/272.43) \times 393.28 \approx 258$, assuming that the radial displacement of the respective node, at which the instability boundary is reached in this mode, is 1 in. = 2.54 cm. Thus the (LCC/weight) ratio of the unoptimized hyperstatic shell-type lattice, under the influence of the imperfection-sensitivity, can be expected to be around 60% to 65% of the theoretical value.

3.7 Direct Evaluation of the Most Degrading Buckling Mode in Equilibrium Using the Total Potential Energy Hypersurfaces

The concept of a total potential energy hypersurface, as it is used here, was first introduced in Ref. 6 and Ref. 7. In this context the total potential energy is defined as the sum of the elastic energy of the lattice and the potential energy of the displaced applied loads. Actually, we consider a change of this energy relative to the initial position or the initial configuration of the lattice. The change in the potential energy is usually negative relative to this configuration. This is so, because the energy required to displace and to drive the system into the buckling process is derived from the field of gravity at the expense of the potential energy of the loads. Since the external loads in a space lattice are invariably applied (centrally) at the nodes, these displacements are also the displacements of the nodes. In the case of flexible elastic members, only (or mainly) their flexural contractions contribute to these displacements. Therefore, we can say that when the contraction of the flexurally shortening members and the elastic constants of the material of which these are made are defined, also the change in the elastic energy absorbed under these deformations can be calculated. Similarly, the change in the potential energy of the applied loads under conservative loading conditions can be readily evaluated, as this was shown in Refs. 7, 8 and Ref. 15 in Chapter 3, pp. 234–240. In the case of three-dimensional lattices, isostatic or redundant, the flexural contractions of the buckling members are defined by the vector ($\mathbf{e}_b$), viz. ($\mathbf{e}_c$); $b \leq c$, and by the vector of the flexural contractions ($\mathbf{e}_\gamma$) of the redundant buckling members.

Let us first consider how the most degrading buckling mode can be determined in states of statical equilibrium. We understand, it is that mode which, under the proportionate applied loads may be maintained in statical equilibrium when these are correctly adjusted, but which is allowed to develop in an unstable motion under the constant critical external loads. In that case, there is a correlation between the change in the total potential energy under equilibrium conditions, which is equal to

$$\Delta V = v = \frac{1}{2} \sum_{i=1,2,\ldots} \delta \mathbf{W}_i^T \, \delta \mathbf{s}_i + \cdots \tag{3.7.1}$$

as explained in Section 1.2 and, more elaborately in Ref. 15, Section 1.7, Eq. (7.20), and the change, or the converted amount of kinetic energy (ΔT), which in this case is equal to

$$\Delta T = -\Delta V = -v = -\frac{1}{2} \sum_{i=1,2,\ldots} \delta \mathbf{W}_i^T \, \delta \mathbf{s}_i + \cdots \tag{3.7.2}$$

($\delta \mathbf{W}_i$) represent the changes in the Cartesian external load components in the three global coordinate directions (x, y, z) at the node or load (i) and ($\delta \mathbf{s}_i$) the corresponding displacement components of the same load. This is so, since, in a conservative mechanical system ($\Delta E = \Delta V + \Delta T = 0$), where ($E$) is the total mechanical energy of the system.

Then, in our notation, $(\delta \mathbf{W}_i)$ are the elements of the external load vector $(\delta \mathbf{W})$ at node (i), assembled in the same sequence as before (following the numbering sequence of the nodes), which we denote by the vector $(\boldsymbol{\beta}' \, \delta w)$. Similarly, $(\delta \mathbf{s}_i = \mathbf{u}_i)$ represent the elements of the nodal displacement vector $(\mathbf{u})$ at joint (i) entered in the same sequence. In this notation Eq. (3.7.1) may be rewritten

$$\Delta V = v = \tfrac{1}{2} \mathbf{u}^T \boldsymbol{\beta}' \, \delta w + \cdots = \tfrac{1}{2} \boldsymbol{\beta}'^T \mathbf{u} \, \delta w + \cdots \tag{3.7.3}$$

where $(\Delta V = v)$ represents the change in the total potential energy of the system. Obviously $(\Delta V = v)$ must be *negative*, in order that a conversion of the total potential into the kinetic energy can take place. The largest amount of conversion into kinetic energy takes place in that mode of elastic deformation in which, under the prescribed loading and the vector $(\boldsymbol{\beta}')$, the negative change in the total potential energy is the greatest. Here, the restriction is made relative to the prescribed proportionate modifiable loads, under which equilibrium (usually unstable equilibrium) is possible, if the unstable motion in the same mode under constant critical loads is not allowed to take place. However, such unstable equilibrium can only be maintained under controlled laboratory conditions using a stabilizing loading device, as explained in Ref. 15, Chapter 5.

We show next that there is a correlation between the equilibrium equations (3.6.21) or (3.6.24), derived in the last section, and the change in the total potential energy of the system in an equilibrium mode of elastic deformation, defined by the vector $(\mathbf{e}_b)$, $(b \le c)$ where (c) is the number of critically stressed isostatic compressive members, referred to the last isostatic subsystem before the ultimate critical state is reached. From the *same* state an onset of unstable motion and the dynamic collapse of the system in flexural buckling becomes possible. Then the applied loads are *not* adjusted according to the Eqs. (3.6.21) or (3.6.24). Rewriting Eq. (3.6.24)

$$\mathbf{E}_b \, \mathbf{e}_b = \mathbf{C}_{1b}'^T \boldsymbol{\beta}' \, \delta w; \qquad b \le c \tag{3.6.24)bis}$$

Multiplying by $(\mathbf{e}_b^T)$ we get,

$$\mathbf{e}_b^T \mathbf{E}_b \, \mathbf{e}_b = \mathbf{e}_b^T \mathbf{C}_{1b}'^T \boldsymbol{\beta}' \, \delta w$$

But, from the kinematic relation (3.3.2) we conclude that

$$\mathbf{e}_b^T \mathbf{C}_{1b}'^T = \mathbf{u}^T \tag{3.7.4}$$

where $(b \le c)$. Thus, Eq. (3.6.24), reduces to

$$\mathbf{e}_b^T \mathbf{E}_b \, \mathbf{e}_b = \mathbf{u}^T \boldsymbol{\beta}' \, \delta w = \mathbf{u}^T \, \delta \mathbf{W} \tag{3.7.5}$$

But, from (3.7.3) it follows that,

$$\tfrac{1}{2} \mathbf{u}^T \boldsymbol{\beta}' \, \delta w = v = \Delta V \tag{3.7.5a}$$

and therefore, that

$$\Delta V = v = \tfrac{1}{2} \mathbf{e}_b^T \mathbf{E}_b \, \mathbf{e}_b = \tfrac{1}{2} \mathbf{u}^T \boldsymbol{\beta}' \, \delta w; \qquad b \le c \tag{3.7.6}$$

Thus, the *stability matrix* $(\mathbf{E}_b)$ plays an important part in the determination of the total potential energy change of the system and, therefore, in its statical stability. Then, according to *Theorem 5*, Ref. 15, p. 74 or *Theorem* 1, Section 1.2, p. 13, if

$$v = \tfrac{1}{2}\mathbf{e}_b^T \mathbf{E}_b \mathbf{e}_b < 0; \qquad b \le c$$

is negative in at least *one* adjacent equilibrium position, then the statical equilibrium in the ultimate critical state is certainly unstable in the *dynamic* sense, as in that case a conversion of the total potential into the kinetic energy may take place in an unstable dynamic collapse by motion. This conversion takes place only if the system is *induced* to a displacement from its initial position. This is accomplished in practice by a *perturbation* of the ideal geometry, through the initial imperfections, which cause the system to "reach" the total potential energy levels (hypersurface) in a configuration or position, *adjacent* to the theoretical ultimate critical state. The expression

$$v = \tfrac{1}{2}\mathbf{e}_b^T \mathbf{E}_b \mathbf{e}_b; \qquad b \le c \tag{3.7.6)bis}$$

represents a quadratic form in the subspace $(v, \mathbf{e}_b)$. If (v) were plotted in this coordinate space, it would represent a *hypersurface* in a $(b + 1)$-dimensional space, $(b \le c)$. Therefore, there exists for each isostatic or hyperstatic system of this type (structure and its loading) in the field of gravity, a *total potential energy hypersurface*, which can be used to probe the stability characteristics of the system and its tendency to collapse by unstable buckling. Of interest here are those regions of this hypersurface, which, near the ultimate critical state, reflect the *largest negative* values in $(\Delta V = v)$, where a conversion of the total potential into the kinetic energy can take place. To prove the stability of the system in the ultimate critical state the expression

$$v = \tfrac{1}{2}\mathbf{e}_b^T \mathbf{E}_b \mathbf{e}_b > 0; \qquad b \le c$$

must be greater than zero (positive) in the *entire* adjacent physical coordinate space defined by the deformation vector $(\mathbf{e}_b)$ and the kinematic restraints imposed in Eq. (3.3.6) where $(b \le c)$, regardless of the magnitudes of the elements $(e_i,\ i = 1, 2, \ldots, b \le c)$. This is generally possible only if the stability matrix $(\mathbf{E}_b)$ is *positive definite*. The test for positive definiteness is that all the principal minors (principal determinants consisting of elements symmetrical about the main diagonal) be *positive*.

An *alternative* procedure to test the positive definiteness of $(\mathbf{E}_b)$ is to introduce *orthogonal* coordinates $(\boldsymbol{\varepsilon}_b)$ linearly related to the flexural contractions $(\mathbf{e}_b)$, i.e.

$$\mathbf{e}_b = \mathbf{B}\boldsymbol{\varepsilon}_b \tag{3.7.7}$$

Substituting in (3.7.6), gives

$$2v = \mathbf{e}_b^T \mathbf{E}_b \mathbf{e}_b = \boldsymbol{\varepsilon}_b^T \mathbf{B}^T \mathbf{E}_b \mathbf{B} \boldsymbol{\varepsilon}_b = \boldsymbol{\varepsilon}_b^T \boldsymbol{\Gamma} \boldsymbol{\varepsilon}_b$$

where

$$\boldsymbol{\Gamma} = \mathbf{B}^T \mathbf{E}_b \mathbf{B} \tag{3.7.8}$$

If $(\boldsymbol{\Gamma})$ is made diagonal, this requires that the column vectors $(\mathbf{b})$ of the linear transformation matrix $(\mathbf{B})$ of order $(b \times b)$ be determined from

$$\mathbf{b}_i^T \mathbf{E}_b \mathbf{b}_j = \delta_{ij}\gamma_j \qquad (\delta_{ij} - \text{Kronecker delta}) \tag{3.7.9}$$

where (γ_j) represent the diagonal elements of $(\boldsymbol{\Gamma})$ and where $(i, j = 1, 2, \ldots, b \leq c)$. If the column vectors $(\mathbf{b}_j)$ are made orthogonal to each other by imposing the orthogonality conditions

$$\mathbf{b}_i^T \mathbf{b}_j = 0; \qquad j \neq i \tag{3.7.10a}$$

$$\mathbf{b}_i^T \mathbf{b}_j = 1; \qquad j = i \tag{3.7.10b}$$

which is identical to the statement that

$$\mathbf{B}^T\mathbf{B} = \mathbf{I} \tag{3.7.11}$$

where $(\mathbf{I})$ is a $(b \times b)$ unit matrix, or that

$$\mathbf{B}^T = \mathbf{B}^{-1} \tag{3.7.12}$$

then, the *inverse* of $(\mathbf{B})$ is equal to the transpose of $(\mathbf{B})$. From (3.7.9) it follows that

$$\mathbf{E}_b \mathbf{B} - (\mathbf{B}^T)^{-1}\boldsymbol{\Gamma} = \mathbf{0} \tag{3.7.13}$$

or, in view of (3.7.12), that

$$\mathbf{E}_b \mathbf{B} - \mathbf{B}\boldsymbol{\Gamma} = \mathbf{0} \tag{3.7.13a}$$

In the second term the post-multiplication by $(\boldsymbol{\Gamma})$ implies that each column (j) of $(\mathbf{B})$ is multiplied by the corresponding diagonal element (γ_j) of $(\boldsymbol{\Gamma})$. Thus, the first column is $(\mathbf{b}_1\gamma_1)$, the second $(\mathbf{b}_2\gamma_2)$ etc. The last result may then be written

$$(\mathbf{E}_b \mathbf{b}_1 + \mathbf{E}_b \mathbf{b}_2 + \cdots + \mathbf{E}_b \mathbf{b}_b) - (\gamma_1 \mathbf{b}_1 + \gamma_2 \mathbf{b}_2 + \cdots + \gamma_b \mathbf{b}_b) = \mathbf{0} \tag{3.7.14}$$

or, generally, since the vectors $(\mathbf{b}_j, j = 1, 2, \ldots, b \leq c)$, are independent from one another,

$$(\mathbf{E}_b - \gamma_j \mathbf{I})\mathbf{b}_j = \mathbf{0}; \qquad j = 1, 2, \ldots, b \leq c \tag{3.7.15}$$

But, for $(\mathbf{b}_j)$ to be different from zero, the following determinant must vanish

$$|\mathbf{E}_b - \gamma_j \mathbf{I}| = 0 \tag{3.7.16}$$

This equation has generally $(b \leq c)$ distinct roots which are all real in this case, since $(\mathbf{E}_b)$ is *symmetrical* about its main diagonal, as we showed earlier Section 3.7, Eq. (3.6.19) (for proof see Ref. 15, p. 33). These roots represent the (b) eigenvalues of the matrix $(\mathbf{E}_b)$. Thus, the expression for the total potential energy change (v) of the buckled deformed system in the vicinity of the ultimate critical state at small but *finite* deformations characterized by the vector $(\mathbf{e}_b)$, equals

$$2v = \mathbf{e}_b^T \mathbf{E}_b \mathbf{e}_b = \boldsymbol{\varepsilon}_b^T \mathbf{B}^T \mathbf{E}_b \mathbf{B} \boldsymbol{\varepsilon}_b = \boldsymbol{\varepsilon}_b^T \boldsymbol{\Gamma} \boldsymbol{\varepsilon}_b = \gamma_1 \varepsilon_1^2 + \gamma_2 \varepsilon_2^2 + \cdots + \gamma_b \varepsilon_b^2 \tag{3.7.17}$$

Now, the independent orthogonal coordinates (ε_i) appear as squares only. Therefore, the eigenvalues (γ_i) determine the stability of the system in the ultimate critical state. If all the eigenvalues (γ_i) are *positive*, the system is certainly stable in the ultimate critical state, since then in all adjacent physically possible positions (configurations)

$$2v = \mathbf{e}_b^T \mathbf{E}_b \mathbf{e}_b > 0 \tag{3.7.18}$$

If *at least one* of these eigenvalues is *negative* the system is unstable because then, along the independent orthogonal coordinate associated with this negative eigenvalue, the total potential energy decreases and conversion into the kinetic energy along that coordinate of deformation may take place, all other independent coordinates being zero. This proof is then sufficient to prove the instability of the system. The (*b*) flexural contractions ($\mathbf{e}_b$) govern, in this case, also the flexural contractions ($\mathbf{e}_\gamma$) of the redundant buckling members. Therefore, if the kinematic condition (3.3.6b) is satisfied, any such post-buckling mode or mechanism is also *kinematically* admissible and it may take place in unstable motion under *constant* applied loads ($\mathbf{W}$). This completes the proof that under the stated conditions the buckling process is either stable or unstable in the *dynamic* sense. In other words, the system is stable if all the eigenvalues of the stability matrix ($\mathbf{E}_b$) are positive. If, at least, one eigenvalue is negative, the system is unstable. Eq. (3.7.6) suggests that *equilibrium* is possible in any one such mode, defined by the deformation vector ($\mathbf{e}_b$), under the *modified* applied loads defined by the load vector

$$(\mathbf{W}^0 + \boldsymbol{\beta}' \, \delta w) \tag{3.7.19}$$

where ($\boldsymbol{\beta}'$) is *arbitrary*, yet such that in any one mode ($\mathbf{e}_b$) is a statically and kinematically admissible solution of Eq. (3.6.24) and Eq. (3.7.6). To find the regions of the largest negative values of ($\Delta V = v$) (global minimum) in the entire physically possible space of elastic deformation characterized by the flexural contraction vector ($\mathbf{e}_b$; $b \leq c$) and by the kinematic restraints (3.3.6) i.e. for the *positive* values of the vectors ($\mathbf{e}_b$) and ($\mathbf{e}_\gamma$), where (c) is the number of the critical members in the last isostatic mechanism before the onset of unrestrained buckling. We proceed by transforming further the Eq. (3.7.6).

The scalar quantity ($\mathbf{u}^T\boldsymbol{\beta}'$) represents the weighted sum of the joint displacements ($\mathbf{u}$) multiplied (weighted) by the load-proportionality constants of the applied loads *after* the onset of buckling, contained in the vector ($\boldsymbol{\beta}'$). We define this weighted sum by the scalar quantity ($\delta\bar{s}$), which has the dimension of *length*, assuming that ($\boldsymbol{\beta}'$) is nondimensional, i.e.

$$\delta\bar{s} = \mathbf{u}^T\boldsymbol{\beta}' \tag{3.7.20}$$

On dividing Eq. (3.7.6) by $(\delta\bar{s})^2$, one gets

$$2\frac{v}{(\delta\bar{s})^2} = \frac{\mathbf{e}_b^T \mathbf{E}_b \mathbf{e}_b}{(\mathbf{u}^T\boldsymbol{\beta}')^2} = \frac{(\mathbf{u}^T\boldsymbol{\beta}')\,\delta w}{(\mathbf{u}^T\boldsymbol{\beta}')^2} = \frac{\delta w}{\delta\bar{s}}; \qquad b \leq c \tag{3.7.21}$$

Here, (δw) is considered to have the dimension of force, so that the right side has the dimension of force/length, i.e. the same as the gradient of the load-displacement equilibrium path. ($\delta\bar{s}$) may be referred to as the weighted total displacement of the nodes. Consider now the problem in which the total potential energy in a hyperstatic lattice under the conservative constant loads ($\mathbf{W}^0$) is defined near the ultimate critical state by the function

$$v = \tfrac{1}{2}\mathbf{x}^T\mathbf{E}(\mathbf{W}^0)\mathbf{x} \tag{3.7.22}$$

where $(\mathbf{x})$ represents an arbitrary vector of flexural contractions $(\mathbf{e}_b)$ based on any last isostatic subsystem of critical bars in the lattice and $(\mathbf{E})$ the corresponding *stability* or *post-buckling stiffness matrix* $(\mathbf{E}_b)$. Then, (v) represents a hypersurface in the coordinate space $(x_i;\ i = 1, 2, 3, \ldots, b \leq \mathbf{c})$. Suppose this hypersurface is bounded by another hypersurface or contour, defined by the quadratic form

$$\tfrac{1}{2}\mathbf{x}^T\mathbf{Q}\mathbf{x} = C \qquad \text{(const.)} \tag{3.7.23}$$

in which (C) may be equated to unity, i.e. $(C = 1)$. The symmetrical matrix $(\mathbf{Q})$ is as yet unspecified. The idea of "bounding" the total potential energy hypersurface comes about, because the magnitude of the vector $(\mathbf{e}_b)$ or $(\mathbf{x})$, in this case, is not constrained by the kinematic or any other conditions. In other words, if the vector $(\mathbf{e}_b = \mathbf{e}_b^*)$ or $(\mathbf{x} = \mathbf{x}^*)$ is the solution, which corresponds to an admissible buckling mode resulting in a total potential energy change $(v = v^*)$, then the vector $(k\mathbf{e}_b)$ is also a solution, where (k) is an arbitrary constant, because $(\mathbf{e}_b)$ is not bounded. Then, the corresponding value of (v) changes to a new value given by $(k^2 v^*)$, since (v) is a quadratic function in $(\mathbf{e}_b)$. Thus, if (v^*) represents a minimum of (v), for example, it is necessary to specify the "contour" tied to the magnitude of the vector $(\mathbf{e}_b)$ or $(\mathbf{x})$, on which this minimum is evaluated. Therefore, the vector $(\mathbf{e}_b)$ or $(\mathbf{x})$ must be bounded. The simplest way to bound this vector might be to specify that its norm (i.e. the sum of the squares of its components) be equal to unity, i.e.

$$\mathbf{e}_b^T\mathbf{e}_b = 1 \tag{3.7.23a}$$

which arbitrarily constrains the vector $(\mathbf{e}_b)$ to the contour $(\mathbf{e}_b^T\mathbf{e}_b = 1)$. A more general way to constrain the vector $(\mathbf{e}_b)$ is that introduced by a general quadratic form in Eq. (3.7.23) in which, for the time being, $(\mathbf{Q})$ is an arbitrary symmetrical matrix.

Suppose it is desired to find the points of the most negative values of (v) on this contour, such that at these points the system may be maintained in statical equilibrium by the *modified* external loads $(\mathbf{W} = \mathbf{W}^0 + \boldsymbol{\beta}'\,\delta w)$, where $(\boldsymbol{\beta}')$ is an *arbitrary* vector of load proportionality constants in the post-buckling range of the system. This problem may be restated, on introducing a set of the Lagrangean multipliers (λ), as follows:

Minimize the function (v) on the same contour subject to the restraints defined by Eq. (3.7.23). This then is identical to minimizing the function $(\bar{F})$, where

$$\bar{F} = \tfrac{1}{2}\mathbf{x}^T\mathbf{E}\mathbf{x} - \lambda(\tfrac{1}{2}\mathbf{x}^T\mathbf{Q}\mathbf{x} - C) \tag{3.7.24}$$

on the same contour. Therefore, the stationary values of (v) on this contour are found by setting the gradient vector of $(\bar{F})$ equal to zero. This operation results in the following matrix equation

$$\nabla\bar{F} = (\mathbf{E} - \lambda\mathbf{Q})\mathbf{x} = \mathbf{0} \tag{3.7.25}$$

where (**E**) and (**Q**) are symmetrical matrices. For ($\mathbf{x} \neq \mathbf{0}$) these equations have a solution only if

$$|\mathbf{E} - \lambda \mathbf{Q}| = 0 \tag{3.7.26}$$

This condition furnishes the required set of Lagrangean multipliers (λ_i, $i = 1, 2, 3, \ldots, b \leq c$) which turn out to be the eigenvalues of (**E**) and (**Q**). Since (**E**) and (**Q**) are symmetrical matrices, all the eigenvalues must be real (for proof see Ref. 15, p. 33). Then for each eigenvalue (λ_i) a corresponding eigenvector ($\mathbf{x}_i$) can be found from Eq. (3.7.25) (short of an undetermined proportionality constant). Putting now ($\lambda = \lambda_i$) and ($\mathbf{x} = \mathbf{x}_i$) in Eq. (3.7.25), we get

$$(\mathbf{E} - \lambda_i \mathbf{Q})\mathbf{x}_i = \mathbf{0} \tag{3.7.26a}$$

Then for each eigenvalue (λ_i) the eigenvector ($\mathbf{x}_i$) is determined by the last equation. The length of the eigenvector ($\mathbf{x}_i$) is then bounded by the condition (3.7.23). In this way the stationary points of the total potential energy change (v) can be evaluated on the contour along the eigenvector directions, which are the solution of Eq. (3.7.26a). Premultiplying now Eq. (3.7.26a) by $(\mathbf{x}_i)^T$, gives

$$\mathbf{x}_i^T \mathbf{E} \mathbf{x}_i - \lambda_i \mathbf{x}_i^T \mathbf{Q} \mathbf{x}_i = \mathbf{0}$$

or

$$\lambda_i = \frac{\mathbf{x}_i^T \mathbf{E} \mathbf{x}_i}{\mathbf{x}_i^T \mathbf{Q} \mathbf{x}_i} = \frac{v_i}{C} \tag{3.7.27}$$

If the matrix (**Q**) of the quadratic form in (3.7.23) is known, the eigenvalues (λ_i) can be found from Eq. (3.7.26). Then ($v_i = C\lambda_i$, $i = 1, 2, \ldots, b \leq c$) represent the values of (v) at the stationary points in the subspace of the contour determined by the quadratic form (3.7.23), where the gradient of the total potential energy hypersurface is zero. In the particular case, when ($C = 1$)

$$v_i = C\lambda_i = \lambda_i = \tfrac{1}{2}\mathbf{x}_i^T \mathbf{E} \mathbf{x}_i \tag{3.7.28}$$

Thus, the eigenvectors ($\mathbf{x}_i$) in Eq. (3.7.26a) define the directions in the (**x**) coordinate space, where the total potential energy has either a minimum or a maximum. These minima or maxima are then given by the ($b \leq c$) eigenvalues (λ_i) obtained from the Eq. (3.7.26). If we now identify the vector (**x**) with the vector ($\mathbf{e}_b$) in an equilibrium mode, defined by the equation (3.6.24), and the matrix (**E**), with the corresponding matrix ($\mathbf{E}_b$), then the eigenvectors ($\mathbf{x}_i$) correspond to the eigenvectors ($\mathbf{e}_{bi}$), where

$$(\mathbf{E}_b - \lambda_i \mathbf{Q})\mathbf{e}_{bi} = \mathbf{0} \tag{3.7.29}$$

The eigenvalues (λ_i) can now be determined if the matrix (**Q**), defining the quadratic form in (3.7.23) and the coordinate space of the contour, is specified or known.

In the case of an *equilibrium* mode (i.e. a collapse mode in which the system may be maintained in statical equilibrium by the modified *applied* proportionate loads ($\mathbf{W}^0 + \boldsymbol{\beta}'\,\delta w$)) the matrix (**Q**) can be easily identified. If the load-proportionality vector ($\boldsymbol{\beta}'$) in post-buckling equilibrium is *arbitrary*, such that the system in the mode defined by ($\mathbf{e}_{bi}$) can be maintained in statical equilibrium, then it follows from (3.7.21), on putting ($\lambda = \lambda_i$) and ($\mathbf{x}_i = \mathbf{e}_{bi}$), that

$$\frac{\delta w}{\delta \bar{s}_i} = \frac{\mathbf{e}_{bi}^T \mathbf{E}_b \mathbf{e}_{bi}}{(\mathbf{u}_i^T \boldsymbol{\beta}')^2} \tag{3.7.30}$$

The denominator of this expression may be rewritten $(\mathbf{u}_i^T \boldsymbol{\beta}')^2 = \mathbf{u}_i^T \boldsymbol{\beta}' \boldsymbol{\beta}'^T \mathbf{u}_i$ and transformed by the kinematic relations (3.3.2) to

$$(\mathbf{u}_i^T \boldsymbol{\beta}')^2 = \mathbf{e}_{bi}^T \mathbf{C}_{1b}'^T \boldsymbol{\beta}' \boldsymbol{\beta}'^T \mathbf{C}_{1b}' \mathbf{e}_{bi} = \mathbf{e}_{bi}^T \mathbf{B}_b \mathbf{e}_{bi} \tag{3.7.31}$$

where the ($b \times b$) matrix ($\mathbf{B}_b$) is equal to

$$\mathbf{B}_b = \mathbf{C}_{1b}'^T \boldsymbol{\beta}' \boldsymbol{\beta}'^T \mathbf{C}_{1b}' \tag{3.7.31a}$$

Eq. (3.7.30) now becomes

$$\frac{\delta w}{\delta \bar{s}_i} = \frac{\mathbf{e}_{bi}^T \mathbf{E}_b \mathbf{e}_{bi}}{(\mathbf{u}_i^T \boldsymbol{\beta}')^2} = \frac{\mathbf{e}_{bi}^T \mathbf{E}_b \mathbf{e}_{bi}}{\mathbf{e}_{bi}^T \mathbf{B}_b \mathbf{e}_{bi}} \tag{3.7.32}$$

If next the matrix ($\mathbf{Q}$) in Eq. (3.7.23) is set equal to the matrix ($\mathbf{B}_b$), defined by Eq. (3.7.31a), the matrix ($\mathbf{E}$) and the eigenvectors ($\mathbf{x}_i$) replaced by ($\mathbf{E}_b$) and the eigenvectors ($\mathbf{e}_{bi}$), respectively, which are also the solutions of the equilibrium equations (3.6.24), the right side of Eq. (3.7.32) becomes identical to the right side of Eq. (3.7.27). Thus, ($\delta w/\delta \bar{s}_i = \lambda_i$), in this case, become the eigenvalues of the matrices ($\mathbf{E}_b$) and the matrix ($\mathbf{Q} = \mathbf{B}_b$) and ($\mathbf{e}_{bi}$) their eigenvectors, where

$$\mathbf{e}_{bi}^T \mathbf{Q} \mathbf{e}_{bi} = \mathbf{e}_{bi}^T \mathbf{B}_b \mathbf{e}_{bi} = \mathbf{u}_i^T \boldsymbol{\beta}' \boldsymbol{\beta}'^T \mathbf{u}_i = \mathbf{e}_{bi}^T \mathbf{C}_{1b}'^T \boldsymbol{\beta}' \boldsymbol{\beta}'^T \mathbf{C}_{1b}' \mathbf{e}_{bi} = 2C \tag{3.7.32a}$$

where the last identity follows from Eq. (3.7.23). Hence,

$$\mathbf{Q}_b = \mathbf{B}_b = \mathbf{C}_{1b}'^T \boldsymbol{\beta}' \boldsymbol{\beta}'^T \mathbf{C}_{1b}' \tag{3.7.33}$$

($\mathbf{Q}$) is then determined in an equilibrium mode defined by the vector ($\mathbf{e}_b$), which is the solution of the equilibrium Eqs. (3.6.21) or (3.6.24). In this mode the system is maintained in statical equilibrium by the applied modified proportionate loads ($\mathbf{W}^0 + \boldsymbol{\beta}'\, \delta w$), where ($\boldsymbol{\beta}'$) is essentially arbitrary. The same system, however may be in unstable motion in the same mode, provided the applied loads are *not* modified but maintained at constant critical values ($\mathbf{W}^0$) and the total potential energy change in the same mode is negative. If, ($\mathbf{e}_{bi}$) is a finite non-zero eigenvector which satisfies the Eqs. (3.7.29) in which ($\mathbf{Q} = \mathbf{B}_b$) is defined by Eq. (3.7.33) and it satisfies also the modified equilibrium equations (3.7.30) derived from the original equilibrium conditions (3.6.24), then, under a prescribed load-proportionality vector ($\boldsymbol{\beta}'$), the eigenvalues (λ_i) of the matrices ($\mathbf{E}_b$) and ($\mathbf{Q} = \mathbf{B}_b$) can be calculated from the condition that

$$|\mathbf{E}_b - \lambda_i \mathbf{Q}| = |\mathbf{E}_b - \lambda_i(\mathbf{C}_{1b}'^T \boldsymbol{\beta}' \boldsymbol{\beta}'^T \mathbf{C}_{1b}')| = 0; \qquad i = 1, 2, \ldots, b \le c \tag{3.7.34}$$

Generally, (b) distinct real eigenvalues (λ_i) can be extracted from this condition, as both matrices ($\mathbf{E}_b$) and ($\mathbf{Q}$) are symmetric in this case. The corresponding eigenvectors ($\mathbf{e}_{bi}$) are found from the Eqs. (3.7.29) and the normalizing condition (3.7.23) on setting ($\mathbf{x}_i \equiv \mathbf{e}_{bi}$). The minimum and maximum values (v_i) of the total potential energy function (v) at the stationary points on the contour defined by Eq. (3.7.23) can be found from Eq. (3.7.27) and these are proportional to the eigenvalues (λ_i), i.e. ($v_i = C\lambda_i$). These stationary points are then fixed along the eigenvector directions by the magnitudes of the eigenvectors ($\mathbf{e}_{bi}$; $i = 1, 2, 3, \ldots, b \le c$). Usually, the points of the most negative eigenvalues are of most practical interest.

Putting, ($\mathbf{x}_i \equiv \mathbf{e}_{bi}$), ($\mathbf{E} = \mathbf{E}_b$) and ($\mathbf{Q} = \mathbf{B}_b = \mathbf{C}_{1b}'^T \boldsymbol{\beta}' \boldsymbol{\beta}'^T \mathbf{C}_{1b}$) we deduce from Eqs. (3.7.27) and (3.7.32) that

$$v_i = v(\mathbf{e}_{bi}) = C\lambda_i = C\frac{\mathbf{e}_{bi}^T \mathbf{E}_b \mathbf{e}_{bi}}{\mathbf{e}_{bi}^T \mathbf{B}_b \mathbf{e}_{bi}} = C\frac{\delta w}{\delta \bar{s}_i}; \qquad i = 1, 2, \ldots, b \le c \tag{3.7.35}$$

and that

$$\frac{\delta w}{\delta \bar{s}_i} = \lambda_i \tag{3.7.35a}$$

It follows from this result that the change (δw) in the load-parameter under the load-proportionality vector ($\boldsymbol{\beta}'$), necessary to maintain the buckled system in statical equilibrium at these stationary points is equal to

$$\delta w = \delta w_i = \lambda_i \delta \bar{s}_i; \qquad i = 1, 2, 3, \ldots, b \le c \tag{3.7.36}$$

We show in Eq. (3.7.37) that on the bounding contour defined by (3.7.33), ($\delta \bar{s}_i = \sqrt{2C}$). It follows then from (3.7.36) that

$$\delta w_i = \lambda_i \delta \bar{s}_i = \lambda_i \sqrt{2C} = \sqrt{\frac{2}{C}}\, v_i; \qquad i = 1, 2, \ldots, b \le c \tag{3.7.36a}$$

so that the necessary changes (δw_i) in the load-parameter for the system to be maintained in statical equilibrium at the stationary points of (v) are proportional to the eigenvalues ($v_i = C\lambda_i$) at these points.

From this we conclude that, given a prescribed load-vector ($\boldsymbol{\beta}'$) and the matrix ($\mathbf{Q} = \mathbf{B}_b = \mathbf{C}_{1b}'^T \boldsymbol{\beta}' \boldsymbol{\beta}'^T \mathbf{C}_{1b}'$), where ($\mathbf{C}_{1b}'$) is derived from the last isostatic subsystem in the evaluation of the ultimate critical state of the system just before this state is reached, the eigenvalues ($\lambda_i = v_i/C$) can be found from the Eq. (3.7.34) and the corresponding eigenvectors from Eq. (3.7.29) and the normalizing condition (3.7.23). The required changes (δw_i), such that the system may be maintained in statical equilibrium by the modified loads ($\mathbf{W}^0 + \boldsymbol{\beta}' \delta w_i$) at these stationary points can then be calculated from Eq. (3.7.36) or (3.7.36a). The prescribed load-proportionality vector ($\boldsymbol{\beta}'$) is somewhat arbitrary, yet it must be sufficiently general, having the essential components at the nodal points of load application in the directions of global coordinate freedoms of the system, so that under arbitrary distortions characterized by the eigenvectors ($\mathbf{e}_{bi}$), equilibrium of the system under the load vector ($\mathbf{W}^0 + \boldsymbol{\beta}' \delta w_i$) may be possible.

Here the eigenvalues (λ_i) are based on an arbitrary last isostatic subsystem, to which the matrix ($\mathbf{C}_{1b}'$) is referred, in a sequence of isostatic subsystems, used in the evaluation of the ultimate critical state, and on the flexural contractions ($\mathbf{e}_b$)

of the buckling critical members related to this subsystem. These eigenvalues obviously depend on the kinematic submatrices $(\mathbf{C}'_{1c})$ and $(\mathbf{C}'_{1b})$ as the matrix $(\mathbf{Q})$ depends on these submatrices. They also depend on the load vector $(\mathbf{W}^0)$, because the matrices $(\mathbf{E}_c)$ or $(\mathbf{E}_b)$, depend on it. Now, $(\mathbf{C}'_{1c})$ obviously depends on the choice of the first and, consequently, the last isostatic subsystem. It was shown by **Davister** (Ref. 27) that these eigenvalues are the *same* regardless of the choice of the isostatic subsystems. This conclusion is important, as the choice of the isostatic subsystems is essentially arbitrary, in any one sequence in which the ultimate critical state is uniquely determined. We see after this discussion that we can treat the ratio $(\delta w/\delta \bar{s})$ as a function of the deformation vector $(\mathbf{e}_b)$. This follows directly from Eq. (3.7.21), i.e.

$$\frac{\delta w}{\delta \bar{s}} = \frac{\mathbf{e}_b^T \mathbf{E}_b \mathbf{e}_b}{(\mathbf{u}^T \boldsymbol{\beta}')^2} = 2\frac{v}{(\delta \bar{s})^2} \tag{3.7.21)bis}$$

where $(\boldsymbol{\beta}')$ is arbitrary, but such that under the external loads $(\mathbf{W}^0 + \boldsymbol{\beta}'\,\delta w)$ the buckled system may be maintained in statical equilibrium. Since

$$(\delta \bar{s})^2 = (\mathbf{u}^T \boldsymbol{\beta}')^2 = \mathbf{u}^T \boldsymbol{\beta}' \boldsymbol{\beta}'^T \mathbf{u} = \mathbf{e}_b^T \mathbf{C}'^T_{1b} \boldsymbol{\beta}' \boldsymbol{\beta}'^T \mathbf{C}'_{1b} \mathbf{e}_b = \mathbf{e}_b^T \mathbf{Q} \mathbf{e}_b = 2C \tag{3.7.37}$$

where

$$\mathbf{Q} = \mathbf{C}'^T_{1b} \boldsymbol{\beta}' \boldsymbol{\beta}'^T \mathbf{C}_{1b}$$

we find

$$\frac{\delta w}{\delta \bar{s}} = \frac{\mathbf{e}_b^T \mathbf{E}_b \mathbf{e}_b}{\mathbf{e}_b^T \mathbf{Q} \mathbf{e}_b} \tag{3.7.38}$$

Then, from the Eqs. (3.7.35), (3.7.36) and the last result we conclude that (λ_i) are also the eigenvalues of the function ratio $(\delta w/\delta \bar{s})$, when $(\mathbf{e}_b)$ is equal to the eigenvector $(\mathbf{e}_{bi})$. This statement implies, however, that the vector $(\boldsymbol{\beta}')$ exists or that it can be prescribed, so that equilibrium is possible under the load parameter change (δw). Since, on the bounding contour the quadratic form $(\mathbf{e}_b^T \mathbf{Q} \mathbf{e}_b = 2C)$ is constant, it follows also from Eq. (3.7.37) that in the subspace of this contour the average weighted displacement $(\delta \bar{s})$,

$$\delta \bar{s} = \sqrt{2C} \tag{3.7.39}$$

is constant. This also holds at the points of stationary values of (v), where $(\delta \bar{s} = \delta \bar{s}_i)$. This suggests that the function $\bar{F}(\mathbf{e}_b)$, where $\bar{F}$ is defined by

$$\bar{F} = \frac{\frac{1}{2}\mathbf{e}_b^T \mathbf{E}_b \mathbf{e}_b}{\frac{1}{2}\mathbf{e}_b^T \mathbf{Q} \mathbf{e}_b} = \frac{v(\mathbf{e}_b)}{C} \tag{3.7.40}$$

may be treated as a function of $(\mathbf{e}_b)$, subject to the constraint

$$\mathbf{e}_b^T \mathbf{Q} \mathbf{e}_b = 2C \tag{3.7.40a}$$

where $\mathbf{Q}$ is defined as before and where $(\mathbf{C}'_b)$ in $\mathbf{Q}$ is computed with respect to an arbitrary last isostatic subsystem. The reader may verify, using the method of Lagrangean multipliers, that the stationary values of $(\bar{F})$ are again determined by the condition

$$\nabla \bar{F} = \nabla\left(\frac{\delta w}{\delta \bar{s}}\right) = \frac{1}{C}(\mathbf{E}_b - \lambda \mathbf{Q})\mathbf{e}_b = \mathbf{0} \tag{3.7.41}$$

and by the same determinantal equation

$$|\mathbf{E}_b - \lambda\mathbf{Q}| = 0 \tag{3.7.41a}$$

so that the eigenvalues ($\lambda = \lambda_i$; $i = 1, 2, 3, \ldots, b \leq c$) are the same as before.

Clearly, the most negative region of the total potential energy (v) or of the function $\bar{F}$ is of most interest, because in that region the largest conversion of the total potential into kinetic energy may be generated. This is where the most degrading collapse mechanisms are most likely to occur. But, the most negative value of (v) is proportional to the largest negative eigenvalue (λ_i), which we denote by (λ_M), which represents the *unrestrained* global minimum of (v) or ($\bar{F}$) in the entire positive orthant of this space defined by ($\mathbf{e}_b \geq \mathbf{0}$). The *most degrading* buckling mode is then determined by the associated eigenvector ($\mathbf{e}_M$). However, not the entire coordinate space ($\mathbf{e}_b \geq \mathbf{0}$) in this region may be kinematically admissible. The admissible regions are governed by the *kinematic restraints* imposed by the kinematic conditions (3.3.6), i.e.

$$\begin{aligned} &\mathbf{e}_b \geq \mathbf{0} && \text{(a)}\\ &\mathbf{e}_\gamma = \mathbf{C}_\gamma \mathbf{C}'_{1b}\mathbf{e}_b \geq \mathbf{0} && \text{(b)}\\ &\mathbf{e}_\sigma = \mathbf{C}_\sigma \mathbf{C}'_{1b}\mathbf{e}_b \geq \mathbf{0} && \text{(c)}\\ &\mathbf{e}_u = \mathbf{C}_u \mathbf{C}'_{1b}\mathbf{e}_b \geq \mathbf{0} && \text{(d)} \end{aligned} \tag{3.3.6)bis}$$

Then, under these constraints, the most negative value of (v) or (F) is not necessarily determinated by an eigenvector ($\mathbf{e}_b$), which corresponds to a stationary point, but by the *constrained global minimum* determined jointly by the form of (v) or ($\bar{F}$) and the kinematic constraints (3.3.6). Let the vector ($\mathbf{e}_b^*$) be the solution of ($\mathbf{e}_b$) which defines the point of this *constrained minimum*. Thus the unconstrained minimum, defined by the largest negative eigenvalue (λ_M), i.e.

$$v(\mathbf{e}_{bM}) \leq v(\mathbf{e}_b^*) < 0 \tag{3.7.42}$$

Alternatively, this minimization problem may be formulated: Minimize the function

$$\bar{F}(\mathbf{x}_b) = \frac{\mathbf{x}_b^T \mathbf{E}_b \mathbf{x}_b}{\mathbf{x}_b^T \mathbf{Q} \mathbf{x}_b} \tag{3.7.43}$$

subject to the linear *kinematic constraints*

$$\begin{aligned} &\mathbf{x}_b \geq \mathbf{0} && \text{(a)}\\ &\mathbf{x}_\gamma = \mathbf{C}_\gamma \mathbf{C}'_{1b}\mathbf{x}_b \geq \mathbf{0} && \text{(b)}\\ &\mathbf{x}_\sigma = \mathbf{C}_\sigma \mathbf{C}'_{1b}\mathbf{x}_b \geq \mathbf{0} && \text{(c)}\\ &\mathbf{x}_u = \mathbf{C}_u \mathbf{C}'_{1b}\mathbf{x}_b \geq \mathbf{0} && \text{(d)} \end{aligned}$$

where the kinematic submatrices ($\mathbf{C}'_{1b}$), ($\mathbf{C}_\gamma$), ($\mathbf{C}_\sigma$) and ($\mathbf{C}_u$) were defined in Eqs. (3.3.2), (3.3.5) and (3.3.6). In this form the minimization problem is generalized and this generalization applies to a *greater variety* of bounding contours, which can be used to bound the region of the total potential energy hypersurface or the hypersurface of the function ($\bar{F}$), adjacent to the ultimate critical state or point. This minimization is then carried out on a contour defined by the normalizing condition

$$\mathbf{x}_b^T \mathbf{Q} \mathbf{x}_b = 2C$$

in which (**Q**) and the constant (C) are essentially arbitrary and where (C) may be set equal to ($C = 1/2$). However one must realize that on such a generalized contour the equilibrium of the system is *not necessarily possible* under the modified applied loads ($\mathbf{W}^0 + \boldsymbol{\beta}' \, \delta w$), as the symmetrical matrix (**Q**) does not generally satisfy the equilibrium condition imposed by the Eq. (3.7.33). The result of this optimisation procedure is the localisation of the global constrained minimum of the function $\bar{F}(\mathbf{x}_b)$ subject to the above constraints and the evaluation of the vector ($\mathbf{x}_b$) which defines this minimum, as well as of the most degrading buckling mode of the system under investigation.

Two independent numerical methods were used to evaluate the most degrading modes under the kinematic restraints (3.3.6) for the unoptimized and the two optimized models of the reticulated shell described in the last section. The first method was the conjugate gradient descent method. In the second method the descent direction, associated with an eigenvector direction obtained from Eq. (3.7.29) in combination with Eq. (3.7.23) was used to start the search for the constrained global minimum. It is referred to as the eigenvector or eigenvalue method. Both methods give practically identical results and both are briefly described in Section 3.10.

3.8 Comparison of the Numerical Results Characterizing the Post-Buckling Equilibrium of Three Model Reticulated Shells for Underwater Applications[4]

(i) The Unoptimized Model Shell (Model No. 1)

The structure of the unoptimized model shell-lattice was already described in detail in Section 3.6. Also the identification of the members and the joints, as well as the status of the members in the three layers of the lattice in the ultimate critical state, were given in Figs. 3.6.1, 3.6.2, 3.6.3 and 3.6.4 respectively in that Section. Two buckling modes marked 16 and 31 were also illustrated in Figs. 3.6.5, 3.6.6 and the superposition of the 8 admissible modes is characterized and displayed in Fig. 3.6.7. The characteristic quantities for these 8 modes may be found in Table 2 in Section 3.6. Here, a summary of the characteristic figures for this case is reproduced. We note that no cover shell, supported at the outer

[4] The numerical computations and the graphical representations of the results in this section were carried out by **M.D. Davister**, Refs. 20 and 22.

peripheral nodes, which would permit the structure to be used for underwater applications, is considered in this study.

The theoretical ultimate critical depth was found to be 272.43 ft. = 83.0366 m, measured from the water surface (sea level) to the mid-height of the shell (fixed at 5.527 ft. = 1.685 m below the crown). This is equivalent to a pressure of 117.866 psi = 8.287 kg*/cm^2 (* implies kg weight). The total pressure area attributable to each peripheral joint on the outer layer of the shell-lattice amounted to some 20 sq.ft. = 1.858 m^2. The average critical load, supported by such a joint was approximately 340,000 lb. = 154,221.41 kg*. In the ultimate critical state 200 compressive members were found to be on the point of buckling (critically stressed) 5 of these being part of the last isostatic subsystem just before this state was reached. Thus 195 of these members represented the redundant bars. From the 31 possible post-buckling modes only 8 were found to be kinematically and statically admissible, conforming to the kinematic conditions (3.3.6) in Section 3.3. These modes are numbered 16, 17, 20, 26, 27, 28, 29 and 31. Only the modes 16 and 31 are displayed graphically in the Figs. 3.6.5 and 3.6.6. The slopes to the load radial-displacement equilibrium curves after buckling were evaluated at each outer (peripheral) loaded joint. Three such typical slopes are displayed in each figure 3.6.5 and 3.6.6 referring to the nodes 9, 34 and 36. The steepest slopes (gradients) tend to diminish at the joints closer to the middle of the shell. Their magnitudes varied considerably and they were all found to be negative (some of these having fairly large negative values). This indicates that the unoptimized shell would certainly collapse by unstable motion at this depth.

These gradients can be meaningfully interpreted if they are related to the actual *total* critical loads or hydrostatic heads at the joints. The steepness of these gradients is then only meaningful in relation to the total critical load. Typical gradients are shown plotted for selected joints against the radial displacements of these joints in Figs. 3.6.5 and 3.6.6. A radial displacement of 10 in. = 25.4 cm was assumed to represent a sufficiently large distortion of the shell lattice, so that the imperfection-sensitivity of the shell could be judged in this range of elastic deformation.

The 8 different modes which were investigated did not reveal very much difference in these slopes at the respective joints. The ranges of these correspond to the typical gradients $(\delta W_i/\delta s_i)$, where (i) denotes a typical peripheral node of load application, such as were obtained in the post-buckling range for general isostatic and hyperstatic pin-jointed lattices and frames discussed in the theory developed in Chapters 1, 2 and 3.

A summary of the post-buckling slopes at all the outer peripheral loaded joints for the unoptimized Model 1 is given in Fig. 3.6.7. These slopes indicate that different degrees of imperfection-sensitivity may be expected in different locations of the shell-lattice and this sensitivity appears to be enhanced towards the boundaries of the shell. There were 195 *redundant* buckling and 5 tensile-yield members, Fig. 3.6.4a, in each of the 8 admissible buckling modes and the number of isostatic buckling members varied from 3 to 5, as indicated in Figs. 3.6.5 and 3.6.6. Altogether 506 members were involved in each buckling mechanism, all of which were, therefore, global. The load-carrying capacity to weight ratio defined by

$$\text{LCC/Weight} = \frac{w^0 \sum_{i=1}^{J} \beta_{zi}}{\text{weight of 506 members}}$$

was found to be equal to 393.28, where (β_{zi}) represent the vertical load proportionality constants (along the z-axis) of the vector $(\boldsymbol{\beta})$ in Eqs. (2.5.4) and (3.2.1) for the unoptimized model reticulated shell based on the ultimate theoretical critical load.

(ii) The First Optimized Model Shell (Model No. 2)

An attempt was made to reduce the absolute values of the post-buckling equilibrium slopes and, therefore, to improve the imperfection-sensitivity of the first optimized model-shell compared to the unoptimized model. This was done by stiffening certain bars adjacent to the shell boundaries which buckled in the unoptimized model-shell and, which seemed to be responsible for the steep negative gradients of the local post-buckling equilibrium paths of the unoptimized shell model. Only the middle layer bars 482 to 506 (i.e. 25 bars) were stiffened. The exact locations of these bars are shown in Figs. 3.6.3a, 3.6.3b, 3.6.3c and in Fig. 3.8.1a and Fig. 3.8.2. These bars are parts of the octahedron and tetrahedron space elements located near the shell boundary.

After this stiffening only *one* of the 31 possible post-buckling modes was found to be kinematically and statically admissible, according to the restraints defined by Eqs. (3.3.6). The resulting post-buckling slopes in the admissible equilibrium mode at the top joint 1 and at the joints 9 and 20, situated midway radially and close to the boundary, are shown plotted in Fig. 3.8.4. These slopes are, indeed, less steep than the comparable slopes for the unoptimized shell. (These slopes were evaluated at the joints closest to those used before. The previous joints, namely 32 to 61, suffered no displacements in this case.) This seems to confirm the prediction that a better post-buckling performance of the lattice showing less steep initial post-buckling equilibrium slopes and, therefore, a somewhat reduced imperfection sensitivity may be achieved by the local stiffening of its elements. However, not any type of stiffening is acceptable. Stiffening the elements in some locations may even produce the opposite effect.

The steepest slope or gradient in this first optimized lattice had the value of approximately 1/3 of the steepest slope for the unoptimized model. Also, the average critical hydrostatic head, referred to the mid-height of the shell, is increased in this case by approximately 8 ft. = 2.438 m of sea-water. The ranges of the post-buckling slopes at all the 51 outer peripheral loaded joints were now found to vary between −11,745 lb./in. = −2,097.42 kg*/cm or in feet or meters of water, from −9.41 ft./in. = −1.13 m/cm to −2.58 ft./in. = −0.31 m/cm of sea-water head. The lay-out of the buckling bars in the one admissible mode, referred to the three parts of the lattice, is shown in Fig. 3.8.4 and the superposition of the gradients for all the loaded peripheral joints, with the deflected shape of the outer layer, is displayed in Fig. 3.8.5. Since the shell has 51 outer peripheral joints and the joints 32 to 61 suffered no displacements in this case, which reduces the number of displaced joints in this case by 30, several of the remaining

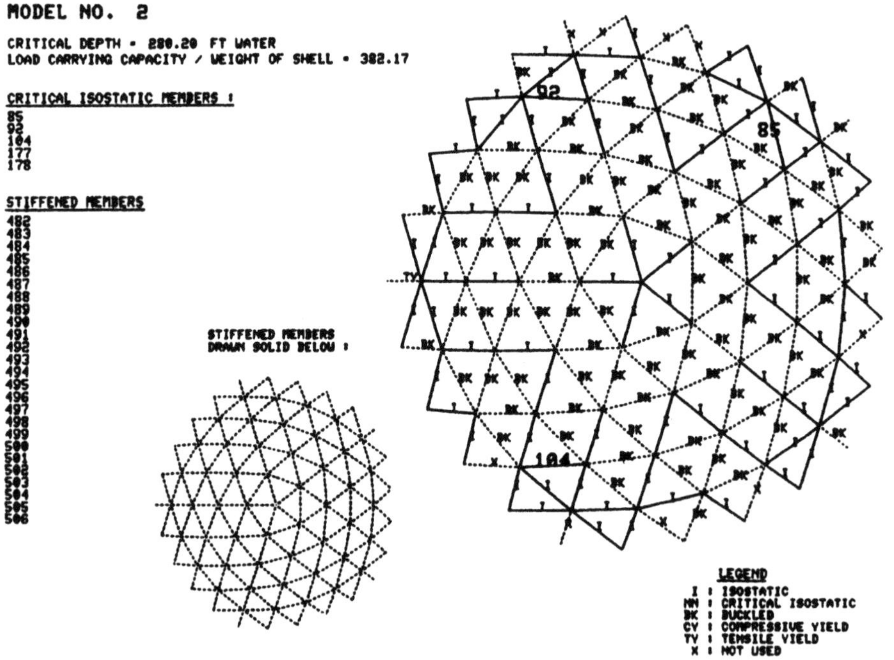

Figure 3.8.1
Status of the members at the critical state (outer layer of members)

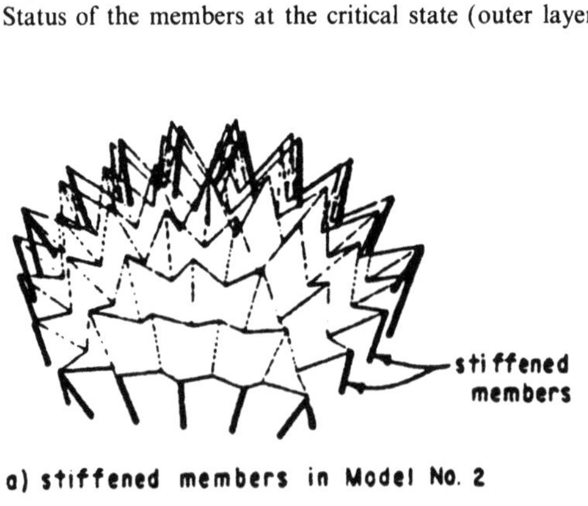

Figure 3.8.1a

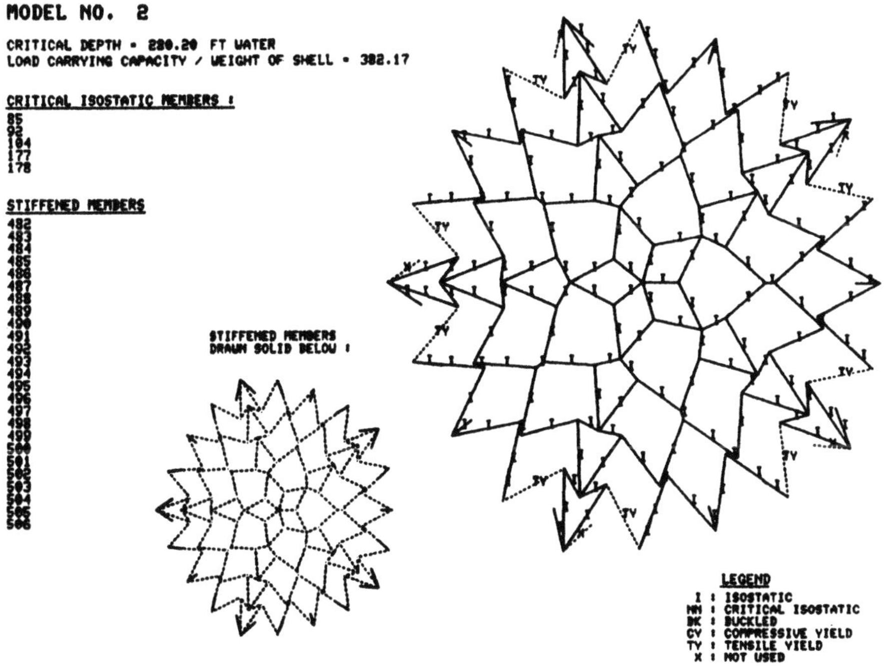

Figure 3.8.2
Status of the members at the critical state (middle layer of members)

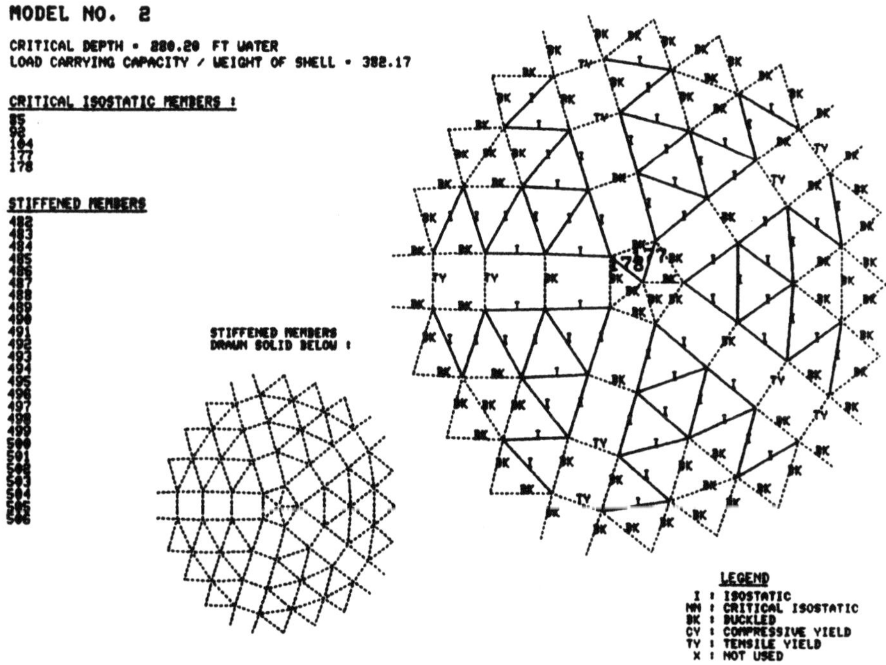

Figure 3.8.3
Status of the members at the critical state (inner layer of members)

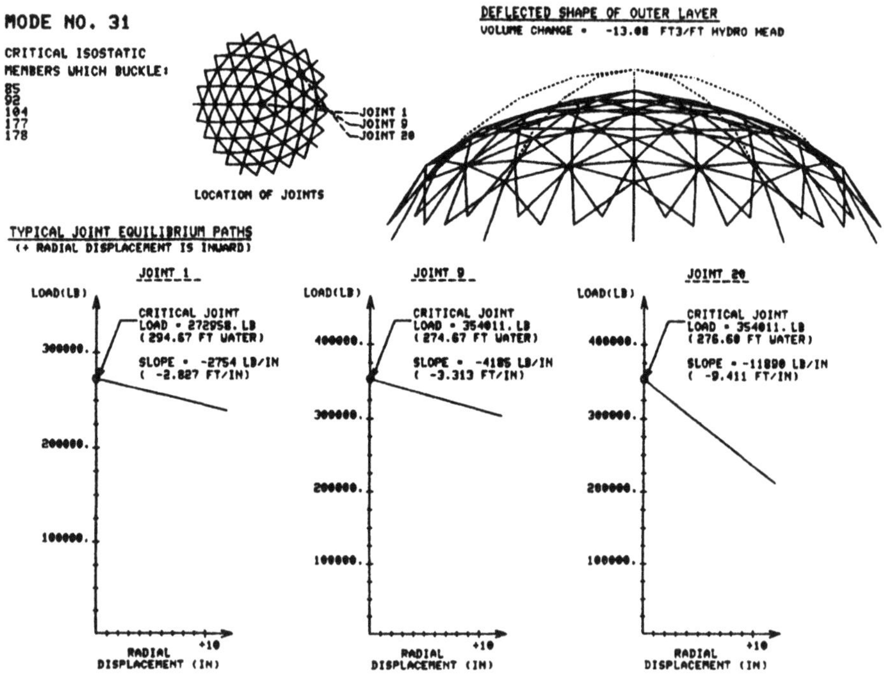

Figure 3.8.4
Plot of the deflected shell and of typical post-buckling equilibrium paths

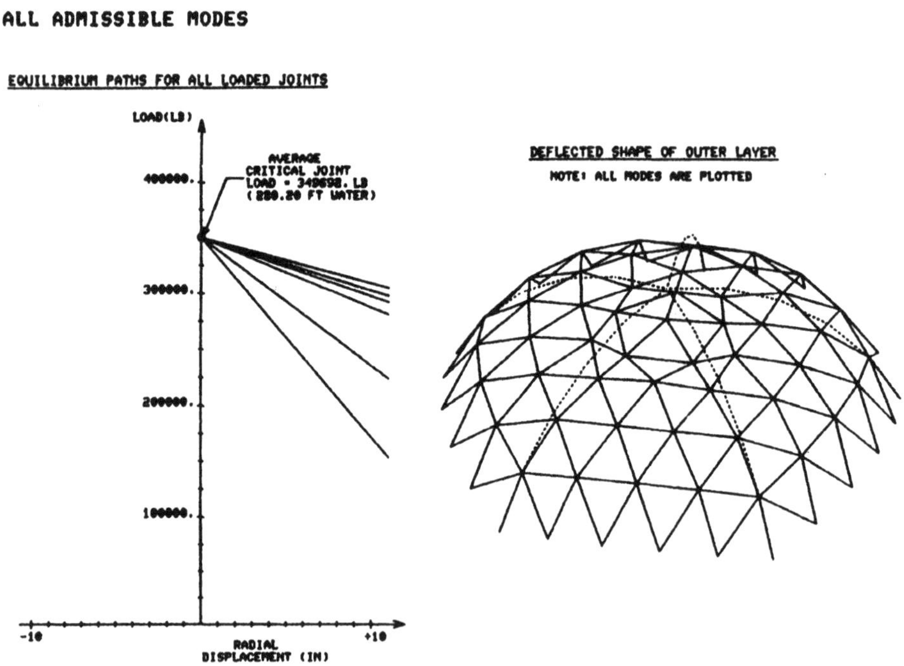

Figure 3.8.5
Superposition of all modes displaying all post-buckling slopes and the deflected outer layer

21 peripheral modes displayed only narrowly different gradients, as only about 6 different groups of slopes are visible in the computer-plotted chart in Fig. 3.8.5. A similar situation occurred in the unoptimized model in which only about 7 distinct groups of gradients were recorded for the 51 peripheral loaded joints, which are displayed in Fig. 3.6.7 in Section 3.6. The shell is clearly unstable at the critical depth, as all the gradients are negative in this case.

The theoretical load-carrying capacity to weight ratio of the first optimized model was found to be equal to about 382. This ratio can be made higher, this depending on the amount of stiffening of the members. However, the actual ratio depends on the peak load under the influence of initial imperfections. In view of the less steep theoretical initial gradients, this peak-load may be higher in this case percentage-wise than in the case of the unoptimized Model No. 1, where this ratio, under the imperfection-sensitivity of the structure, was estimated to be around 60% to 65% of the theoretical value.

(iii) The Second Optimized Model (Model No. 3)

In the second optimization of the shell 25 members in the unoptimized lattice situated along the meridians between the five icosahedron spherical triangles on the outer layer Fig. 3.8.6a were stiffened and prevented from buckling. The respective locations of these members are identified by the members 1 to 5, 12, 15, 18, 21, 24, 37, 42, 47, 52, 57, 77, 84, 91, 98, 105, 132, 141, 150, 159 and 168, Fig. 3.6.3a in Section 3.6. This stiffening had two important consequences:

(i) the ultimate critical state of the lattice was raised to 417.89 ft. = 127.37 m of sea water, which corresponds to 181.10 psi = 12.73 kg*/cm^2 (* denotes weight of 1 kg), as opposed to 272.43 ft. of water (83.04 m of water) and 117.87 psi = 8.287 kg*/cm^2 for the unoptimized shell.
(ii) the number of possible, kinematically and statically admissible post-buckling equilibrium modes is considerably higher in this case.

In fact, 1023 possible primary post-buckling modes based on a single last isostatic subsystem, (immediately before the ultimate critical state was reached), were discovered. But, only 116 of these were found statically and kinematically admissible according to the kinematic constraints specified by the Eqs. (3.3.6). The reason for such a sudden increase in modes was the increase in the number of compressive critical members in the last isostatic subsystem from 5 in the unoptimized model, to up to 10 members. These 10 members, together with the 147 redundant compressive critical bars, 50 tensile bars (stressed to the yield limit) and 3 unused (stress-free) bars were involved in these buckling mechanisms of the lattice. The effective number of compressive critical bars capable of generating new mechanisms was 147 + 10 = 157, which resulted in 1023 mode combinations based on the 10 critical isostatic members (see Eq. (3.6.33), p. 183) in the last subsystem obtained in the evaluation of the ultimate critical state with only 116 of these being statically and kinematically admissible. Since the total number of members in the lattice was 506 and only 157 + 3 + 50 = 210 members

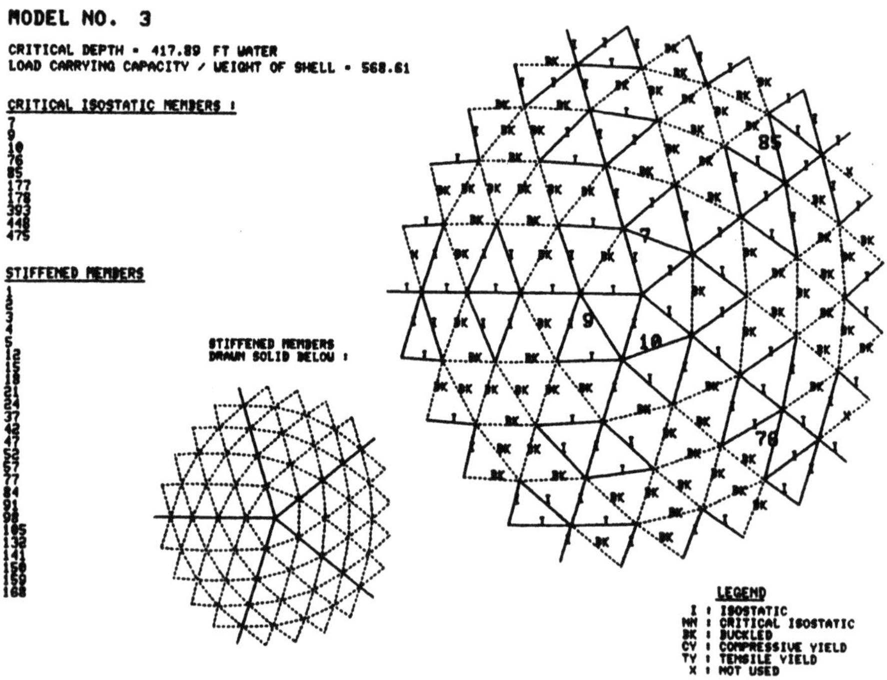

Figure 3.8.6
Status of the members at the critical state (outer layer of members)

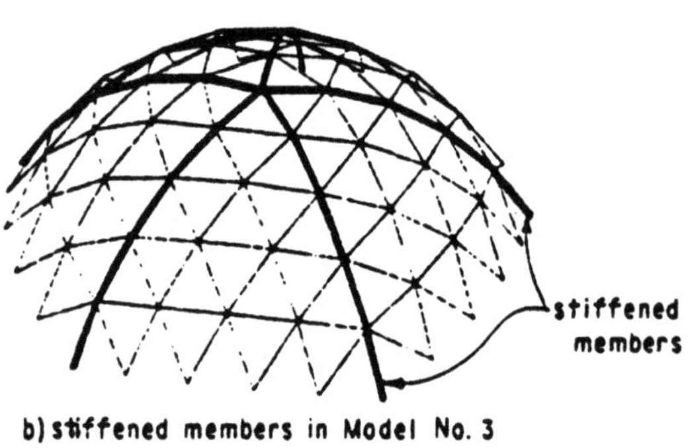

Figure 3.8.6a
Stiffened meridional members in Model No. 3

were involved in the buckling mechanism this suggests that not all the bars contributed to the post-buckling distortions of the structure. The status of the critical state in the outer, middle and inner layers of the shell lattice is presented in Figs. 3.8.6, 3.8.7 and 3.8.8.

The output data were scanned and 10 out of the 116 admissible modes were chosen for presentation in the original publication, Ref. 20. Only 2 typical, out of these 10 modes, are presented in Figs. 3.8.9 and 3.8.10, which are numbered 143 and 1023 (the member numbers in these figures refer to the numbers on the member identification charts in Figs. 3.6.3a to 3.6.3c). In the mode 143 there are only 3 isostatic critical bars and in mode 1023 there are 10 such bars. The slopes

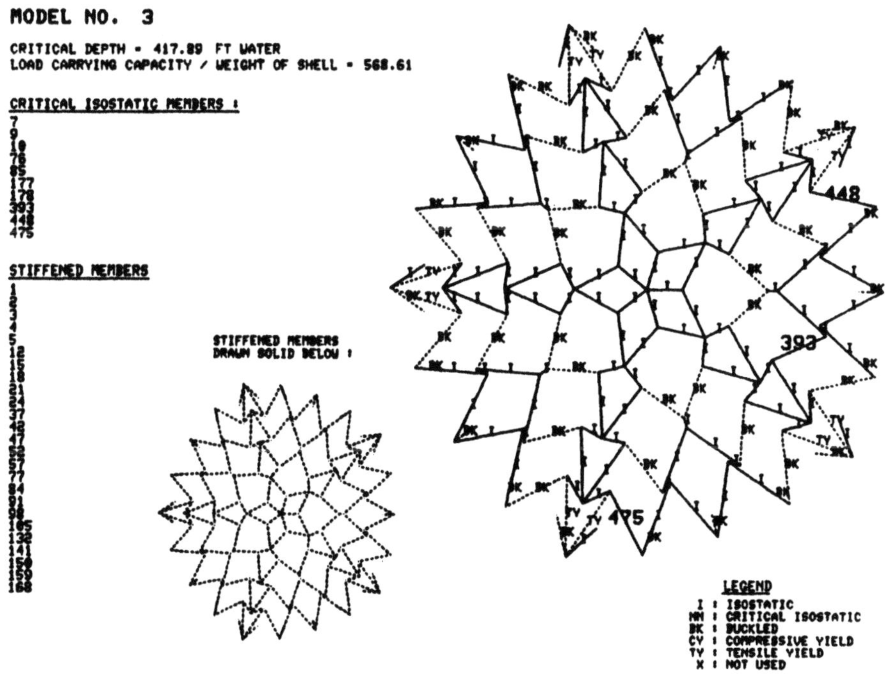

Figure 3.8.7
Status of the members at the critical state (middle layer of members)

(gradients) of the computed post-buckling equilibrium paths in each of these modes were plotted at three different joints on the outer shell layer, i.e. at the joints 2, 7 and 17 which are located on the meridian dividing two adjacent spherical triangles of the icosahedron. These are indicated in Figs. 3.8.9 and 3.8.10. The lay-out of the buckling members is shown in Figs. 3.8.6 to 3.8.8. These slopes show little difference in the 10 different modes originally studied. This suggests that the sensitivity of the shell is not strongly biased towards a particular group of modes. The superimposed gradients at the peripheral outer joints for all the 116 modes, together with the deformed shapes of the outer layer of the shell, are shown plotted in Fig. 3.8.11. The load-carrying capacity to weight (LCC/Wt) ratio of the lattice was markedly affected by the stiffening of the 25 meridional bars and this value jumped from 393.28 for the unoptimized Model No. 1 to 568.61 based on the theoretical loads for the optimized Model No. 3, resulting in a considerably more efficient lattice.

The statical stability of the shell in the ultimate critical state is determined by the change in the total potential energy of the system in an adjacent configuration of the deformed lattice, assuming the system is conservative. We showed in Section 1.2, Eq. 1.2.6 and by the Theorem 1 in that Section that this change can be expressed as a sum of the post-buckling gradients of all the load-displacement equilibrium paths at the ultimate critical state. Generally, it suffices to find only one mode for which this change is negative to prove the instability of the shell in the critical state. If all the slopes in at least one mode are negative, as in this

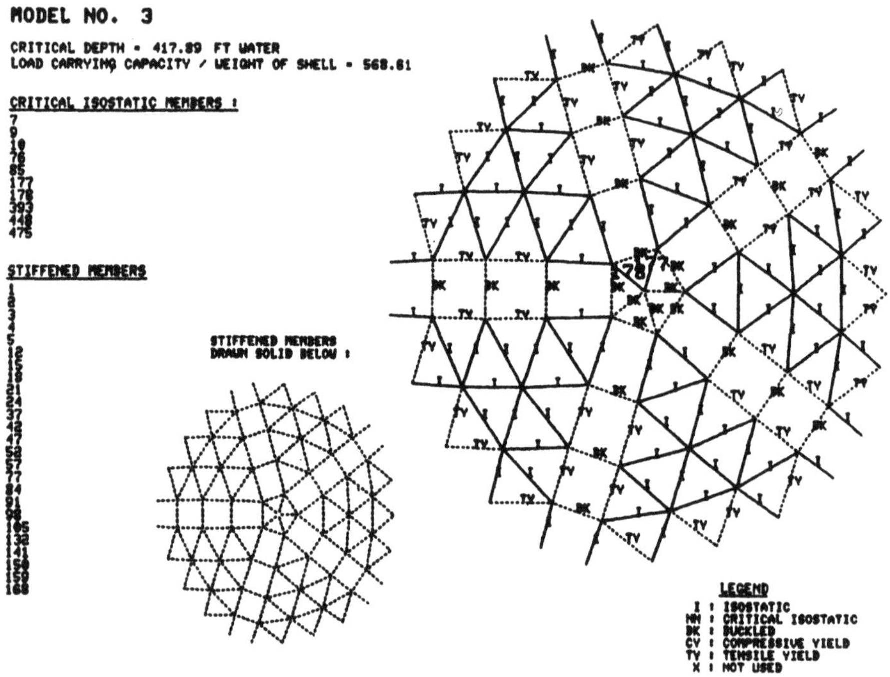

Figure 3.8.8
Status of the members at the critical state (inner layer of members)

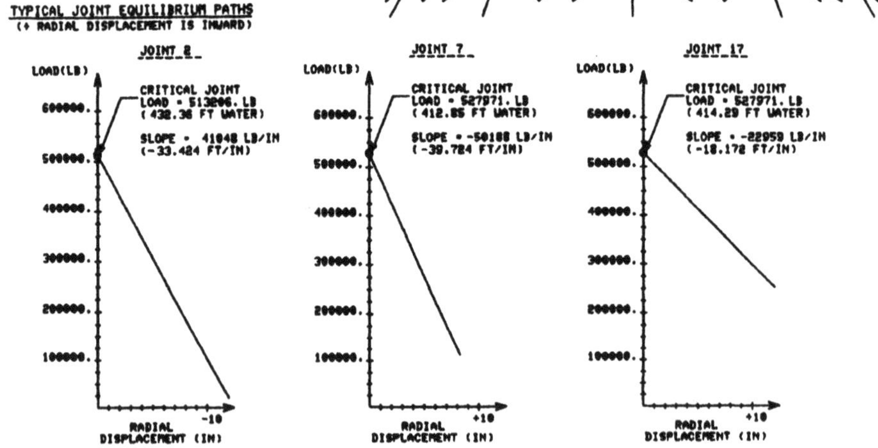

Figure 3.8.9
Plot of the deflected shell and of typical postbuckling equilibrium paths

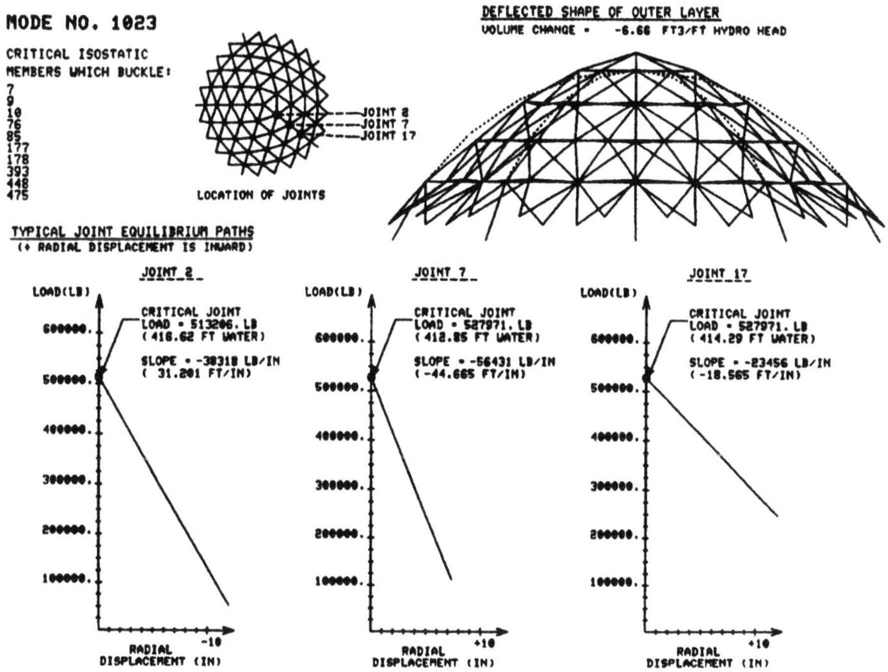

Figure 3.8.10
Plot of the deflected shell and of typical postbuckling equilibrium paths

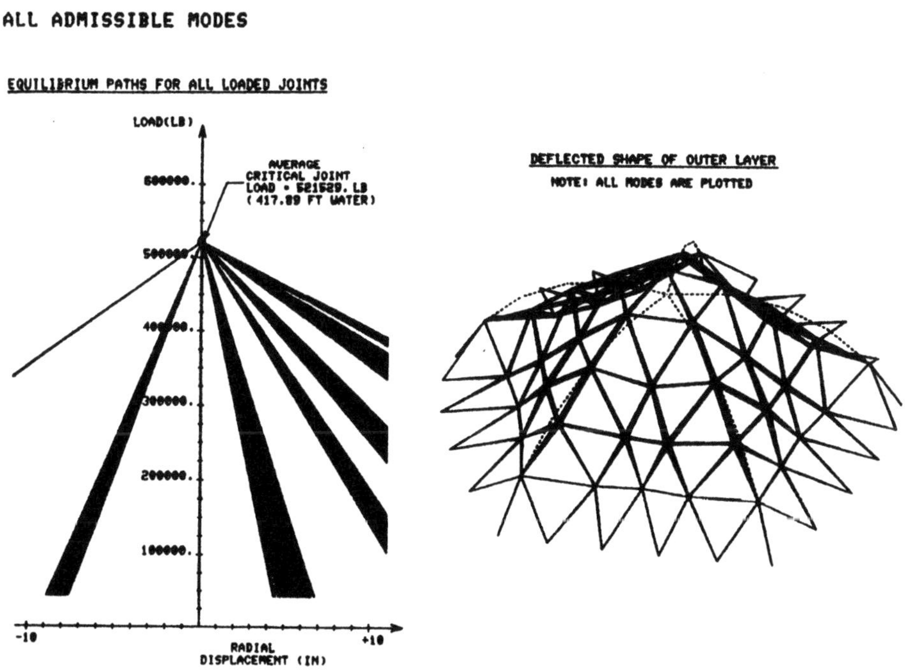

Figure 3.8.11
Superposition of all modes displaying all postbuckling slopes and the deflected outer layer

case, the ultimate critical state is certainly unstable and the structure fails in unstable motion if not artificially stabilized in equilibrium.

The changes in the total potential energy of each system, measured from the ultimate critical state at constant critical loads, were evaluated for all the modes for the three models considered. These values are relative and they correspond to the distorted lattice of each model in each mode, such that the structure in each case can be maintained in statical equilibrium by the modificed hydrostatic pressure (head), which is by 1 ft. = 30.48 cm smaller than that of the respective ultimate critical pressure or head. These results show that all the total potential energy changes are negative and that all the three models are statically unstable in their respective ultimate critical states and that no stable modes exist. These total potential energy changes are listed in Tables 3a to 3c. The conclusions which can be drawn from these studies indicate that in some cases the stiffening of certain elements in the lattice does indeed produce less steep post-buckling equilibrium gradients (Model No. 2) and that a better performance and less imperfection-sensitivity of the structure can be expected in the critical region of loading, compared to the unoptimized model shell. The results obtained for the first optimized model (Model No. 2) tend to confirm this conclusion.

In the first optimized model shell the gradients of the post-buckling equilibrium paths were less steep, but no appreciable increase in the ultimate critical pressure or head was obtained. Also, the theoretical LCC/Wt ratio was not raised in relation to that of the unoptimized model. On the other hand, in the second

Tables 3a, 3b, 3c
Relative potential energy changes for the three model shells in postbuckling equilibrium under a head 1 ft. less than the respective critical head

Table 3a
Total potential energy change for the modes of Model 1

Mode	Total potential energy change	Mode	Total potential energy change	Mode	Total potential energy change	Mode	Total potential energy change
16	−0.64609082	17	−0.64449603	20	−0.64449603	26	−0.64610877
27	−0.64610877	28	−0.64449697	29	−0.64449697	31	−0.64629737

Table 3b
Total potential energy change for the modes of Model 2

Mode	Total potential energy change	Mode	Total potential energy change	Mode	Total potential energy change	Mode	Total potential energy change
31	−0.40817072	0	0.00000000	0	0.00000000	0	0.00000000

Table 3c
Total potential energy change for the modes of Model 3

Mode	Total potential energy change	Mode	Total potential energy change	Mode	Total potential energy change	Mode	Total potential energy change
143	−0.20465689	227	−0.20485118	318	−0.20602493	352	−0.20495026
355	−0.20496335	358	−0.20477255	359	−0.20485603	444	−0.20622587
478	−0.20517795	481	−0.20514544	484	−0.20509768	485	−0.20508690
486	−0.20513718	514	−0.20626872	554	−0.20578978	555	−0.20608688
583	−0.20627746	586	−0.20631062	589	−0.20614755	590	−0.20614599
617	−0.20552630	620	−0.20506528	621	−0.20514917	623	−0.20508264
624	−0.20516577	626	−0.20495428	681	−0.20630048	709	−0.20650980
712	−0.20649954	715	−0.20648544	716	−0.20637592	743	−0.20574990
746	−0.20543539	747	−0.20541652	748	−0.20548553	749	−0.20539295
750	−0.20538350	752	−0.20531611	765	−0.20650980	768	−0.20655538
771	−0.20642365	772	−0.20654108	802	−0.20604411	803	−0.20633694
805	−0.20609406	806	−0,20639101	808	−0.20623829	820	−0.20680561
823	−0.20639936	824	−0.20639936	826	−0.20643679	827	−0.20643468
829	−0.20625438	841	−0.20564625	842	−0.20572983	844	−0.20524683
845	−0.20526724	852	−0.20672642	855	−0.20675042	856	−0.20676002
887	−0.20658331	890	−0.20659213	892	−0.20660525	904	−0.20703391
907	−0.20677962	908	−0.20666316	910	−0.20676000	911	−0.20665196
913	−0.20662095	925	−0.20601362	926	−0.20599317	927	−0.20604433
928	−0.20565623	929	−0.20561361	932	−0.20703391	935	−0.20666316
936	−0.20677962	938	−0.20671439	939	−0.20683281	941	−0.20668094
953	−0.20659919	954	−0.20688974	956	−0.20648670	957	−0.20654626
960	−0.20693224	961	−0.20693224	963	−0.20650703	964	−0.20654626
967	−0.20583186	971	−0.20703088	972	−0.20703088	974	−0.20702665
975	−0.20703930	977	−0.20704937	990	−0.20713258	992	−0.20689776
993	−0.20689776	996	−0.20730966	997	−0.20719228	999	−0.20691806
1000	−0.20689776	1003	−0.20623879	1004	−0.20719228	1005	−0.20730966
1007	−0.20691806	1008	−0.20697643	1011	−0.20704410	1012	−0.20704410
1013	−0.20755731	1014	−0.20755731	1016	−0.20732967	1017	−0.20732967
1020	−0.20745273	1021	−0.20745273	1022	−0.20745273	1023	−0.20786194

optimized model (Model No. 3) the critical load or head was increased from some 272 ft. to 417 ft. which represents an increase of some 53% in relation to the critical head of the unoptimized model, whereas only 25 meridional members needed to be marginally stiffened. The gradients of the post-buckling equilibrium paths became steeper by some 30% compared to those of the unoptimized Model No. 1. This does not mean, however, that under the influence of geometrical imperfections the first optimized model (Model No. 2) would perform better than the second optimized Model No. 3. The modified or "imperfect" equilibrium paths may well exhibit higher peak-values for Model No. 3 than for Model No. 2, as we can infer from the explanations of imperfection-sensitivity given in Chapter 2. Thus, the second optimization process may yield better results than the first.

3.9 The Most Degrading Post-Buckling Modes for the Three Model Reticulated Shells Intended for Underwater Applications

The theory concerned with the direct evaluation of the most degrading buckling modes developed in Section 3.7 was applied to the three model reticulated structures, i.e. Models 1, 2 and 3, studied in the last section, under equilibrium conditions. The numerical procedure adopted in the solution of the problem is described in the following Section. Here, we confine our attentions to the numerical results which were obtained for the three most degrading equilibrium modes for the three models considered. These results are based on the investigations published in Ref. 23 and they are reproduced in Fig. 3.9.1(a), (b) and (c).

The results are shown plotted against the weighted displacement ($\delta\bar{s}$) of the nodes, defined in Eq. (3.7.20). ($\delta\bar{s}$) is measured in feet. The ordinate represents

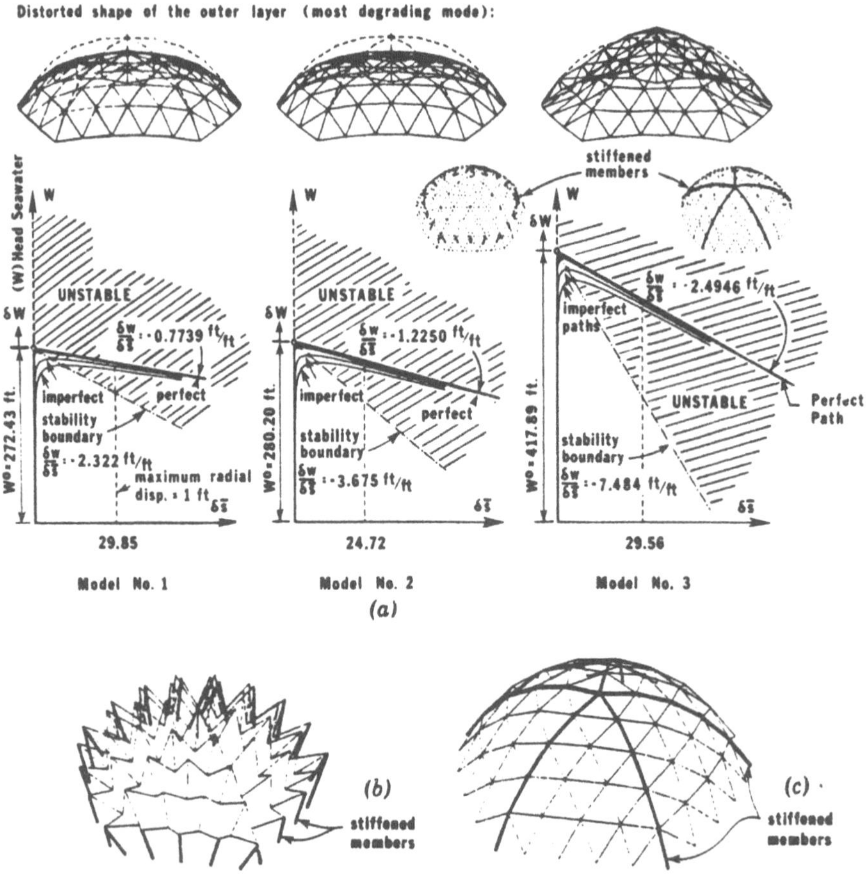

Figure 3.9.1
(a) Perfect and imperfect post-buckling equilibrium paths for most degrading mode with ($\boldsymbol{\beta}' = \boldsymbol{\beta}$) and $\delta\bar{s} = \mathbf{u}^T\boldsymbol{\beta}'$: (b) stiffened members in Model No. 2; (c) stiffened members in Model No. 3

the hydrostatic head (W) in feet of sea water (1 ft. = 30.48 cm). For the unoptimized Model No. 1 the average critical head amounted to some 272 ft. and the steepest negative gradient to the equilibrium path in the most degrading mode to ($\delta W/\delta \bar{s} = -0.7739$ ft./ft.). The computed buckled shape of the outer layer of the shell in this mode is indicated in Fig. 3.9.1(a).

The first optimized model (Model 2) was generated similarly as before, by stiffening 25 members close to the shell boundaries of the unoptimized Model No. 1, as indicated in Fig. 3.9.1(b). The improved average theoretical critical head was thereby only marginally raised to some 280 ft. of sea water. The steepest post-buckling gradient, based on the average weighted displacement ($\delta \bar{s}$), in the most degrading equilibrium mode is now more negative having the value of ($\delta W/\delta \bar{s} = -1.2250$ ft./ft.). The buckled shape of the outer layer for this case is also indicated in Fig. 3.9.1(a).

In the second optimized Model No. 3, 25 meridional bars between the five spherical triangles in the outer layer were stiffened, as explained in the last Section. These stiffened bars are indicated in Fig. 3.9.1(c). Now the critical theoretical head is much higher and it amounts to some 417 ft. of sea water. The steepest negative gradient, based on the average weighted displacement ($\delta \bar{s}$) in the most degrading postbuckling equilibrium mode, is now equal to ($\delta W/\delta \bar{s} = -2.4946$ ft./ft.). The computed buckled shape of the outer layer for this case is also indicated in Fig. 3.9.1(a).

The results, obtained numerically according to the eigenvector method and the conjugate gradient method are presented in Table 4. Thus, the two independent numerical approaches are shown to give practically identical results for the global constrained minimum value of the total potential energy and for the most degrading buckling modes in equilibrium.

The imperfect paths in each case may be evaluated and plotted using the theory in Section 2.14, particularly the form of Eq. (2.14.40), which is

$$W_{\text{impf.}} = W^0 + 2C_{\text{pf.}} \frac{s^{*3/2}}{\sqrt{s}} + C_{\text{pf.}} s \qquad (2.14.40)\text{bis}$$

where ($W_{\text{impf.}}$) represents the ordinate of a typical external load at a node or of the total load in the form of hydrostatic preasure. (s) represents the correspond-

Table 4
Comparison of results

Model Reticulated Shell		Constrained Minimum ($\delta w/\delta \bar{s}$)	
Number (1)	Model type (2)	Eigenvector method (3)	Conjugate gradient method (4)
1	Basic	−0.77389	−0.77389
2	1st optimized	−1.22501	−1.22498
3	2nd optimized	−2.48457	−2.48457

ing displacement of such a typical load and (s^*) the displacement where the imperfect stable-unstable path has its peak, which occurs always on the stability boundary having a gradient 3 times that of the path in the same mode. This holds only for pin-jointed and quasi-pin-jointed elastic systems. The constant ($C_{\text{pf.}}$) represents the gradient of the perfect equilibrium path in the same mode, which can be calculated theoretically, as described by the theory in Chapters 2 and 3.

3.10 Minimization Methods in the Direct Evaluation of the Most Degrading Buckling Modes

It was shown in Section 3.7, Eq. (3.7.6), that the expression for the total potential energy change (v) from the ultimate critical state may be written in the form[5]

$$v = \tfrac{1}{2}\mathbf{e}_b^T \mathbf{E}_b \mathbf{e}_b; \qquad b \leq c \tag{3.10.1}$$

where

$$\mathbf{E}_b = \left(1 + \frac{1}{K}\right)(\mathbf{T}_b + \mathbf{C}_{1b}'^T \mathbf{C}_\gamma^T \mathbf{T}_\gamma \mathbf{C}_\gamma \mathbf{C}_{1b}'^T) + \mathbf{C}_{1b}' \mathbf{C}_\sigma^T \mathbf{T}_\sigma' \mathbf{C}_\sigma \mathbf{C}_{1b}' - \mathbf{C}_{1b}'^T \mathbf{P}_L^0 \mathbf{C}_{1b}' \tag{3.10.2}$$

as given by Eq. (3.6.19), except that the subscript (c) is replaced by (b), which denotes the number of the actually buckling members. The subscript (c) denotes the number of critically *stressed* bars in the last isostatic subsystem used to determine the ultimate critical state and the maximum number of isostatic buckling members as ($b \leq c$). Not all of the (c) bars need shorten flexurally in post-buckling, as some may unload and remain straight in the post-buckling process in which they rotate only bodily.

The expression for (v) may be considered as representing a ($b + 1$)-dimensional hypersurface in a ($b + 1$)-dimensional space defined by the independent variables (e_b); $b \leq c$. This hypersurface contains also the negative energy regions, measured from the ultimate critical state, corresponding to the most degrading post-buckling modes of the hyperstatic or isostatic pin-jointed elastic system under consideration, characterized by the vector ($\mathbf{e}$) composed of the vectors ($\mathbf{e}_1$) and ($\mathbf{e}_2$), defined in the Eqs. (3.2.3), (3.2.4) and (3.2.5) respectively. The vector ($\mathbf{e}_1$) contains the independent isostatic flexural member shotenings represented by the vector ($\mathbf{e}_b$). The vector ($\mathbf{e}_2$) represents the flexural shortenings of the redundant members and, according to the Eqs. (3.2.5), (3.2.7) and (3.3.2), it is equal to

$$\mathbf{e}_2 = \mathbf{C}_2 \mathbf{C}_{1b}' \mathbf{e}_b \geq 0; \qquad b \leq c \tag{3.10.3}$$

Thus, the total potential energy (v) is essentially a function of (b) variables which are the elements of the vector ($\mathbf{e}_b$) which characterize a particular post-buckling mode.

---

[5] Vectors and matrices are printed in bold type.

As flexural contractions may be only positive numbers, the positive orthant (when $b = 2$, the positive quadrant) represents the kinematically admissible region of the v-hypersurface, i.e. for ($\mathbf{e}_b \geq \mathbf{0}$). When the system is hyperstatic, only a restricted region of this orthant is admissible, as a result of the positiveness of the vector ($\mathbf{e}_2$), expressed by the condition (3.10.3).

Before a proper local minimum or the global minimum value of the total potential energy change can be evaluated, the v-hypersurface must be bounded. This is so, because the magnitude of the vector ($\mathbf{e}_b$) is not constrained by the kinematic conditions (3.3.6) in Section 3.3 or, equivalently, by the condition (3.10.3) above. In this form the total potential energy (v) is said to be unbounded. This means that, if the vector ($\mathbf{e}_b = \mathbf{e}_b^*$) is the solution corresponding to the admissible buckling mode resulting in a minimum value of ($v = v^*$), then also the vector ($k\mathbf{e}_b^*$) is an admissible value, where (k) is any positive constant. This is so, since the kinematic constraints in (3.3.6), or in the equivalent form in Eq. (3.10.3) above, are linear. Therefore (v^*) is not a minimum, as a more negative value ($k^2 v^*$) in the admissible region can always be found by selecting a new value for (k).

This implies that before determining a proper minimum on the v-hypersurface the length of the vector ($\mathbf{e}_b$) must be bounded. One way of bounding the length of the vector ($\mathbf{e}_b$) is to arbitrarily impose the requirement or constraint that

$$\mathbf{e}_b^T \mathbf{e}_b = 1 \tag{3.10.4}$$

which limits the magnitude (absolute value) of the vector to unity (normalizing condition).

We have discussed this problem already in Section 3.7, where only the equilibrium states under proportionately modified applied loading of the system in the post-buckling modes of deformation were considered. We saw that, under those conditions, the constraint on the magnitude of the vector ($\mathbf{e}_b$) had to assume a special form given by relation (3.7.23) and (3.7.40a) i.e.

$$\mathbf{e}_b^T \mathbf{Q} \mathbf{e}_b = 2C = \text{constant} \qquad \text{(3.7.40a)bis}$$

where, in the case of post-buckling equilibrium,

$$\mathbf{Q} = \mathbf{C}_{1b}'^T \boldsymbol{\beta}' \boldsymbol{\beta}'^T \mathbf{C}_{1b}' \qquad \text{(3.7.33)bis}$$

in which ($\boldsymbol{\beta}'$) was essentially arbitrary. There, we were able to introduce an arbitrary load vector change ($\boldsymbol{\beta}'\, \delta w$) to minimize the total potential energy of the system in equilibrium in the post-buckling process. The change (δw) could be then evaluated from Eq. (3.7.36) or Eq. (3.7.38) in the form of a function ratio, the denominator of which represented the quadratic form given by Eq. (3.7.40a).

The magnitude of the vector ($\mathbf{e}_b$) may be bounded by other similar conditions. For example, one such bounding may be accomplished by imposing the requirement that the vector embodying the flexural contractions of the isostatic members (vector $\mathbf{e}_b$) and the hyperstatic flexural member shortenings ($\mathbf{e}_\gamma$) be normalized. In that case

$$\mathbf{e}_b^T \mathbf{e}_b + \mathbf{e}_\gamma^T \mathbf{e}_\gamma = 1 \tag{3.10.5}$$

In Section 3.3, Eq. (3.3.6b) we derived that ($\mathbf{e}_\gamma = \mathbf{C}_\gamma \mathbf{C}'_{1c} \mathbf{e}_c$). If, instead of ($\mathbf{e}_c$) the vector ($\mathbf{e}_b$) is introduced, then

$$\mathbf{e}_\gamma = \mathbf{C}_\gamma \mathbf{C}'_{1b} \mathbf{e}_b \tag{3.3.6)bis}$$

and the bounding condition (3.10.5) becomes

$$\mathbf{e}_b^T \mathbf{e}_b + \mathbf{e}_b^T \mathbf{C}'^T_{1b} \mathbf{C}_\gamma^T \mathbf{C}_\gamma \mathbf{C}_{1b} \mathbf{e}_b = 1$$

or

$$\mathbf{e}_b^T \mathbf{e}_b + \mathbf{e}_b^T \mathbf{e} = \mathbf{e}_b^T \mathbf{B}_e \mathbf{e}_s \tag{3.10.6}$$

Substituting in (3.10.5) gives the contour

$$\mathbf{e}_b^T \mathbf{B}_e \mathbf{e}_b = 1 \tag{3.10.7}$$

where

$$\mathbf{B}_e = \mathbf{I} + \mathbf{C}'^T_{1b} \mathbf{C}_\gamma^T \mathbf{C}_\gamma \mathbf{C}'_{1b} \tag{3.10.8}$$

and ($\mathbf{B}_e$) is the matrix which defines the quadratic form in Eq. (3.10.7). The magnitude of the vector ($\mathbf{e}_b$) on the v-hypersurface is not constant now and (v) is evaluated at unequal distances from the origin of the v-hypersurface in the ($b + 1$) dimensional space (v, $\mathbf{e}_b$).

Other bounding hypersurfaces or contours may be conceived similarly. If one defines, by the symbol (U_{ee}), the excess elastic energy change caused by the changes ($\delta \mathbf{P}_b$) in the compressive axial loads in the buckling members, beyond that due to the critical axial loads, then

$$U_{ee} = \tfrac{1}{2} \delta \mathbf{P}_b^T \mathbf{e}_b + \tfrac{1}{2} \delta \mathbf{P}_\gamma^T \mathbf{e}_\gamma \tag{3.10.9}$$

where ($\delta \mathbf{P}_b$) and ($\delta \mathbf{P}_\gamma$) are axial load increments in the isostatic and the redundant buckling members, respectively. Substituting for

$$\delta \mathbf{P}_b = \mathbf{T}_b \mathbf{e}_b ; \qquad b \leq c \tag{3.4.6)bis}$$

and writing similarly, using the Eq. (3.3.6),

$$\delta \mathbf{P}_\gamma = \mathbf{T}_\gamma \mathbf{e}_\gamma = \mathbf{T}_\gamma \mathbf{C}_\gamma \mathbf{C}'_{1b} \mathbf{e}_b \tag{3.10.10}$$

we get

$$2U_{ee} = \mathbf{e}_b^T (\mathbf{T}_b + \mathbf{C}'^T_{1b} \mathbf{C}_\gamma^T \mathbf{T}_\gamma \mathbf{C}_\gamma \mathbf{C}'_{1b}) \mathbf{e}_b \tag{3.10.11}$$

where ($\mathbf{T}_\gamma$) is, clearly, a diagonal matrix so that ($\mathbf{T}_\gamma^T = \mathbf{T}_\gamma$). Putting now

$$\mathbf{S} = \mathbf{T}_b + \mathbf{C}^T_{1b} \mathbf{C}_\gamma^T \mathbf{T}_\gamma \mathbf{C}_\gamma \mathbf{C}'_{1b} \tag{3.10.12}$$

where (**S**) is defined as the "excess strain energy" matrix, one finds

$$U_{ee} = \tfrac{1}{2} \mathbf{e}_b^T \mathbf{S} \mathbf{e}_b \tag{3.10.13}$$

(**S**) is then another matrix, additional to the matrices (**Q**) and ($\mathbf{B}_e$), that may be used to bound the total potential energy hypersurface (v). These and other contours in bounding the total potential energy hypersurfaces of complex hyperstatic lattices and their effects on the numerical results in calculating the local and global minima of such hypersurfaces were studied by **M.D. Davister**, Ref. 27. **Davister** used two more bounding contours referred to as the g-contour and the g_L-contour, (op.cit.).

Before continuing this discussion, we formulate the *constrained minimization problem*, regardless of the physical significance of the matrices involved to bound the total potential energy hypersurface (v) near its origin. Thus, the expression:

$$v = \tfrac{1}{2}\mathbf{e}_b^T \mathbf{E}_c \mathbf{e}_b; \qquad b \leq c$$

is required to be a proper or global constrained minimum, subject to the following inequality and equality constraints:

$$\begin{aligned} &1.\ \mathbf{e}_b \geq 0 \\ &2.\ \mathbf{e}_2 = \mathbf{C}_2\mathbf{C}'_{1b}\mathbf{e}_b \geq 0 \\ &3.\ \tfrac{1}{2}\mathbf{e}_b^T \mathbf{B}\mathbf{e}_b = 1; \qquad b \leq c \end{aligned} \tag{3.10.14}$$

where ($\mathbf{C}_2$) is defined in Eqs. (3.2.3) and (3.6.2) and ($\mathbf{B}$) may be any of the ($b \times b$) matrices, ($\mathbf{Q}$), ($\mathbf{B}_e$) or ($\mathbf{S}$). In the above expression ($\mathbf{e}_b$) is a normalized vector with respect to the matrix ($\mathbf{B}$).

To illustrate this point consider an arbitrary vector ($\mathbf{e}'_b$) which is not normalized with respect to ($\mathbf{B}$) in the quadratic form in Eq. (3.10.14; 3). This form has now the value

$$\tfrac{1}{2}\mathbf{e}'^T_b \mathbf{B}\mathbf{e}'_b = C' \tag{3.10.14a}$$

where (C') is a constant. On dividing the last equation by (C') and comparing the left side with the condition (3.10.14; 3) for all values of ($\mathbf{e}_b$) and ($\mathbf{e}'_b$), one finds that

$$\mathbf{e}_b = \frac{\mathbf{e}'_b}{\sqrt{C'}} \tag{3.10.14b}$$

Substituting now this quantity in the expression for (v) which is constrained by the condition ($\mathbf{e}_b^T \mathbf{B}\mathbf{e}_b = 1$), we get,

$$v = \frac{1}{2}\mathbf{e}_b^T \mathbf{E}_b \mathbf{e}_b = \frac{1}{2}\frac{\mathbf{e}'^T_b \mathbf{E}_b \mathbf{e}'_b}{C'} \tag{3.10.14c}$$

Substituting further for (C') from (3.10.14a), one finds that

$$v = \frac{\mathbf{e}'^T_b \mathbf{E}_b \mathbf{e}'_b}{\mathbf{e}'^T_b \mathbf{B}\mathbf{e}'_b} \tag{3.10.14d}$$

where ($\mathbf{e}'_b$) is an *arbitrary* vector of the flexural contractions not necessarily constrained by the constraint condition 3 in Eq. (3.10.14). Since ($\mathbf{e}'_b$) is arbitrary, the last expression holds also for the special case when ($\mathbf{e}'_b = \mathbf{e}_b$), i.e.

$$v = \frac{\mathbf{e}_b^T \mathbf{E}_b \mathbf{e}_b}{\mathbf{e}_b^T \mathbf{B}\mathbf{e}_b} \tag{3.10.14e}$$

where now ($\mathbf{e}_b$) is normalized with respect to ($\mathbf{B}$) and ($\mathbf{e}_b^T \mathbf{B}\mathbf{e}_b = 1$).

Setting (**B**) equal to (**S**) and ($\mathbf{B}_e$), respectively where (**S**) is given by Eq. (3.10.12) and ($\mathbf{B}_e$) by Eq. (3.10.8), we obtain the corresponding function-ratios

$$v_S = \frac{\mathbf{e}_b^T \mathbf{E}_b \mathbf{e}_b}{\mathbf{e}_b^T \mathbf{S} \mathbf{e}_b} = v_1 \tag{3.10.14f}$$

and

$$v_e = \frac{\mathbf{e}_b^T \mathbf{E}_b \mathbf{e}_b}{\mathbf{e}_b^T \mathbf{B}_e \mathbf{e}_b} = v_2 \tag{3.10.14g}$$

The minimization problem can now be restated in a somewhat simplified form, (**Davister**, op.cit.). To find the minimum value of (v) on the bounded v-hypersurface, minimize

$$v = \frac{\mathbf{e}_b^T \mathbf{E}_b \mathbf{e}_b}{\mathbf{e}_b^T \mathbf{B} \mathbf{e}_b}; \qquad b \leq c \tag{3.10.15}$$

subject to

$$\mathbf{e}_b \geq 0,$$

and

$$\mathbf{e}_2 = \mathbf{C}_2 \mathbf{C}'_{1b} \mathbf{e}_b \geq 0$$

where ($\mathbf{e}_b$) is essentially arbitrary (the prime on ($\mathbf{e}_b$) is now omitted). The matrix (**B**) must be positive definite for (v) not to have any singularities. The matrices (**Q**), ($\mathbf{B}_e$) and (**S**) usually satisfy this condition. Thus, any quadratic form, which is positive definite, can be used to define the ratio of quadratic forms in (3.10.15) which is to be minimized to evaluate the admissible most degrading post-buckling modes, associated with the lowest energy level on a contour of the v-hypersurface. But, *the equilibrium conditions on this contour are not generally satisfied*, except when ($\mathbf{B} \equiv \mathbf{Q}$), where (**Q**) is defined by Eq. (3.7.33). Therefore the buckling mode so determined represents a mode of deformation while the system is in *a state of motion*. If the vector ($\mathbf{e}_b$) corresponds to the constrained minimum (global minimum) of the v-hypersurface such a mode is referred to as the *most degrading dynamic collapse mode*.

In addition to the contour-hypersurfaces defined by the matrices ($\mathbf{B}_e$) and (**S**) in Eqs. (3.10.8) and (3.10.14), two other matrices or contours were used in the mentioned treatments (Ref. 27 and Ref. 28) for the evaluation of the most degrading collapse modes of complex hyperstatic lattices. Following **Davister**, we refer to these as the g- and the g_L-"contours" or hypersurfaces.

The g-Hypersurface

A bounding hypersurface may be defined by considering the displacements (**u**) of the joints of the lattice in the elastic post-buckling process. Similarly, as in the Eq. (3.10.4) the magnitude of the ($\mathbf{e}_b$) vector and in Eq. (3.10.5) the magnitude of the vector ($\mathbf{e}_b^T \mid \mathbf{e}_y^T$) was set equal to unity, now the length of the vector (**u**) is held constant in all directions of the vector ($\mathbf{e}_b$) by putting

$$\mathbf{u}^T\mathbf{u} = 1 \tag{3.10.16}$$

Using Eq. (3.3.2) and replacing the subscript (c) by (b) in that equation, we get

$$\mathbf{e}_b^T \mathbf{C}_{1b}'^T \mathbf{C}_{1b}' \mathbf{e}_b = 1$$

or

$$\mathbf{e}_b^T \mathbf{J} \mathbf{e}_b = 1 \tag{3.10.17}$$

where

$$\mathbf{J} = \mathbf{C}_{1b}'^T \mathbf{C}_{1b}'; \qquad b \le c \tag{3.10.17a}$$

Then the values of the total potential energy (v) on the g-hypersurface are given, similarly as before, by

$$g = \frac{\mathbf{e}_b^T \mathbf{E}_b \mathbf{e}_b}{\mathbf{e}_b^T \mathbf{J} \mathbf{e}_b} = v_3; \qquad b \le c \tag{3.10.18}$$

where ($\mathbf{e}_b$) is an arbitrary vector representing the flexural contractions of the lattice and of the system in one of its post-buckling modes.

The g_L-Hypersurface

The g_L hypersurface or contour is obtained by normalizing only the displacements of the loaded joints, instead of normalizing the displacements of all the joints. If the displacements of the loaded joints are denoted by ($\mathbf{u}_L$) and those of unloaded joints by ($\mathbf{u}_{uL}$) ((uL) standing for "unloaded"), then the bounding surface may be defined by

$$\mathbf{u}_L^T \mathbf{u}_L = \mathbf{u}^T \bar{\mathbf{I}} \mathbf{u} = 1 \tag{3.10.19}$$

where the matrix ($\bar{\mathbf{I}}$) is generally an ($N \times N$) matrix equal to

$$\bar{\mathbf{I}} = \left(\begin{array}{c|c} \mathbf{I} & \mathbf{0} \\ \hline \mathbf{0} & \mathbf{0} \end{array}\right) \tag{3.10.20}$$

and ($\mathbf{I}$) is of the order ($L \times L$), where (L) is the order of ($\mathbf{u}_L$). Thus

$$\mathbf{u}^T \bar{\mathbf{I}} \mathbf{u} = (\mathbf{u}_L^T \mid \mathbf{u}_{uL}^T) \left(\begin{array}{c|c} \mathbf{I} & \mathbf{0} \\ \hline \mathbf{0} & \mathbf{0} \end{array}\right) \left(\begin{array}{c} \mathbf{u}_L \\ \hline \mathbf{u}_{uL} \end{array}\right) = \mathbf{u}_L^T \mathbf{I} \mathbf{u}_L = \mathbf{u}_L^T \mathbf{u}_L = 1 \tag{3.10.21}$$

Thus, the displacements corresponding to the unloaded directions at the joints are omitted in the vector ($\mathbf{u}_L$). These directions are marked by the zero elements in the load vector ($\boldsymbol{\beta}$), already introduced in Eq. (2.5.4) in Chapter 2.

Substituting now for ($\mathbf{u}$) from Eq. (3.3.2) and writing the index (b) instead of (c),

$$\mathbf{u} = \mathbf{C}_{1b}' \mathbf{e}_b; \qquad b \le c \tag{3.3.2)bis}$$

we find

$$\mathbf{u}_L^T \mathbf{u}_L = \mathbf{u}^T \bar{\mathbf{I}} \mathbf{u} = \mathbf{e}_b^T \mathbf{C}_{1b}'^T \bar{\mathbf{I}} \mathbf{C}_{1b}' \mathbf{e}_b = \mathbf{e}_b^T \mathbf{J}_L \mathbf{e}_b = 1$$

where the matrix $\mathbf{J}_L$ is defined by

$$\mathbf{J}_L = \mathbf{C}_{1b}^{\prime T}\bar{\mathbf{I}}\mathbf{C}_{1b}^{\prime} \tag{3.10.22}$$

Then on the bounding surface, defined by (3.10.19) and (3.10.20),

$$v = \mathbf{e}_b^T \mathbf{E}_b \mathbf{e}_b \quad \text{and} \quad \mathbf{e}_b^T \mathbf{J}_L \mathbf{e}_b = 1 \tag{3.10.22a}$$

If now $(\mathbf{e}_b^{\prime})$ is a vector of flexural contractions of the system of arbitrary magnitude in one of its post-buckling modes, then similarly, as shown in the steps between Eqs. (3.10.15) and (3.10.18), in this case

$$v = g_L = \frac{\mathbf{e}_b^{\prime T} \mathbf{E}_b \mathbf{e}_b^{\prime}}{\mathbf{e}_b^{\prime T} \mathbf{J}_L \mathbf{e}_b^{\prime}} \tag{3.10.23}$$

In the special case, when $(\mathbf{e}_b^{\prime} = \mathbf{e}_b)$

$$g_L = \frac{\mathbf{e}_b^T \mathbf{E}_b \mathbf{e}_b}{\mathbf{e}_b^T \mathbf{J}_L \mathbf{e}_b} = v_4 \tag{3.10.23a}$$

where the matrix $(\mathbf{J}_L)$ is defined by (3.10.22) and its quadratic form with respect to vector $(\mathbf{e}_b)$ by Eq. (3.10.22a).

Generalized formulation of the minimization problem

A generalized formulation of the minimization problem can be then restated in terms of the simplified problem defined by the Eqs. (3.10.15). For the convenience of mathematical argument we replace the vectors $(\mathbf{e}_b)$ by $(\mathbf{x})$ and $(\mathbf{e}_2)$ by $(\mathbf{y})$ and the matrix $(\mathbf{E}_b)$ by $(\mathbf{A})$ and the matrix product $(\mathbf{C}_2\mathbf{C}_{1b})$ in Eq. (3.10.14;2) simply by the matrix $(\mathbf{C})$†. Then the generalized minimization problem becomes: minimize

$$F(\mathbf{x}) = \frac{\mathbf{x}^T\mathbf{A}\mathbf{x}}{\mathbf{x}^T\mathbf{B}\mathbf{x}} \tag{3.10.24}$$

subject to the non-equality restraints

$$\mathbf{x} \geq \mathbf{0} \tag{3.10.25a}$$

$$\mathbf{y} = \mathbf{C}\mathbf{x} \geq \mathbf{0} \tag{3.10.25b}$$

where $(\mathbf{A})$ and $(\mathbf{B})$ must be symmetric $(n \times n)$ matrices. $(\mathbf{C})$ is an $(R \times n)$ matrix, where $(R = M - N)$ and (M) is the number of members in the hyperstatic elastic lattice and $(n \leq N)$ the number of its degrees of kinematic freedoms. It is understood that $(\mathbf{x})$ depicts only the $(n \leq N)$ flexural contractions of the slender and buckling isostatic members, which are supposed to be flexible. The variables making up the set associated with this minimization problem can be represented by the vector $(\mathbf{z})$ where

$$\mathbf{z} = \begin{pmatrix} \mathbf{x} \\ \hline \mathbf{y} \end{pmatrix} \tag{3.10.26}$$

and where $(\mathbf{y})$ is given by (3.10.25b).

† Not to be confused with the kinematic matrix $(\mathbf{C})$ in Eq. (3.2.3).

Generally, the methods used to minimize a function of several variables may be classified as gradient and non-gradient methods[6]. A non-gradient method may be a method which, in addition to the gradient, requires the information about higher derivatives of the function or of its curvature at a point $(\mathbf{x})$. Examples of the gradient method include the steepest descent and the conjugate gradient method.

When the minimum value of $F(\mathbf{x})$ is sought, which must satisfy certain constraints, one usually begins the search starting with an initial feasible solution, say $[\mathbf{x}^1, F(\mathbf{x}^1)]$, which satisfies these constraints. Subsequent solutions are computed in a "downhill" direction, such that

$$F(\mathbf{x}^{k+1}) < F(\mathbf{x}^k)$$

where

$$\mathbf{x}^{k+1} = \mathbf{x}^k + \alpha^k \mathbf{p}^k$$

There, $(\mathbf{p}^k)$ is a vector defining this downhill direction and (α^k) is the steplength in the kth step in this direction. Each iteration then exists of computing a direction $(\mathbf{p}^k)$ such that for a small $(\alpha^k = \alpha)$

$$F(\mathbf{x}^k + \alpha \mathbf{p}^k) < F(\mathbf{x}^k) \tag{3.10.27}$$

Before presenting the algorithm used to solve the minimization problem defined by Eqs. (3.10.24) and (3.10.25) it is necessary to discuss the characteristics of this function-ratio, which include its gradient $(\mathbf{g})$ and its second derivatives in the form of the Hessian matrix $(\mathbf{G})$. Following **Davister** (op.cit.), we derive the necessary expressions required in the strategy for finding the constrained minimum of $F(\mathbf{x})$. First let us consider briefly the unconstrained problem.

The stationary points of $F(\mathbf{x})$ are obtained by solving the $(n \leq N)$-conditions

$$\frac{\partial F(\mathbf{x})}{\partial x_i} = 2\frac{\mathbf{x}^T\mathbf{B}\mathbf{x}(\mathbf{a}^i\mathbf{x}) - \mathbf{x}^T\mathbf{A}\mathbf{x}(\mathbf{b}^i\mathbf{x})}{(\mathbf{x}^T\mathbf{B}\mathbf{x})^2} = 0; \qquad i = 1, 2, \ldots, n \leq N \tag{3.10.28}$$

where $(\mathbf{a}^i)$ and $(\mathbf{b}^i)$ are the respective row vectors of $(\mathbf{A})$ and $(\mathbf{B})$, or

$$[\mathbf{A} - F(\mathbf{x})\mathbf{B}]\mathbf{x} = \mathbf{0}; \qquad \text{where } [\mathbf{a}^i] = \mathbf{A},\ [\mathbf{b}^i] = \mathbf{B} \tag{a}$$

For $(\mathbf{x})$ to be non zero, $F(\mathbf{x})$ must be a root of the characteristic equation

$$|\mathbf{A} - F(\mathbf{x})\mathbf{B}| = 0 \tag{b}$$

Then, at the stationary points $F(\mathbf{x})$ takes on the eigenvalues (λ_i) of the last equation. For each eigenvalue, the eigenvector $(\mathbf{x}_i)$ may then be found from Eq. (a). The solution is given then by the $(n \leq N)$ eigenpairs

$$[\lambda_1 = F(\mathbf{x}_1), \mathbf{x}_1], (\lambda_2, \mathbf{x}_2), \ldots, (\lambda_n, \mathbf{x}_n) \tag{c}$$

---

[6] See e.g. R.L. Fox "Optimization Methods for Engineering Design", Addison-Wesley Pub. Co. 1971.

On premultiplying Eq. (a) by $(\mathbf{x}_i^T)$ and eliminating $F(\mathbf{x}_i)$ one may represent the same result in the form

$$\left[\left(\lambda_1 = F(\mathbf{x}_1) = \frac{\mathbf{x}_1^T\mathbf{A}\mathbf{x}_1}{\mathbf{x}_1^T\mathbf{B}\mathbf{x}_1}\right), \mathbf{x}_1\right], \left(\frac{\mathbf{x}_2^T\mathbf{A}\mathbf{x}_2}{\mathbf{x}_2^T\mathbf{B}\mathbf{x}_2}, \mathbf{x}_2\right), \ldots \text{etc.}$$

The global *unconstrained minimum of* $F(\mathbf{x})$ is given by the *most negative eigenvalue* in the set of $(\lambda_i;\ i = 1, \ldots, n \leq N)$. If this eigenvalue is denoted by $[\lambda_1 = F(\mathbf{x}_1)]$, any constrained minimum of $F(\mathbf{x})$ must be necessarily greater than $F(\mathbf{x}_1)$. If the *global constrained* minimum of $F(\mathbf{x})$ under the constraints defined by Eqs. (3.10.25a and b) is denoted by $F(\mathbf{x}^*)$, where $(\mathbf{x}^*)$ denotes the point of this global minimum, then

$$F(\mathbf{x}_1) = \frac{\mathbf{x}_1^T\mathbf{A}\mathbf{x}_1}{\mathbf{x}_1^T\mathbf{B}\mathbf{x}_1} \leq F(\mathbf{x}^*) = \frac{\mathbf{x}^{*T}\mathbf{A}\mathbf{x}^*}{\mathbf{x}^{*T}\mathbf{B}\mathbf{x}^*} \tag{d}$$

The unconstrained minima of (v) on different contours, defined by the matrices $(\mathbf{B} = \mathbf{Q}, \mathbf{S}, \mathbf{B}_e, \mathbf{J}, \mathbf{J}_e$, etc.), are obtained by solving the characteristic equations:

$$\begin{aligned} &|\mathbf{E}_b - v_i\mathbf{B}_e| = 0 \\ &|\mathbf{E}_b - v_i\mathbf{S}| = 0 \\ &|\mathbf{E}_b - v_i\mathbf{J}| = 0 \\ &|\mathbf{E}_b - v_i\mathbf{J}_L| = 0 \\ &\text{etc.} \end{aligned} \tag{e}$$

where (v_i) then represent the corresponding eigenvalues.

We recall that $(\mathbf{x}^T\mathbf{A}\mathbf{x})$ can be considered as representing the value $F(\mathbf{x})$, where $F(\mathbf{x})$ is bounded by a hypersurface or contour $(\mathbf{x}^T\mathbf{B}\mathbf{x} = 1)$ and that, for $(\mathbf{B})$ arbitrary, the system is *not* in statical equilibrium in the post-buckling process, but in an unstable accelerated motion away from the initial configuration of the critical state. In this context at the point $[\mathbf{x}, F(\mathbf{x})]$, the functions $\mathbf{g}(\mathbf{x})$, $\mathbf{G}(\mathbf{x})$ and ΔF, where

$$\Delta F = F(\mathbf{x} + \alpha\mathbf{p}) - F(\mathbf{x}) \tag{3.10.28}$$

are said to be evaluated on this contour. The gradient vector $\mathbf{g}(\mathbf{x})$ of $F(\mathbf{x})$ at any point is defined as

$$\mathbf{g}(\mathbf{x}) = \begin{Bmatrix} g_1(\mathbf{x}) \\ g_2(\mathbf{x}) \\ \cdot \\ g_i(\mathbf{x}) \\ \cdot \\ g_n(\mathbf{x}) \end{Bmatrix} \tag{3.10.29}$$

where, on differentiating Eq. (3.10.24), one gets

$$g_i(\mathbf{x}) = \frac{\partial F(\mathbf{x})}{\partial x_i} = \frac{\mathbf{x}^T\mathbf{B}\mathbf{x}(\mathbf{a}^i + \mathbf{a}_i^T)\mathbf{x} - (\mathbf{x}^T\mathbf{A}\mathbf{x})(\mathbf{b}^i + \mathbf{b}_i^T)\mathbf{x}}{(\mathbf{x}^T\mathbf{B}\mathbf{x})^2} \tag{3.10.30}$$

where $(\mathbf{a}^i)$ represents the ith-row vector of $(\mathbf{A})$ and $(\mathbf{a}_i)$ its ith-column vector. Similarly, $(\mathbf{b}^i)$ and $(\mathbf{b}_i)$ of $(\mathbf{B})$. Since, $(\mathbf{A})$ and $(\mathbf{B})$ must be symmetrical and positive definite matrices, ($\mathbf{a}^i = \mathbf{a}_i^T$). Taking this into consideration and substituting (3.10.30) into (3.10.29), one finds on dividing by ($\mathbf{x}^T\mathbf{B}\mathbf{x}$), that

$$\mathbf{g}(\mathbf{x}) = 2\frac{\mathbf{A}\mathbf{x} - \dfrac{\mathbf{x}^T\mathbf{A}\mathbf{x}}{\mathbf{x}^T\mathbf{B}\mathbf{x}}\mathbf{B}\mathbf{x}}{\mathbf{x}^T\mathbf{B}\mathbf{x}}$$

Substituting further for $F(\mathbf{x})$ from (3.10.24) this becomes

$$\mathbf{g}(\mathbf{x}) = 2\frac{\mathbf{A}\mathbf{x} - F(\mathbf{x})\mathbf{B}\mathbf{x}}{\mathbf{x}^T\mathbf{B}\mathbf{x}} \tag{3.10.31}$$

If also the vector ($\mathbf{x}$) is normalized with respect to the matrix ($\mathbf{B}$) by putting

$$\mathbf{x}^T\mathbf{B}\mathbf{x} = 1 \tag{3.10.32}$$

the last result simplifies to

$$\mathbf{g}(\mathbf{x}) = 2[\mathbf{A} - F(\mathbf{x})\mathbf{B}]\mathbf{x} \tag{3.10.33}$$

Here, $\mathbf{g}(\mathbf{x})$ is said to be the gradient vector of ($\mathbf{x}^T\mathbf{A}\mathbf{x}$) on the contour defined by Eq. (3.10.32).

The Hessian ($n \times n$) matrix $\mathbf{G}(\mathbf{x})$ is now defined by

$$\mathbf{G}(\mathbf{x}) = \left[\frac{\partial^2 F(\mathbf{x})}{\partial x_i\, \partial x_j}\right]; \qquad i, j = 1, 2, \ldots, n \leq N \tag{3.10.34}$$

On dividing the numerator and the denominator in (3.10.30) by ($\mathbf{x}^T\mathbf{B}\mathbf{x}$) and considering the symmetry of ($\mathbf{A}$) and ($\mathbf{B}$), one gets

$$g_i = \frac{\partial F(\mathbf{x})}{\partial x_i} = \frac{2(\mathbf{a}^i - F(\mathbf{x})\mathbf{b}^i)\mathbf{x}}{\mathbf{x}^T\mathbf{B}\mathbf{x}} \tag{3.10.35}$$

Setting the numerator in (3.10.35) equal to the scalar quantity (z) and differentiating with respect to (x_j), one finds

$$\frac{\partial^2 F(\mathbf{x})}{\partial x_i\, \partial x_j} = \frac{(\mathbf{x}^T\mathbf{B}\mathbf{x})\dfrac{\partial z}{\partial x_j} - z\dfrac{\partial(\mathbf{x}^T\mathbf{B}\mathbf{x})}{\partial x_j}}{(\mathbf{x}^T\mathbf{B}\mathbf{x})^2} \tag{3.10.36}$$

where

$$\frac{\partial z}{\partial x_j} = 2[a_{ij} - F(\mathbf{x})b_{ij}] - 2\mathbf{b}^i\mathbf{x}\frac{\partial F(\mathbf{x})}{\partial x_j} \tag{3.10.36a}$$

and

$$\frac{\partial(\mathbf{x}^T\mathbf{B}\mathbf{x})}{\partial x_j} = (\mathbf{b}^j + \mathbf{b}_j^T)\mathbf{x} = 2\mathbf{b}^j\mathbf{x} \tag{3.10.36b}$$

The last identity follows from the symmetry of ($\mathbf{B}$). Since $\mathbf{G}(\mathbf{x})$ is to be evaluated on the bounding hypersurface or contour ($\mathbf{x}^T\mathbf{B}\mathbf{x} = \text{constant}$) the partial derivation of ($\mathbf{x}^T\mathbf{B}\mathbf{x}$) on this contour are zero and Eq. (3.10.36a) simplifies to:

$$\frac{\partial z}{\partial x_j} = 2[a_{ij} + F(\mathbf{x})b_{ij}]$$

so that (3.10.36) reduces to

$$\frac{\partial^2 F(\mathbf{x})}{\partial x_i\, \partial x_j} = 2[a_{ij} - F(\mathbf{x})b_{ij}] \tag{3.10.37}$$

where ($\mathbf{x}^T\mathbf{B}\mathbf{x} = 1$). Here, ($a_{ij}$) and ($b_{ij}$) are the $(i, j)^{\text{th}}$ elements of ($\mathbf{A}$) and ($\mathbf{B}$), respectively. By substituting this result into (3.10.34), the Hessian matrix $\mathbf{G}(\mathbf{x})$ of $F(\mathbf{x})$ becomes

$$\mathbf{G}(\mathbf{x}) = 2[\mathbf{A} - F(\mathbf{x})\mathbf{B}] \tag{3.10.38}$$

Therefore, from (3.10.33),

$$\mathbf{g}(\mathbf{x}) = \mathbf{G}(\mathbf{x})\mathbf{x} \tag{3.10.39}$$

where ($\mathbf{x}$) is normalized with respect to ($\mathbf{B}$), according to Eq. (3.10.32).

Consider next the change (ΔF) in $F(\mathbf{x})$ as one moves from the point ($\mathbf{x}$) to ($\mathbf{x} + \alpha\mathbf{p}$) on the $(n + 1)$-dimensional hypersurface [$F(\mathbf{x}) = \mathbf{x}^T\mathbf{A}\mathbf{x}$]. The new point ($\mathbf{x} + \alpha\mathbf{p}$) is assumed to lie on the contour ($\mathbf{x}^T\mathbf{B}\mathbf{x} = 1$) and the vector ($\mathbf{x} + \alpha\mathbf{p}$), as well as ($\mathbf{x}$), must be normalized with respect to matrix ($\mathbf{B}$). ($\mathbf{p}$) is arbitrary in so far as it points in a "downhill" direction and (α) is an arbitrary small steplength. To calculate (ΔF), the vector ($\mathbf{x} + \alpha\mathbf{p}$) is substituted in Eq. (3.10.28), where $F(\mathbf{x})$ is given by (3.10.24). This gives, after matrix multiplication,

$$\Delta F = \frac{\mathbf{x}^T\mathbf{A}\mathbf{x} + 2\alpha\mathbf{x}^T\mathbf{A}\mathbf{p} + \alpha^2\mathbf{p}^T\mathbf{A}\mathbf{p}}{\mathbf{x}^T\mathbf{B}\mathbf{x} + 2\alpha\mathbf{x}^T\mathbf{B}\mathbf{p} + \alpha^2\mathbf{p}^T\mathbf{B}\mathbf{p}} - \frac{\mathbf{x}^T\mathbf{A}\mathbf{x}}{\mathbf{x}^T\mathbf{B}\mathbf{x}} \tag{3.10.40}$$

After bringing this to the common denominator, the result for (ΔF) becomes

$$\Delta F = \frac{2\alpha\mathbf{x}^T[\mathbf{A} - F(\mathbf{x})\mathbf{B}]\mathbf{p} + \alpha^2\mathbf{p}^T[\mathbf{A} - F(\mathbf{x})\mathbf{B}]\mathbf{p}}{\mathbf{x}^T\mathbf{B}\mathbf{x} + 2\alpha\mathbf{x}^T\mathbf{B}\mathbf{p} + \alpha^2\mathbf{p}^T\mathbf{B}\mathbf{p}} \tag{3.10.41}$$

and by (3.10.38), on factorizing the denominator,

$$\Delta F = \frac{\alpha\mathbf{x}^T\mathbf{G}(\mathbf{x})\mathbf{p} + \frac{1}{2}\alpha^2\mathbf{p}^T\mathbf{G}(\mathbf{x})\mathbf{p}}{(\mathbf{x} + \alpha\mathbf{p})^T\mathbf{B}(\mathbf{x} + \alpha\mathbf{p})} \tag{3.10.42}$$

By (3.10.39) this simplifies finally to

$$\Delta F = \frac{\alpha\mathbf{g}^T(\mathbf{x})\mathbf{p} + \frac{1}{2}\alpha^2\mathbf{p}^T\mathbf{G}(\mathbf{x})\mathbf{p}}{(\mathbf{x} + \alpha\mathbf{p})^T\mathbf{B}(\mathbf{x} + \alpha\mathbf{p})} \tag{3.10.43}$$

where, in the transposition of the matrix $\mathbf{G}(\mathbf{x})$ its symmetry has been considered.

The quantity (ΔF) in (3.10.43) must be negative if ($\mathbf{p}$) points in a "downhill" direction within the bounding contour on the $F(\mathbf{x})$-hypersurface. Since the bounding contour given by (3.10.32) must be a positive definite quadratic form, and, therefore, the denominator in (3.10.43) is positive, then the condition that ($\mathbf{p}$) points in a "downhill" direction is that the numerator of (3.10.43) be negative, i.e.

$$\alpha\mathbf{g}^T(\mathbf{x})\mathbf{p} + \tfrac{1}{2}\alpha^2\mathbf{p}^T\mathbf{G}(\mathbf{x})\mathbf{p} < 0 \tag{3.10.44}$$

A strategy for finding the constrained global minimum of $F(\mathbf{x})$ may be undertaken from an initial feasible solution at the point $[\mathbf{x}^0, F(\mathbf{x}^0)]$, this solution then satisfies the contour equation ($\mathbf{x}^T\mathbf{B}\mathbf{x} = 1$) and the constraints specified by the Eqs. (3.10.25a and b). All subsequent intermediate solutions $[\mathbf{x}^1, F(\mathbf{x}^1)]$, $[\mathbf{x}^2, F(\mathbf{x}^2)], \ldots, [\mathbf{x}^k, F(\mathbf{x}^k)]$, etc., are obtained by moving an amount (α^k) in some "downhill" direction ($\mathbf{p}^k$) from the current point $[\mathbf{x}^k, F(\mathbf{x}^k)]$. All of these solutions satisfy the constraints in Eq. (3.10.25), then

$$\mathbf{x}^{k+1} = \mathbf{x}^k + \alpha^k\mathbf{p}^k \qquad \text{for } k = 1, 2, \ldots$$

and

$$F(\mathbf{x}^{k+1}) < F(\mathbf{x}^k)$$

The series ends in the global constrained minimum when ($k = m$) and ($\mathbf{x}^m = \mathbf{x}^*$) and $F(\mathbf{x}^m) = F(\mathbf{x}^*)$, which is reached in ($m$) steps. An optimal selection of (α^k) and ($\mathbf{p}^k$) is required, so that (ΔF^k) may be sufficiently large in each iteration and the method may converge rapidly. Since, (ΔF) is a function of the step-length (α) defined by (3.10.43), the stationary points, where (ΔF) is the minimum (most negative value), are of interest in this search. At this minimum the curvature ($\partial^2\Delta F/\partial\alpha^2$) with respect to the step-length (α) must be positive. To obtain this value we consider the stationary values of (ΔF) determined by

$$\frac{\partial \Delta F}{\partial \alpha} = 0 \tag{3.10.45}$$

where the index (k) is omitted. This partial derivative can be found from Eq. (3.10.43). Thus

$$\begin{aligned}\frac{\partial \Delta F}{\partial \alpha} = \{&[\mathbf{x}^T\mathbf{B}\mathbf{x} + 2\alpha\mathbf{x}^T\mathbf{B}\mathbf{p} + \alpha^2\mathbf{p}^T\mathbf{B}\mathbf{p}][\mathbf{g}^T\mathbf{p} + \alpha\mathbf{p}\mathbf{G}\mathbf{p}] \\ &- [\alpha\mathbf{g}^T\mathbf{p} + \tfrac{1}{2}\alpha^2\mathbf{p}^T\mathbf{G}\mathbf{p}][2\mathbf{x}^T\mathbf{B}\mathbf{p} + 2\alpha\mathbf{p}^T\mathbf{B}\mathbf{p}]\}/ \\ &[\mathbf{x}^T\mathbf{B}\mathbf{x} + 2\alpha\mathbf{x}^T\mathbf{B}\mathbf{p} + \alpha^2\mathbf{p}^T\mathbf{B}\mathbf{p}]^2\end{aligned} \tag{3.10.46}$$

After carrying out the multiplication in the numerator and considering that the point $[\mathbf{x}, F(\mathbf{x})]$ is on the contour ($\mathbf{x}^T\mathbf{B}\mathbf{x} = 1$) and that the vector ($\mathbf{p}$) is arbitrary and may also be normalized by setting ($\mathbf{p}^T\mathbf{B}\mathbf{p} = 1$), the last result reduces to

$$\frac{\partial \Delta F}{\partial \alpha} = \frac{\mathbf{g}^T\mathbf{p} + \mathbf{p}^T\mathbf{G}\mathbf{p}\alpha + [\mathbf{x}^T\mathbf{B}\mathbf{p}(\mathbf{p}^T\mathbf{G}\mathbf{p}) - \mathbf{g}^T\mathbf{p}]\alpha^2}{[1 - 2\alpha\mathbf{x}^T\mathbf{B}\mathbf{p} + \alpha^2]^2} \tag{3.10.47}$$

The stationary points of (ΔF) may now be evaluated upon equating the numerator in the last result to zero according to the condition (3.10.45), i.e.

$$\mathbf{g}^T\mathbf{p} + \mathbf{p}^T\mathbf{G}\mathbf{p}\alpha + [\mathbf{x}^T\mathbf{B}\mathbf{p}(\mathbf{p}^T\mathbf{G}\mathbf{p}) - \mathbf{g}^T\mathbf{p}]\alpha^2 = 0 \tag{3.10.48}$$

Two roots, say (α^{1*}) and (α^{2*}) in the direction of ($\mathbf{p}$), from the point ($\mathbf{x}$), which correspond to the maximum and minimum of (ΔF), may be found on solving this quadratic equation. The plot of (ΔF) with respect to (α) is shown in Fig. 3.10.1, including the minimum and maximum points.

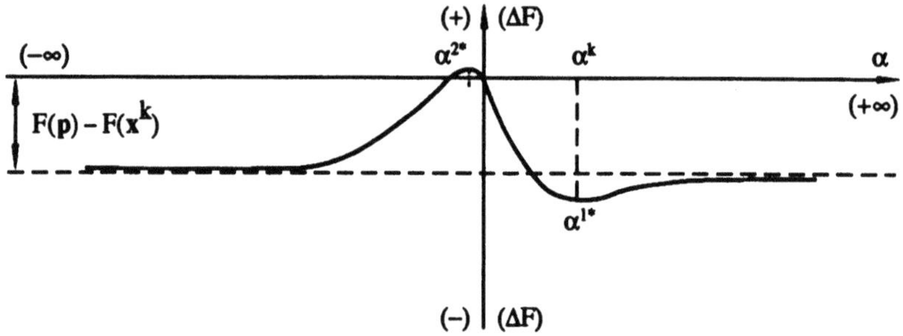

Figure 3.10.1
The variation of (ΔF) with (α) in an arbitrary direction $(\mathbf{p})$ at the point $(\mathbf{x})$, plotted according to **Davister** (op.cit.)

On dividing the numerator and denominator in Eq. (3.10.40) by (α^2) and letting $(\alpha \to \pm\infty)$, one obtains the values of (ΔF) as (α) approaches infinity in the form

$$\lim_{\alpha \to \pm\infty} \Delta F = \frac{\mathbf{p}^T\mathbf{A}\mathbf{p}}{\mathbf{p}^T\mathbf{B}\mathbf{p}} - \frac{\mathbf{x}^T\mathbf{A}\mathbf{x}}{\mathbf{x}^T\mathbf{B}\mathbf{x}} = F(\mathbf{p}) - F(\mathbf{x}) \tag{3.10.49}$$

which are indicated in Fig. 3.10.1. The variation of (ΔF) with (α) in an arbitrary direction $(\mathbf{p})$ is then as shown in the figure. The optimal steplength (α^{1*}) corresponds to the minimum value of (ΔF). This is the steplength that may be used in any step if the minimization problem is without constraints, or if (α^{1*}) does not violate any constraints in a constrained minimization problem.

After substituting for $(\mathbf{G})$ from Eq. (3.10.38), the expression $(\mathbf{p}^T\mathbf{G}\mathbf{p})$ becomes

$$\mathbf{p}^T\mathbf{G}\mathbf{p} = 2[\mathbf{p}^T\mathbf{A}\mathbf{p} - F(\mathbf{x})\mathbf{p}^T\mathbf{B}\mathbf{p}] = 2[F(\mathbf{p}) - F(\mathbf{x})] \tag{3.10.50}$$

in which both $(\mathbf{x})$ and $(\mathbf{p})$ are normalized with respect to the matrix $(\mathbf{B})$ according to Eq. (3.10.32). Similarly, after substituting for $[\mathbf{g}(\mathbf{x})]$ from (3.10.33), we get that

$$\mathbf{g}^T\mathbf{p} = 2[\mathbf{x}^T\mathbf{A}\mathbf{p} - F(\mathbf{x})\mathbf{x}^T\mathbf{B}\mathbf{p}] \tag{3.10.50a}$$

Then the third term in Eq. (3.10.48) transforms to

$$\mathbf{x}^T\mathbf{B}\mathbf{p}(\mathbf{p}^T\mathbf{G}\mathbf{p}) - \mathbf{g}^T\mathbf{p} = -2\mathbf{x}^T[\mathbf{A} - F(\mathbf{p})\mathbf{B}]\mathbf{p} \tag{3.10.51}$$

Minimization of $F(\mathbf{x})$ Using the Eigenvector Search Directions

If now $(\mathbf{p})$ is made to be an *eigenvector* of

$$F(\mathbf{x}) = \frac{\mathbf{x}^T\mathbf{A}\mathbf{x}}{\mathbf{x}^T\mathbf{B}\mathbf{x}} = \mathbf{x}^T\mathbf{A}\mathbf{x} \tag{3.10.52}$$

then, putting $(\mathbf{x} = \mathbf{p})$,

$$[\mathbf{A} - F(\mathbf{p}_i)\mathbf{B}]\mathbf{p}_i = \mathbf{0} \tag{3.10.53}$$

where $F(\mathbf{p}_i)$ is the ith eigenvalue and $(\mathbf{p}_i)$ the ith eigenvector for $(i = 1, 2, \ldots, n \le N)$. (On premultiplying the last result by $(\mathbf{p}^T)$ and using the normalizing condition (3.10.32), the result in (3.10.52) may be recovered. Thus, the condition (3.10.53) is acceptable.)

By virtue of (3.10.53) we conclude from (3.10.51) that, if ($\mathbf{p}$) *is an eigenvector* of $F(\mathbf{x})$, and

$$\mathbf{x}^T\mathbf{B}\mathbf{p}(\mathbf{p}^T\mathbf{G}\mathbf{p}) - \mathbf{g}^T\mathbf{p} = 0$$

the expression (3.10.47) for ($\partial \Delta F/\partial \alpha$) at the point ($\mathbf{x}$) simplifies to

$$\frac{\partial \Delta F}{\partial \alpha} = \frac{\mathbf{g}^T\mathbf{p} + (\mathbf{p}^T\mathbf{G}\mathbf{p})\alpha}{[1 + 2\alpha(\mathbf{x}^T\mathbf{B}\mathbf{p}) + \alpha^2]^2}$$

then, the stationary point (α^{**}) in an *eigenvector* direction at ($\mathbf{x}$) is obtained by setting the last expression equal to zero, i.e. when

$$\alpha = \alpha^{**} = -\frac{\mathbf{g}^T\mathbf{p}}{\mathbf{p}^T\mathbf{G}\mathbf{p}} \tag{3.10.54}$$

On substituting for ($\mathbf{g}^T\mathbf{p}$) from (3.10.50a) and for ($\mathbf{p}^T\mathbf{G}\mathbf{p}$) from (3.10.50) and introducing from (3.10.53) the expression

$$\mathbf{x}^T\mathbf{A}\mathbf{p} = F(\mathbf{p})\mathbf{x}^T\mathbf{B}\mathbf{p},$$

which may be obtained by premultiplying (3.10.53) by ($\mathbf{x}^T$), the expression for a *single* stationary point (α^{**}) in an eigenvector direction at the point ($\mathbf{x}$) reduces to the very simple form

$$\alpha^{**} = -\mathbf{x}^T\mathbf{B}\mathbf{p} \tag{3.10.55}$$

This result was first derived by Davister (op.cit.). Davister has also shown (op.cit., p. 93) that (α^{**}) corresponds to a *maximum* value of (ΔF) rather than to a minimum, when $F(\mathbf{p}) < F(\mathbf{x})$. Consequently, ($\alpha^{**}$) *cannot* be used as an optimal steplength (α^*) in the step ($\mathbf{p}$) in the present minimization problem. In this case, the variation of (ΔF) with respect to (α) is as shown in Fig. 3.10.2. As ($\alpha \to \pm\infty$) the value of (ΔF) at the point ($\mathbf{x}$) tends to $[F(\mathbf{p}) - F(\mathbf{x})]$, as it may be verified from Eq. (3.10.40) or (3.10.49). Using Eq. (3.10.50a) and the result that

$$\mathbf{x}^T\mathbf{A}\mathbf{p} = F(\mathbf{p})\mathbf{x}^T\mathbf{B}\mathbf{p} \tag{3.10.56}$$

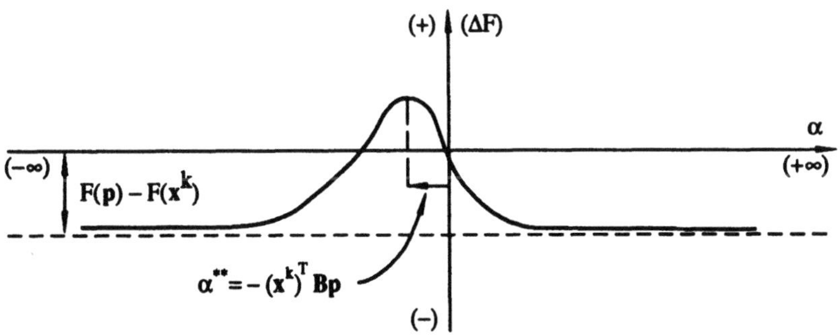

Figure 3.10.2
The variation of (ΔF) with (α) in an eigenvector direction (according to **Davister** op.cit.)

which is obtained by premultiplying (3.10.53) by ($\mathbf{x}^T$), one finds first, on substituting this into (3.10.50a), that

$$\mathbf{g}^T\mathbf{p} = 2[F(\mathbf{p}) - F(\mathbf{x})]\mathbf{x}^T\mathbf{B}\mathbf{p} \tag{3.10.57}$$

and then, on substituting the last result and Eq. (3.10.50) into the expression (3.10.43) for (ΔF) and carrying out the multiplications, that

$$\Delta F = [F(\mathbf{p}) - F(\mathbf{x})]\frac{2\alpha\mathbf{x}^T\mathbf{B}\mathbf{p} + \alpha^2}{1 + 2\alpha\mathbf{x}^T\mathbf{B}\mathbf{p} + \alpha^2} \tag{3.10.58}$$

When ($\alpha \to \pm\infty$), $\Delta F \to [F(\mathbf{p}) - F(\mathbf{x})]$, as stated before.

Then, clearly, the strategy in the present minimization problem, when ($\mathbf{p}$) points in an eigenvector direction, *is to select the steplength* (α) *as large as possible*, so that no constraint is violated. Based on this discussion it is possible to present a strategy originally proposed by **Davister** (op.cit.), which allows a selection of the vector ($\mathbf{p}^k$) and the steplength (α^k) at a feasible point. If the constraints

$$\mathbf{x}^k > \mathbf{0} \tag{3.10.59a}$$

$$\mathbf{y}^k = \mathbf{C}\mathbf{x}^k > \mathbf{0} \tag{3.10.59b}$$

are sufficiently positive, this ensures that a sufficiently small steplength ($\Delta\alpha$) may be taken in any direction ($\mathbf{p}$) without causing any of the variables ($\mathbf{x}$) and ($\mathbf{y}$) to become negative, i.e. that,

$$\mathbf{x}^k + \Delta\alpha\mathbf{p}^k \geq \mathbf{0} \tag{3.10.59c}$$

$$\mathbf{C}(\mathbf{x}^k + \Delta\alpha\mathbf{p}^k) \geq \mathbf{0} \tag{3.10.59d}$$

Since (ΔF) is to be minimized at each step (k), it follows from (3.10.58) that $[F(\mathbf{p}) - F(\mathbf{x})]$ is a minimum when ($\mathbf{p}^k$) is *the eigenvector corresponding to the most negative eigenvalue* of $[F(\mathbf{x}) = \mathbf{x}^T\mathbf{A}\mathbf{x}/\mathbf{x}^T\mathbf{B}\mathbf{x}]$, which we denoted earlier by ($\mathbf{x}_1$) [see Eq. (c), following (3.10.27)], i.e. when

$$\mathbf{p}^k = \mathbf{x}_1 \tag{3.10.60}$$

Thus, the vector ($\mathbf{p}^k$) *is chosen as the eigenvector of the eigenpair* $[\mathbf{x}_1, F(\mathbf{x}_1)]$, *where* $F(\mathbf{x}_1)$ *is the most negative eigenvalue of* $[F(\mathbf{x}) = \mathbf{x}^T\mathbf{A}\mathbf{x}/\mathbf{x}^T\mathbf{B}\mathbf{x}]$ *of the unconstrained minimization problem*. Davister (op.cit.) also showed that, when one or more of the constraint variables ($\mathbf{x}^k$) and ($\mathbf{y}^k$) are zero, ($\mathbf{x}_1$) corresponds to an eigenvector of an *abridged* quadratic function ratio, where the abridged form is obtained by deleting the rows and columns in the matrices ($\mathbf{A}$) and ($\mathbf{B}$) or their linear combinations, corresponding to the zero elements of the unabridged vector ($\mathbf{x}_1$).

In the case when ($\mathbf{x}_1$) is an admissible solution and

$$\mathbf{x}_1 \geq \mathbf{0} \tag{3.10.61a}$$

$$\mathbf{y}_1 = \mathbf{C}\mathbf{x}_1 \geq \mathbf{0} \tag{3.10.61b}$$

the optimal steplength (α^*) is obviously equal to (∞). In such a case no constraint hyperplanes are encountered as one moves in the direction ($\mathbf{x}_1$).

Whenever the last conditions are not satisfied and one or more elements of the vectors $(\mathbf{x}_1)$ or $(\mathbf{y}_1)$ are *negative*, the optimal solution denoted by $[\mathbf{x}^*, F(\mathbf{x}^*)]$ must be *greater* than $[\mathbf{x}_1, F(\mathbf{x}_1)]$, as a constrained minimum can never be less than the global *unconstrained minimum* of $F(\mathbf{x})$.

Consider now a point $(\mathbf{x}^k)$ which may be in the positive direction as one moves along the vector $(\mathbf{x}_1)$ when $(\alpha > 0)$, or in the negative direction when $(\alpha < 0)$, before encountering a constraint hyperplane, when

$$x_i = 0; \quad i, j = 1, 2, \ldots, n \leq N \tag{3.10.62a}$$

or

$$y_j = C_{jk} x_k; \quad \text{sum on } k = 1, 2, \ldots, n \leq N \tag{3.10.62b}$$

The steplengths in the direction $(\mathbf{p} = \mathbf{x}_1)$, where the first hyperplanes are encountered are denoted by (α_1) for $(\alpha > 0)$ and (α_2) for $(\alpha < 0)$, respectively. These are indicated in Fig. 3.10.3. Thus the strategy to determine the optimal constrained steplength, becomes to evaluate (ΔF) at (α_1) and (α_2) and select the steplength associated with the more negative value either at (ΔF_1) or (ΔF_2).

Although (ΔF) increases initially as one moves towards the maximum value of $[\alpha^{**} = -\mathbf{x}^T\mathbf{Bp}]$ given by (3.10.55), (ΔF) may become negative *beyond* this value of (α) when, according to (3.10.58),

$$\alpha(2\mathbf{x}^T\mathbf{Bp} + \alpha) = 0 \tag{3.10.63a}$$

$$\alpha = -2\mathbf{x}^T\mathbf{Bp} \tag{3.10.63b}$$

Obviously, the possibility also exists that a constraint hyperplane will be encountered before (α) reaches the value $[-2(\mathbf{x}^k)^T\mathbf{Bp}]$. In this case, the opposite positive direction is a feasible downhill direction. So, the rule in the step (k) in which (α) takes the value of either (α_1^k) or (α_2^k) drives either one of the $(n \leq N)$ independent variables $(\mathbf{x})$ or one of the (R) dependent constraint variables $(\mathbf{y})$ to zero. To make the procedure easier we combine the two sets of variables $(\mathbf{x})$ and $(\mathbf{y})$ into a single $(M' \times 1)$ vector $(\mathbf{z})$, where $(M' = n + R)$, according to (3.10.26). Then,

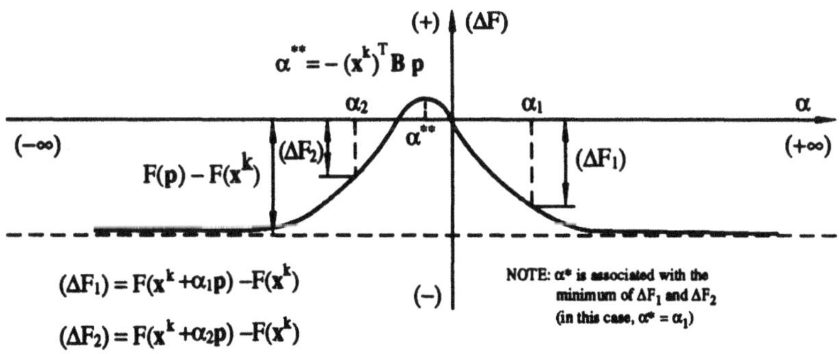

Figure 3.10.3
The optimal steplength (α^*) in an eigenvector direction under a set of linear constraints plotted according to **Davister** (op.cit., p. 99)

the constraint condition may be expressed as

$$\mathbf{z} = \left(\frac{\mathbf{I}}{\mathbf{C}}\right)\mathbf{x} \geq \mathbf{0} \tag{3.10.64}$$

where (**I**) is a $(n \times n)$ unit matrix and (**C**) is the $(R \times n)$ matrix in Eq. (3.10.25b). Each of the variables in the vector (**z**) when equated to zero or to another constant, represents a constraint hyperplane in the n-dimensional space defined by the contour $(\mathbf{x}^T\mathbf{B}\mathbf{x} = 1)$, which bounds the $(n+1)$-dimensional hypersurface defined by the function $[F(\mathbf{x}) = \mathbf{x}^T\mathbf{A}\mathbf{x}]$. The equation of the ith constraint hyperplane is then

$$z_i = 0 \tag{3.10.65}$$

The set of (M') constraint hyperplanes which the vector (**z**) admits may be written

$$\mathbf{z} = \left(\frac{\mathbf{I}}{\mathbf{C}}\right)\mathbf{x} = \mathbf{0} \tag{3.10.66}$$

If the difference in the vector (**z**) in the kth step per unit steplength $(\alpha = 1)$ is defined by the vector $(\bar{\mathbf{z}}^k)$ in the direction of $(\mathbf{p}^k)$, then

$$\bar{\mathbf{z}}^k = \left(\frac{\mathbf{I}}{\mathbf{C}}\right)\mathbf{p}^k \tag{3.10.66a}$$

and for an arbitrary (α) in the direction $(\mathbf{p}^k)$,

$$\mathbf{z} = \mathbf{z}^k + \alpha\bar{\mathbf{z}}^k \tag{3.10.67}$$

A single element (z_i) of this vector becomes

$$z_i = z_i^k + \alpha\bar{z}_i^k \tag{3.10.67a}$$

The steplength $(\alpha = \alpha_i)$ from the current point $(\mathbf{z}^k)$ to the first constraint hyperplane $(z_i = 0)$ in the direction $(\mathbf{p}^k)$ through $(\mathbf{z}^k)$ may now be computed by equating the last result to zero, i.e.

$$z_i = z_i^k + \alpha_i\bar{z}_i^k = 0 \tag{3.10.67b}$$

where (i) is any number between 1 and (M'). Hence,

$$\alpha_i = -\frac{z_i^k}{\bar{z}_i^k} \qquad \text{for any } i = 1, 2, \ldots, M'; \qquad \bar{z}_i^k \neq 0 \tag{3.10.68}$$

If $(\bar{z}_i^k = 0)$ then $(z_i = z_i^k)$ while (α) may be nonzero.

To explain in which direction the vector $(\mathbf{p}^k)$ points if $(\bar{z}_i^k = 1)$ in the kth step, consider the following example. Suppose the (i^{th}) element of $(\mathbf{z}_i)$ is taken from the dependent constraints of

$$\mathbf{y} = \mathbf{C}\mathbf{x} \geq \mathbf{0}$$

in Eq. (3.10.66) and suppose the (i^{th}) element of (**z**) corresponds to the jth element in (**y**). Then from

$$\mathbf{z} = \left(\frac{\mathbf{x}}{\mathbf{y}}\right) = \left(\frac{\mathbf{I}}{\mathbf{C}}\right)\mathbf{x} \geq \mathbf{0} \tag{3.10.64)bis}$$

the equation of this i^{th}-constraint hyperplane is

$$z_i = y_j = C_{j1}x_1 + C_{j2}x_2 + \cdots + C_{jn}x_n = 0 \tag{3.10.69}$$

where (C_{j1}, $j = 1, \ldots, R$), etc. are the elements of the matrix ($\mathbf{C}$).

Let the move now be made in step (k) from the point ($\mathbf{z}^k$) to an adjacent point ($\mathbf{z}^{k+1}$), when the point defined by the vector ($\mathbf{z}^k$) is *not* on this constraint hyperplane. In that case

$$z_i^k = y_j^k = C_{j1}x_1^k + C_{j2}x_2^k + \cdots + C_{jn}x_n^k = C_1 \neq 0$$

The new point ($\mathbf{z}^{k+1}$) is defined by the vector

$$\mathbf{z}^{k+1} = \mathbf{z}^k + \Delta\mathbf{z}^k = \left(\frac{\mathbf{x}^{k+1}}{\mathbf{y}^{k+1}}\right) = \left(\frac{\mathbf{I}}{\mathbf{C}}\right)(\mathbf{x}^k + \alpha\mathbf{p}^k) = \mathbf{z}^k + \alpha\bar{\mathbf{z}}^k \tag{3.10.70}$$

where

$$\Delta\mathbf{z}^k = \alpha\bar{\mathbf{z}}^k = \alpha\left(\frac{\mathbf{I}}{\mathbf{C}}\right)\mathbf{p}^k \tag{3.10.70a}$$

The ith component or element of this vector is

$$\Delta z_i^k = \alpha\bar{z}_i^k = \alpha(C_{j1}p_1^k + C_{j2}p_2^k + \cdots + C_{jn}p_n^k) \tag{3.10.70b}$$

where ($p_1, p_2, \ldots, p_n$) are the components of the vector ($\mathbf{p}$). If ($\bar{z}_i^k = 0$), as assumed, then

$$C_{j1}p_1^k + C_{j2}p_2^k + \cdots + C_{jn}p_n^k = 0 \tag{3.10.71}$$

This equation is identical to (3.10.69), except that the coordinates are the components of the vector ($\mathbf{p}$). But, this means that the coordinates of the vector ($\mathbf{p}$) satisfy the equation of the ith-constraint hyperplane. Therefore, if in the kth-iteration ($\bar{z}_i^k = 0$) and, if ($z_i = 0$) represents the equation of the ith-constraint hyperplane, the vector ($\mathbf{p}$) lies in a plane which is *parallel* to this hyperplane, and so the kth-move is parallel to the corresponding constraint hyperplane ($z_i = 0$).

If the ith-element of ($\mathbf{z}$) is taken from the first ($n \leq N$) constraints ($\mathbf{x} \geq \mathbf{0}$) in (3.10.64), say, from the lth-element, in that case the corresponding element (p_l) of the vector ($\mathbf{p}$) is zero, i.e.

$$p_l = 0 \tag{3.10.71a}$$

In that case $\bar{z}_i^k = p_l^k = 0$ and the move along the vector ($\mathbf{p}$) is parallel to the coordinate plane ($x_l = 0$).

Now, (α_i) may be computed from (3.10.68) so that the point ($\mathbf{z}^{k+1}$) lies on the ith-constraint hyperplane ($z_i = 0$), while the point ($\mathbf{z}^k$) is a feasible point which satisfies the constraints (3.10.64) and it is not on that hyperplane. We defined by (α_1) and (α_2) the value of (α_i) associated with the *first* constraint hyperplane encountered by moving in the direction ($\mathbf{p}^k$) through the feasible point (x^k), viz. ($\mathbf{z}^k$) in the positive, ($\alpha > 0$), and negative, ($\alpha < 0$), directions respectively. Therefore, (α_1) may now be determined from Eq. (3.10.68) as the *minimum of all positive values of* ($-z_i^k/\bar{z}_i^k$) and (α_2) as the *maximum of all negative values*

$(-z_i^k/\bar{z}_i^k)$ for ($i = 1$ to M'), provided that ($\bar{z}_i^k \neq 0$).

$$\begin{aligned} \alpha_1 &= \text{minimum positive value of } (-z_i^k/\bar{z}_i^k) \\ \alpha_2 &= \text{maximum negative value of } (-z_i^k/\bar{z}_i^k) \end{aligned} \tag{3.10.72}$$

If ($z_i^k = 0$), in this case also ($\alpha_i = 0$) and the point ($\mathbf{x}^k$) viz. ($\mathbf{z}^k$) lies on the intersection of the ith-constraint hyperplane with the ($n + 1$)-dimensional hypersurface ($\mathbf{x}^T\mathbf{A}\mathbf{x}$).

The move along the direction of the vector ($\mathbf{p}$) must be such that one always remains in the feasible region on the hypersurface ($\mathbf{x}^T\mathbf{A}\mathbf{x}$), where the constraints

$$z_i \geq 0; \qquad \text{for all values of } i = 1, 2, \ldots, M'$$

For ($\mathbf{z}^k$) to be feasible, (z_i^k) must be greater than zero. If ($\bar{z}_i^k > 0$), ($z_i = z_i^{k+1} = z_i^k + \alpha\bar{z}_i$) is greater than zero for all positive (α). Then, one moves always within the feasible region of the ($\mathbf{x}^T\mathbf{A}\mathbf{x}$)-hypersurface in the direction of ($\mathbf{p}^k$) when

$$x_i = p_i > 0; \qquad i = 1, 2, \ldots, n \leq N \tag{3.10.72a}$$

and

$$C_{j1}p_1^k + C_{j2}p_2^k + \cdots + C_{jn}p_n^k = C_j > 0; \quad j = N + 1, \ldots, M \tag{3.10.72b}$$

and one moves *parallel* to a particular constraint hyperplane when the coordinates of the vector ($\mathbf{p}$) are such that the corresponding hyperplane constraint condition is always equal to zero, as already shown in (3.10.71). On the other hand, ($z_i = z_i^{k+1} < 0$) for all negative (α), when ($\bar{z}_i^k > 0$). Moving in the negative direction gets the point into the *inadmissible* region of the hypersurface ($\mathbf{x}^T\mathbf{A}\mathbf{x}$).

Davister (op.cit.) classified all the variables of the vector ($\mathbf{z}$), which in the kth iteration satisfy the condition

$$z_i = 0 \qquad \text{for any } i = 1 \text{ to } M' \tag{3.10.73}$$

as *active constraints*. Then any feasible move from the point ($\mathbf{x}^k$) under a set of active constraints must be such that the direction of movement along the vector ($\mathbf{p}^k$) is *either parallel to the active constraint hyperplanes or directed into the feasible region away from these active constraint hyperplanes*. Feasible move directions under a set of active constraints are generated by driving one of the independent variables ($\mathbf{x}$) or one of the dependent variables ($\mathbf{y}$) to zero, as (α^k) is computed either as (α_1) or (α_2). The variables driven to zero are added to the set of active constraints and any subsequent direction of movement ($\mathbf{p}$) must be such that ($\mathbf{p}$) is *parallel to all active constraint hyperplanes*, the added ones included.

If the number of the active constraints is denoted by (a_c) and if these are associated with the vector ($\mathbf{x}$) in the kth iteration, then ($\mathbf{x}^k$) contains at least (a_c^k) zero elements and ($n - a_c^k$) nonzero elements. Any move parallel to the (a_c^k) active constraint hyperplanes must be such, that the a_c^k-zero elements in ($\mathbf{x}^k$) remain zero as one moves continuously in the ($\mathbf{p}^k$) direction.

The direction of $(\mathbf{p})$ which minimizes $\Delta F(\mathbf{p})$ at a point $(\mathbf{x})$ in Eq. (3.10.58) and ultimately the function $F(\mathbf{x})$ at the point $(\mathbf{x}=\mathbf{x}^*)$, and which satisfies the parallel-move criteria may be determined by solving the *reduced* minimization problem as follows. Minimize the function ratio

$$F(\mathbf{x}) = \frac{\mathbf{x}^T\mathbf{A}\mathbf{x}}{\mathbf{x}^T\mathbf{B}\mathbf{x}} \tag{3.10.74}$$

in which $(\mathbf{x})$ is partitioned according to the (a_c) zero elements and $(n - a_c)$ nonzero elements $(\bar{\mathbf{x}})$ i.e.

$$\mathbf{x} = \begin{pmatrix} \bar{\mathbf{x}} \\ \hline \mathbf{0} \end{pmatrix} \tag{3.10.74a}$$

If $(\mathbf{A})$ and $(\mathbf{B})$ are now partitioned accordingly and the leading submatrices are denoted by $(\bar{\mathbf{A}})$ and $(\bar{\mathbf{B}})$ respectively, then

$$\mathbf{x}^T\mathbf{A}\mathbf{x} = (\bar{\mathbf{x}}^T \mid \mathbf{0})\left(\begin{array}{c|c} \bar{\mathbf{A}} & \cdot \\ \hline \cdot & \cdot \end{array}\right)\begin{pmatrix} \bar{\mathbf{x}} \\ \hline \mathbf{0} \end{pmatrix} = \bar{\mathbf{x}}^T\bar{\mathbf{A}}\bar{\mathbf{x}}$$

where a dot denotes the corresponding nonzero or zero submatrices. Similarly, $(\mathbf{x}^T\mathbf{B}\mathbf{x} = \bar{\mathbf{x}}^T\bar{\mathbf{B}}\bar{\mathbf{x}})$ and the reduced function ratio becomes

$$F(\mathbf{x}) = F(\bar{\mathbf{x}}) = \frac{\bar{\mathbf{x}}^T\bar{\mathbf{A}}\bar{\mathbf{x}}}{\bar{\mathbf{x}}^T\bar{\mathbf{B}}\bar{\mathbf{x}}} \tag{3.10.74b}$$

$F(\mathbf{x})$ is now a ratio of quadratic forms of the order $(n - a_c) \times (n - a_c)$. Thus, this ratio and the order of the quadratic forms is *reduced by one dimension for each encountered active constraint.*

Denoting the eigenvector associated with the minimum eigenvalue of this ratio by $(\bar{\mathbf{x}}_1)$ the original $(n \times 1)$ vector $(\mathbf{p}^k)$ in the kth step or iteration, parallel to the set of (a_c^k) active constraint hyperplanes in this step may be written

$$\mathbf{p}^k = \begin{pmatrix} \bar{\mathbf{x}}_1 \\ \hline \mathbf{0} \end{pmatrix} \tag{3.10.75}$$

where $(\mathbf{0})$ denotes an $(a_c \times 1)$ null vector. Thus, the problem is relatively simple when the active constraints are associated with the *independent* set of variables $(\mathbf{x})$. The ratio of the reduced quadratic forms may be referred to as the *abridged* quadratic function ratio. When in the kth-iteration step an active constraint is associated with one of the dependent variables of the vector $(\mathbf{y})$ say, (y_i), a transformation of variables may be performed and the vector $(\mathbf{p}^k)$ evaluated from the reduced problem as described. This involves interchanging the dependent variable (y_i) with one of the nonzero independent variables, say (x_j). Now, (y_i) is treated as an independent variable, which is entered into the modified vector $(\mathbf{x})$ in place of (x_j) which is eliminated from the equation for (y_i) and entered into the modified vector $(\mathbf{y})$ of dependent variables in lieu of (y_i). The modified vectors are treated as before and the modified reduced problem is solved for the most negative eigenvalue and the associated eigenvector of an abridged quadratic form. **Davister** (op.cit., p. 107) has developed the analysis further to incorporate active constraints associated with both independent and dependent variables *to replace the dependent variables* (y_i) in the vector $(\mathbf{y})$ in the problem and has shown that under a linear transformation of coordinates the stationary values (eigenvalues) of $F(\mathbf{x})$ expressed by the modified abridged quadratic forms under this transformation are *identical* to those obtained earlier by performing the minimization of the unconstrained problem, defined by the Eqs. (3.10.24), (3.10.25a) and (3.10.25b), (op.cit., p. 110). In this treatment all active constraints are then associated with the independent variables of the abridged problem in each iteration step (k) in which the vector $(\mathbf{p}^k)$ is determined. Then, the optimal move direction $(\mathbf{p}^k)$ under a set of such active constraints corresponds to that of the eigenvector of the most negative eigenvalue of an abridged quadratic form. The order of the abridged quadratic forms is decreased by one for each constraint added to the previous set of the active constraints. For further details relating to this method the reader is referred to Ref. 27, where also the flow of calculations to determine the optimal constrained minimum of $F(\mathbf{x})$ in Eq. (3.10.74), using the eigenvector search directions, is presented in a flow-diagram (p. 128, op.cit.).

Since each active constraint is associated with the independent variables $(\mathbf{x})$ or $(\bar{\mathbf{x}})$, adding and deleting an active constraint is synonymous with adding or deleting an independent variable from a set of constraint variables in the vector $(\mathbf{x})$. To determine if a particular active constraint variable, say (x_a), is to be deleted, it is necessary to investigate if (ΔF) decreases as one moves normal to but away from the constraint hyperplane $(x_a = 0)$ into the *feasible* region. If (ΔF) decreases in this direction, then the active constraint variable is deleted from the set of active constraints, i.e. (x_a) becomes nonzero. It can be shown (op.cit., p. 114) that (ΔF) decreases in the direction normal to the constraint hyperplane $(x_a = 0)$, if the component of the gradient vector $g(\mathbf{x}^k)$ of $F(\mathbf{x}^k)$ in the positive coordinate direction (x_a) (i.e. normal to the coordinate plane $x_a = 0$) is *negative*. Any move in an iteration step (k) in the direction parallel to the coordinate axis of the independent variable (x_a) is feasible, if (x_a) becomes positive for a *sufficiently small* positive value of the steplength (α^k). Since the direction of $(x_a > 0)$ is not an eigenvector direction, the variation of (ΔF) with (α) is as shown in Fig. 3.10.1, where (ΔF) decreases for positive (α) (see Eq. (3.10.47)), as long as (α) is small and less than (α^{1*}). Thus, (x_a) can be deleted from the set of active constraints as (x_a) becomes nonzero, by moving to a new admissible point on the (x_a)-coordinate axis, provided this move does not drive any of the other variables to zero. In this way, active constraints may be deleted and the abridged quadratic form is increased by one dimension for each deleted constraint.

The *necessary conditions for a constrained minimum of* $F(\mathbf{x})$ must be examined next. Suppose that at some iteration (k), the vector $(\mathbf{x}_1)$ of the abridged vector $(\mathbf{x})$, corresponds to the eigenvector pointing to the most negative stationary point (most negative eigenvalue) of $F(\mathbf{x})$ defined by (3.10.74b) and that all of its elements are positive. The corresponding n-dimensional vector $(\mathbf{p}^k)$ is then given by (3.10.75) where $(\mathbf{x}_1)$ is augmented by (a_c^k) zeros. Suppose that also the $(M' \times 1)$ vector $(\bar{\mathbf{z}}^k)$ in Eq. (3.10.67) is found to contain only positive elements, where $(\bar{\mathbf{z}}^k)$ and $(\mathbf{p}^k)$ are related by Eq. (3.10.70a). When this is the case, none of the kinematic constraint inequalities (3.10.64) can be driven to zero in this k^{th}-iteration step by increasing (α). This means that no constraint hyperplanes can be encountered as one moves in the direction $(\mathbf{p}^k)$ through the point $(\mathbf{x}^k)$ or $(\mathbf{z}^k)$ from the origin $(\alpha = 0)$ to very large values when $(\alpha \to +\infty)$. In that case

$$\lim_{\alpha \to +\infty} \Delta F = F(\mathbf{p}^k) - F(\mathbf{x}^k) \tag{3.10.76}$$

as already shown in Fig. 3.10.3. Then the value of $F(\mathbf{x})$ at the new point $(\mathbf{x}^{k+1})$ becomes in the limit as $(\alpha \to +\infty)$,

$$F(\mathbf{x}^{k+1}) = F(\mathbf{x}^k) + \lim_{\alpha \to +\infty} \Delta F = F(\mathbf{p}^k) \tag{3.10.77}$$

Since, in this case, $(\mathbf{p}^k)$ is given by the vector $(\mathbf{x}_1)$ augmented by a $(a_c \times 1)$ null subvector according to (3.10.75), it can be easily verified on substituting $(\mathbf{p}^k)$ into $F(\mathbf{p}^k)$ according to (3.10.74), where $(\mathbf{x})$ is replaced by $(\mathbf{p}^k)$, that

$$F(\mathbf{p}^k) = F(\bar{\mathbf{x}}_1) = F(\mathbf{x}^{k+1}) \tag{3.10.78}$$

Davister (op.cit., pp. 117 to 126) has shown that if the components of the gradient vector $\mathbf{g}(\mathbf{x}^{k+1})$ in the positive coordinate directions, normal to the constraint coordinate hyperplanes, on deleting these constraints in the sense explained before, are all *positive*, then the point $[\mathbf{p}^k, F(\mathbf{p}^k)]$ satisfying these conditions qualifies as the *optimal constrained minimum solution* $[\mathbf{x}^*, F(\mathbf{x}^*)]$, as it satisfies the necessary condition for a local constrained minimum of $F(\mathbf{x})$ viz. for the local *unconstrained* minimum of $F(\bar{\mathbf{x}})$ in the abridged coordinate space $(\bar{\mathbf{x}})$.

A computer program based on this theory was developed to solve the constrained minimization of $F(\mathbf{x})$ under the corresponding inequality constraints defined in Eqs. (3.10.24) and (3.10.25). The calculations always begin with an initial feasible solution $[\mathbf{x}^0, F(\mathbf{x}^0)]$ and the optimal constrained solution $[\mathbf{x}^*, F(\mathbf{x}^*)]$ is determined in an iteration process according to the procedure just described. For further details relating to this method and the flow of calculations the reader is referred to Ref. 27, p. 128 (op.cit.).

Verifications of this method were carried out on different problems in evaluating the most degrading post-buckling modes for different model structures using different starting conditions, in which the vector $(\mathbf{x})$ represents the vector $(\mathbf{e}_b;\ b \le c)$ of the flexural shortenings of the critical isostatic members in the last isostatic subsystem, used to determine the ultimate critical state of a hyperstatic pin-jointed system of flexible elastic members, as discussed in the preceding sections. The vector $(\bar{\mathbf{x}})$ is then synonymous with the abridged vector $(\bar{\mathbf{e}}_b)$ of the flexural shortenings of the actually buckling isostatic members.

The results of this verification were checked on several model structures by an independent constrained minimization descent method in which the algorithm was based on the conjugate gradient method. The numerical results obtained by these two independent approaches for different model structures were practically *identical*, as it will be shown later.

Constrained Minimization Based on the Conjugate Gradient Method

The gradient $\mathbf{g}(\mathbf{x})$ of $F(\mathbf{x})$ was calculated earlier and it is given by Eq. (3.10.33) and, in terms of the Hessian matrix $\mathbf{G}(\mathbf{x})$, in Eq. (3.10.39). The simplest gradient method is the method of *steepest descent*[7] in which the search direction in a kth iteration defined by the vector $(\mathbf{p}^k)$ is set equal to $[-\mathbf{g}(\mathbf{x}^k)]$. In practical cases this method does not converge sufficiently rapidly for most functions $F(\mathbf{x})$. A modification of this method is the conjugate gradient method[8,9], which usually converges more rapidly. The steps which make up the conjugate gradient method when there are no constraints on the function $F(\mathbf{x})$, are as follows:

For a feasible initial point $(\mathbf{x}^1)$,

$$\mathbf{p}^1 = -\mathbf{g}(\mathbf{x}^1) \tag{3.10.79}$$

the subsequent point $(\mathbf{x}^{k+1})$ in the kth-iteration is computed from

$$\mathbf{x}^{k+1} = \mathbf{x}^k + \alpha^k \mathbf{p}^k; \quad k = 1, 2, \ldots \tag{3.10.80}$$

in which $(\mathbf{p}^1)$ is given by (3.10.79). The subsequent search direction vector $(\mathbf{p})$ is determined from

$$\mathbf{p}^{k+1} = -\mathbf{g}(\mathbf{x}^{k+1}) + \beta^k \mathbf{p}^k \tag{3.10.80a}$$

where

$$\beta^k = |\mathbf{g}(\mathbf{x}^{k+1})|^2 / |\mathbf{g}(\mathbf{x}^k)|^2 \tag{3.10.80b}$$

The steplength (α^k) in (3.10.80) is the value of (α) which minimizes $F(\mathbf{x}^k + \alpha \mathbf{p}^k)$ and it is determined as already discussed in the development following Eq. (3.10.48).

In the unconstrained minimization of $F(\mathbf{x})$ when $(\mathbf{p})$ does not point in an eigenvector direction, (α^k) takes either the value of (α^{1*}) or (α^{2*}) depicted in Fig. 3.10.1, which are determined by solving Eq. (3.10.48).

When the minimization is impeded by kinematic constraints, the optimal steplengths (α^{1*}) and (α^{2*}) may not be feasible and (ΔF) must be evaluated at the maximum permissible steplengths (α_1) and (α_2) in the positive and negative direction of $(\mathbf{p}^k)$, as shown in Fig. 3.10.4.

Here, (α_1) and (α_2) represent the minimum *positive* and the maximum *negative* values computed from Eq. (3.10.68), based on the criterion defined in (3.10.72). The optimal value of $(\alpha = \alpha^*)$ is such that

$$-|\alpha_2| \leq \alpha^* \leq |\alpha_1| \tag{3.10.81}$$

---

[7] See for example: Fox, R.L. "Optimization Methods for Engineering Design", Addison-Wesley Publishing Co., pp. 71–78, 1971.

[8] Fox, op.cit., pp. 87–97.

[9] Fletcher, R. & Reeves, C.M. "Function Minimization by Conjugate Gradients", Comp. J., vol. 6(2), pp. 163-168, 1963.

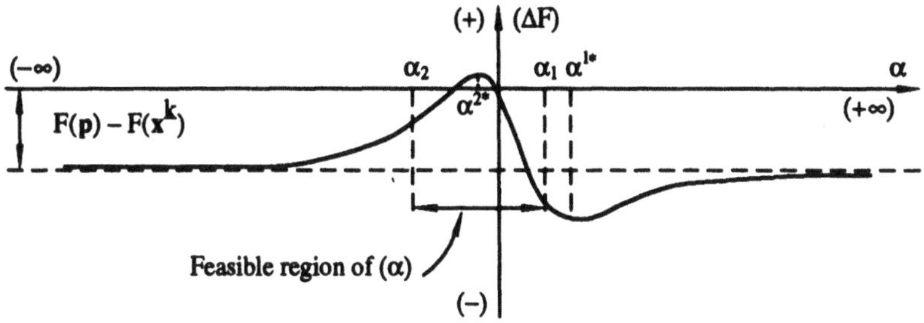

Figure 3.10.4
The optimal steplength (α^*) in an arbitrary direction ($\mathbf{p}^k$) through ($\mathbf{x}^k$) for the conjugate gradient minimization method (**Davister**, op.cit.)

as shown in Fig. 3.10.4. For the case shown (α) would be chosen as (α_1) which gives a more negative value of (ΔF). In an unconstrained problem (α^*) would be chosen as (α^{1*}) which defines the minimum value of (ΔF) associated with its stationary point.

The optimal constrained minimum solution of $F(\mathbf{x})$ in the conjugate gradient method, corresponds, similarly as in the eigenvector method, to a stationary point in an abridged coordinate space of the vector ($\mathbf{x}$), as the independent variables are driven to zero in successive iterations. The solution can be obtained if the constraints are made to correspond to the independent variables and if the elements or components of the gradient vector $\mathbf{g}(\bar{\mathbf{x}}^k)$ in the kth-iteration, associated with the nonzero independent variables of the abridged vector ($\bar{\mathbf{x}}^k$) vanish at the point ($\bar{\mathbf{x}}^k$). As in the eigenvector method, whenever a dependent variable (y_i) is driven to zero, the dependent variable is interchanged with another nonzero independent variable (x_j) using the linear transformations ($\mathbf{y} = \mathbf{C}\mathbf{x}$) in (3.10.64). The result of this procedure is that all active constraint hyperplanes become associated with the independent variables. One then investigates the abridged coordinate space of the nonzero independent variables for a stationary value of $F(\bar{\mathbf{x}})$ at each iteration (k). The condition for a stationary value of $F(\bar{\mathbf{x}})$ at ($\bar{\mathbf{x}}$) in the kth-iteration is that

$$\frac{\partial F(\bar{\mathbf{x}})}{\partial x_i} = 0; \qquad i = 1, 2, \ldots, (n - a_c^k) \tag{3.10.82}$$

where (a_c^k) denotes the number of active constraints at k^{th}-iteration. The last conditions are equivalent to the vanishing of all the components of the gradient vector $\mathbf{g}(\bar{\mathbf{x}})$ of $F(\bar{\mathbf{x}})$, associated with the directions of the nonzero variables ($\bar{\mathbf{x}}$).

In the flow of calculations the iterative process continues until a constraint is encountered. Then the procedure is restarted with the search direction ($\mathbf{p}$) corresponding to the direction of steepest descent by putting [($\mathbf{p}^1 = -\mathbf{g}(\mathbf{x}^1)$] as in Eq. (3.10.79). Every time that the set of nonzero independent variables denoted by ($\bar{\mathbf{x}}$) in Eq. (3.10.74a) is changed, the set of conjugate directions given by (3.10.80a) is reinitialized. This is done to ensure compatibility between the successive gradient vectors $\mathbf{g}(\mathbf{x}^k)$ and $\mathbf{g}(\mathbf{x}^{k+1})$ used to evaluate the scalar (β^k) in (3.10.80b) within the abridged coordinate space that is being searched for a stationary point. Each time the conjugate gradient iteration is restarted the point $[\mathbf{x}^1, F(\mathbf{x}^1)]$ is not arbitrary, but it corresponds to the last point from the preceding iteration step.

The computational methodologies and the flow of calculations in successive iteration steps for both the eigenvector and the conjugate gradient search direction methods of solution of this minimization problem were developed by **Davister** (op.cit.) and the interested reader is referred to this source for further details.

Several studies in evaluating the most degrading post-buckling modes in equilibrium and in dynamic collapse by unstable motion were conducted, in which the function $F(\mathbf{x})$ represents the total potential energy hypersurface $v(\mathbf{e}_b)$; $(b \leq c)$ and the vector $(\mathbf{x})$ is identified with the vector $(\mathbf{e}_b)$ of flexural contractions, using in each case both the eigenvector search direction method and the conjugate gradient method. Of the twenty-four example studies carried out by Davister (op.cit.), there was total agreement between the two methods in twenty-three of these studies. In the remaining case the minimum evaluated by the eigenvector method was 1% less than the minimum obtained by the conjugate gradient method, which indicates that in this particular case the conjugate gradient method may have evaluated a marginally close local minimum rather than the global constrained minimum, as in all the other cases studied.

3.11 Numerical Evaluation of the Most Degrading Dynamic and Static Buckling Modes and the Structural Stability Optimization Strategies in Hyperstatic Pin-Jointed Elastic Systems

We showed in the preceding sections that the most general form of the total potential energy of a hyperstatic or isostatic system in the vicinity of the ultimate critical state, marked by the critical value (w^0) of the parameter of the external loading applied to this system centrally at the joints, is given by

$$v = \tfrac{1}{2}\mathbf{e}_b^T \mathbf{E}_b \mathbf{e}_b; \qquad b \leq c \tag{3.11.1}$$

(c) represents the number of critically stressed compressive elastic bars in the last isostatic subsystem in the evaluation of the ultimate critical state. Then, in the context of the notation in the minimization problem discussed in the preceding section, the vector

$$\mathbf{e}_b \equiv \mathbf{x} \tag{3.11.2}$$

and the post-buckling stiffness matrix or the stability matrix $(\mathbf{E}_b)$ is clearly identical to the matrix $(\mathbf{A})$ in (3.10.24) i.e. $(\mathbf{E}_b \equiv \mathbf{A})$. With this in mind the methodologies discussed in the last section can be applied to any structural system of this type. The solutions ($\bar{\mathbf{x}}_1 = \mathbf{e}_{b1}$) and ($\bar{\mathbf{x}}^* = \mathbf{e}_{b1}^*$) then correspond to the optimal solutions for the global constrained and the unconstrained minima, respectively, of the total potential energy function ($v = \mathbf{e}_b^T \mathbf{E}_b \mathbf{e}_b$) subject to the constraints

$$\mathbf{e}_2 = \mathbf{C}_2 \mathbf{C}'_{1b} \mathbf{e}_b \geq \mathbf{0}; \qquad b \leq c$$

where now

$$\mathbf{e}_2 \equiv \mathbf{y} \tag{3.11.3}$$

and

$$\mathbf{C}_2 \mathbf{C}'_{1b} \equiv \mathbf{C} \tag{3.11.4}$$

as in the notation of the minimization problem in the last section. If an arbitrary vector ($\mathbf{e}_b$) is now bounded by an arbitrary contour

$$\tfrac{1}{2} \mathbf{e}_b^T \mathbf{B} \mathbf{e}_b = C$$

where also (**B**) and the constant (C) are arbitrary and, therefore, the system in the vicinity of the critical state is considered to be in the state of motion and not necessarily in equilibrium, (we recall that only if $\mathbf{B} \equiv \mathbf{Q}$, Eq. (3.7.33), the system may be in a state of post-buckling equilibrium, so that with $\mathbf{x} = \mathbf{e}_b$, $F(\mathbf{x}) = \delta w / \delta \bar{s}$, as shown in Eq. (3.7.38)!), the expression (3.11.1) for (v) then transforms to Eq. (3.10.14d) and the minimization problem discussed in the last section, reduces to that defined in (3.10.15), i.e., minimize

$$v = \frac{\mathbf{e}_b^T \mathbf{E}_b \mathbf{e}_b}{\mathbf{e}_b^T \mathbf{B} \mathbf{e}_b}; \qquad b \leq c \tag{3.10.15)bis}$$

subject to

$$\mathbf{e}_b \geq \mathbf{0} \qquad \text{and} \qquad \mathbf{e}_2 = \mathbf{C}_2 \mathbf{C}'_{1b} \mathbf{e}_b \geq \mathbf{0}$$

The minimization methods discussed in the last section were applied to this problem using several model structures of this type. To illustrate here the results of this minimization in computing the so-called *most degrading* post-buckling modes under different physical conditions, we show these results for the model reticulated shell structures described in Sections 3.6 and 3.8 in the three versions denoted as the unoptimized model (Model No. 1), and as the first and second optimized models (or Models No. 2 and No. 3) respectively. The studies carried out by **Britvec** and **Davister** on the model reticulated shells in this connection, Refs. (23) and (28) involved different contours characterized by different matrices (**B**). Both, the eigenvector search direction method and the conjugate gradient method were employed in the solution of the problems. These contours may be denoted in the context of the present notation by

$$v_{\text{contour}} = \frac{1}{2} \frac{\mathbf{e}_b^T \mathbf{E}_b \mathbf{e}_b}{\mathbf{e}_b^T \mathbf{B} \mathbf{e}_b}; \qquad b \leq c \tag{3.11.5}$$

The detailed description with the numbers and the lay-out of the members and joints of the basic model (Model No. 1) and the modified or optimized models (Models No. 2 and No. 3) is given in Sections (3.6) and (3.8) in this Chapter.

This numerical study is concerned with the evaluation of *ultimate* critical states using the method explained in Section 3.2 in this Chapter, and of the most degrading buckling modes associated with the most negative regions of the total potential energy hypersurfaces, subject to the kinematic constraints defined in (3.10.15). The method employed four different ratios of quadratic forms denoted by (v_1), (v_2), (v_3) and (v_4) according to the Eqs. (3.10.14f), (3.10.14g), (3.10.18) and (3.10.23a) respectively, as explained in the preceding sections.

The numbers, lengths and the outside diameters of the members in the unoptimized model lattice (Model No. 1) are given in Table 1 in Section 3.6 in this Chapter. In this case only members with solid cross-sections were used. All members were designed to buckle in the elastic range and were conceived as high-tensile steel members with a yield point of 50 Kips/in.^{2} (3,515.35 kg*/cm^2). The diameters of the members were [1/20] of their respective lengths and they were computed so that elastic buckling was initiated at 90% of the yield load. Moderately and slightly stressed members can be replaced by hollow tubes in a final design.

The first optimized model (Model No. 2) was obtained from Model No. 1 by stiffening 25 members located at the base of the shell-lattices as shown in Fig. 3.8.1a and these were prevented from buckling elastically and yielding in tension.

The second optimized model (Model No. 3) was obtained by stiffening 25 members along the meridians between the five spherical triangles on the outer layer of Model No. 1, as shown in Fig. 3.8.6a.

The ultimate critical states for the three models were calculated using a computer program based on the theory explained in Section 3.2 in this Chapter. The cross-sectional proportions of the stiffened members are then governed by the axial loads in the respective ultimate critical states.

In practice a shell of this type would be enveloped by a cover-shell, as explained earlier, which would contribute to the overall behavior of the composite structure, this depending on the stiffness and other features of such a cover-shell. The influence of such a cover-shell on the response of the composite structures under loading was not considered in this study. Also, the computed (LCC/Wt) ratios are based on the structural lattices alone, without the weight of a cover-shell.

The unoptimized reticulated shell Model No. 1 was designed for a depth of some 270 ft. of sea-water. The hydrostatic pressure was integrated over the outer shell area and distributed over the peripheral joints according to the area attributable to each joint. In this case the load parameter (w) represents the hydrostatic head of sea-water. The (x, y, z)-components of the radial hydrostatic loads, exerted by 1 ft. of water, then represent the components of the load vector $(\boldsymbol{\beta})$ in Eq. (3.2.1) in Section 3.2. The results of the computer analysis, as computed by **Davister**, Ref. 28, are reproduced for the three models in Table 5.

Table 5
Characteristics of the ultimate critical states for the Models Nos. 1, 2 and 3 computed by **Davister** (op.cit.)

Model No.	No. of members with $e \neq 0$	No. of redundant members with $e = 0$	No. of tensile yield members	No. of critic. compr. members on the point of buckling	No. of isostatic critical members in the *last* isostatic subsystem	average ultim. crit. load param. (w^0) in feet of sea-water	estimated theoretical (LCC/weight) ratio
1	205	0	5	200	5	272.43	320
2	175	30	20	155	5	280.20	330
3	200	5	50	150	10	417.89	255*

The quantities of critically stressed members on the point of buckling, the critical isostatic members and the tensile members on the point of yielding plastically are given in the table. The ultimate critical states of the models are listed in feet of sea-water. The critical load for Model No. 3 is significantly larger than those of Models No. 1 and 2. However, in Model No. 3 the critical loads in the 25 stiffened meridional members become considerable and these were found to be in the region of 3000 Kips (1,360.8 tons). To accommodate such axial loads it is necessary to resort to very heavy sections. However, if advanced lighter materials are used, such as fibre reinforced composite materials, to fabricate such special members from glass or carbon fibre reinforced composites of high strength, a much better performance may be obtained in heavily stressed bars. The (LCC/Wt) ratio quoted in Table 5 for Model No. 3 is based on very heavy meridional steel members, which would *not* be used in practice, as an alternative design would be pursued. Therefore, these figures should be regarded as tentative and very conservative. Earlier estimates of stiffened reticulated shells of the same type, Ref. 20, produced higher theoretical figures for the (LCC/Wt) ratios. Thus, for the unoptimized and the first optimized models these ratios were in the region of 380 to 390 and for the second optimized model (Model No. 3) much higher than the conservative figure* in Table 5. If heavy member sections are avoided and such special members are designed from lighter composite materials of comparable superior strength, the (LCC/Wt) ratio may be drastically improved. Earlier theoretical estimates showed that the ratio for Model No. 3 could be increased to 500 or above, this depending also on the overall design and the estimated weight of the connecting nodal spheres and the matching connectors and other fittings. However, to be on the safe side and not to overestimate the (LCC/Wt) ratios, the low figures are quoted in Table 5. It should be noted that the average theoretical ultimate critical load for Model No. 3 is very much higher than the loads for the other two model structures. It should also be borne in mind that such theoretical estimates are reduced in practice by the imperfection-sensitivity of the lattice under near-critical loads, as already explained in Section 3.8.

We saw earlier in discussing the analysis of hyperstatic pin-jointed lattices in Sections 3.2 and 3.3 that not all the critically stressed redundant members necessarily buckle, as some of these may be constrained from buckling by other members or they may contract axially in plastic deformation. Such members do not buckle if they are located in a part of the lattice that is displaced as a rigid body or that remains stationary. In that case a buckling mechanism is local or partial and not global. Such non-buckling bars are readily identified using the kinematic condition [3.3.6(b)] in Section 3.3 in this Chapter, namely,

$$\mathbf{e}_\gamma = \mathbf{C}_\gamma \mathbf{C}'_{1b} \mathbf{e}_b \geq 0; \qquad b \geq c \tag{3.3.6b)bis}$$

where the subscript (c) is replaced by (b). If any row in the $(\gamma \times b)$ product matrix $(\mathbf{C}_\gamma \mathbf{C}'_{1b})$ contains only zero elements, the associated redundant member flexural shortening is zero, regardless of the vector $(\mathbf{e}_b)$, i.e. for all modes. A similar conclusion follows for tensile yielding members. Thus the criterion by which a post-buckling mechanism is local or global depends also on the redundant member-shortenings, associated with the zero coefficients in the above product matrix.

The number of critical isostatic and the redundant members which may shorten flexurally in each model is given in Table 5, as well as the number of redundant members associated with the zero rows in the above product matrix, which cannot buckle. Therefore, the buckling mechanisms associated with Model No. 1 are global and those in Models No. 2 and 3 may be only partial.

The status of the critically stressed members which buckle for Model No. 1 are given in Figs. 3.6.4a to 3.6.4c, for Model No. 2 in Figs. 3.8.1, 3.8.2 and 3.8.3 and for Model No. 3 in Figs. 3.8.6, 3.8.7 and 3.8.8. Buckling members are denoted by the letters (BK) and those which yield, by (YD), as shown in the diagrams.

The most likely buckling mode to develop under dynamic or static conditions in post-buckling is *that which is associated with the largest conversion of the total potential energy of the elastic system into the kinetic energy*. Many tests on model structures of all types conducted by the author in a longstanding research over many years confirm this tendency. We therefore adopt this criterion as a measure of "degradation" of the structural behavior and refer to such modes as *the most degrading buckling modes*. These are then invariably *associated with the most negative regions of the total potential energy changes* of the system measured from the respective ultimate critical equilibrium state, where the onset of dynamic buckling may freely develop. If collapse takes place in a mode in accelerated motion and not under artificially stabilized equilibrium conditions, which are only possible in a laboratory under stabilized equilibrium using some sort of stabilizing loading device‡, such modes are referred to as the *most degrading dynamic modes* as described in Refs. (8), (9), (15). The most degrading *equilibrium* modes also exist under a system of prescribed external loads, when

‡ For example, See Ref. 15, p. 286

($\mathbf{B} = \mathbf{Q} = \mathbf{B}_b$) as determined by Eq. (3.7.33), as we saw in Sections 3.8 and 3.9, but these may be *very different* from the most degrading dynamic modes, which are always associated with that deformation vector ($\mathbf{e}_b = \mathbf{e}_b^*$), which leads to the largest conversion of the total potential into kinetic energy of the system, subject to its kinematic constraints as discussed.

This minimum value of (v) is determined by performing a constrained minimization of the ratio of (v) and the bounding contour given by ($\mathbf{e}_b^T \mathbf{B} \mathbf{e}_b = 1$), which we now denote by (v_i, $i = 1, 2, 3, 4, \ldots$), where the matrix ($\mathbf{B}$) is chosen from the matrices ($\mathbf{S}$), ($\mathbf{B}_e$), ($\mathbf{J}$), ($\mathbf{J}_L$) etc. which we defined before. Four such contours were used in the evaluation of the most degrading modes for the mentioned reticulated shell intended for subsea applications. These were,

$$v_1 = \frac{\mathbf{e}_b^T \mathbf{E}_b \mathbf{e}_b}{\mathbf{e}_b^T \mathbf{S} \mathbf{e}_b}; \qquad \mathbf{S} \text{ (Eq. 3.10.12)} \tag{3.11.6}$$

$$v_2 = \frac{\mathbf{e}_b^T \mathbf{E}_b \mathbf{e}_b}{\mathbf{e}_b^T \mathbf{B}_e \mathbf{e}_b}; \qquad \mathbf{B}_e \text{ (Eq. 3.10.8)} \tag{3.11.7}$$

$$v_3 = \frac{\mathbf{e}_b^T \mathbf{E}_b \mathbf{e}_b}{\mathbf{e}_b^T \mathbf{J} \mathbf{e}_b}; \qquad \mathbf{J} \text{ (Eq. 3.10.17a)} \tag{3.11.8}$$

$$v_4 = \frac{\mathbf{e}_b^T \mathbf{E}_b \mathbf{e}_b}{\mathbf{e}_b^T \mathbf{J}_L \mathbf{e}_b}; \qquad \mathbf{J}_L \text{ (Eq. 3.10.22)} \tag{3.11.9}$$

Each function (v_i; $i = 1, 2, 3, 4$) is associated with a different contour on the same total potential energy hypersurface given by

$$v = \tfrac{1}{2} \mathbf{e}_b^T \mathbf{E}_b \mathbf{e}_b \tag{3.11.10}$$

But, equilibrium under the proportionately variable prescribed loading is now *not* possible, since the matrices ($\mathbf{B} = \mathbf{S}$, $\mathbf{B}_e$, $\mathbf{J}$ and $\mathbf{J}_L$) are now different from the matrix ($\mathbf{Q}$) defined in Eq. (3.7.33). The denominators in the four function-ratios may be written as follows

$$\mathbf{e}_b^T \mathbf{S} \mathbf{e}_b = \sum_{i=1}^{b+\gamma} \delta P_i e_i = \sum_{i=1}^{b+\gamma} \frac{P_i^0}{2L_i} (e_i)^2; \qquad b \le c \tag{3.11.11}$$

$$\mathbf{e}_b^T \mathbf{B}_e \mathbf{e}_b = \sum_{i=1}^{b+\gamma} e_i^2; \qquad b \le c \tag{3.11.12}$$

$$\mathbf{e}_b^T \mathbf{J} \mathbf{e}_b = \sum_{j=1}^{J} \mathbf{u}_j^T \mathbf{u}_j = \mathbf{u}^T \mathbf{u} \tag{3.11.13}$$

$$\mathbf{e}_b^T \mathbf{J}_L \mathbf{e}_b = \sum_{j=1}^{J_L} \mathbf{u}_j^T \mathbf{u}_j \tag{3.11.14}$$

As before, (b) denotes the number of buckling members and (c) the number of compressive critically stressed members in the last isostatic subsystem used in the determination of the ultimate critical state, (γ) the number of critical redundant members in the lattice, (J) the number of joints of the lattice and (J_L) the number of loaded joints. (δP_i) are the axial load increments (positive compressive) measured from the critical loads (P_i^0). (L_i) is the initial length of member (i) and

(e_i) its flexural shortening. (u_j) denote the magnitudes of the component displacements of the node or pin-connected joint (j). This defines the four contours related to different measures of elastic distortion of the lattice in its buckled form.

Model No. 1

The characteristic computer results obtained from the analysis in the determination of the most degrading buckling modes for Model No. 1, Refs. (27), (28) under hydrostatic pressure are reproduced in Table 6. For this model structure subjected to hydrostatic loading there are five critical isostatic members, i.e. ($c = 5$) in the last isostatic subsystem, before reaching the ultimate critical state at some 272 ft. of sea water. In the case of (v_1) and (v_2) none of the constraints were activated at the minimum points, which means that the constrained and unconstrained minima in these two cases are identical. Also, the minimum values of the function-ratios (v_i) are not appreciably different from the initial feasible values, which indicates that these functions are fairly "flat" and insensitive to the vectors ($\mathbf{e}_b$) throughout this region. Only (v_4) produced very large negative values, which were caused by the singularity of the matrix ($\mathbf{J}_L$) at the minimum point. The four buckling modes corresponding to the four contours were calculated by evaluating the vectors ($\mathbf{e}^*_{bi}$; $i = 1, 2, 3, 4$) and the corresponding vectors ($\mathbf{e}^*_{yi}$) from Eq. 3.3.6(b) (on replacing the index (c) by (b)) for each most degrading mode. The relative values of these four vectors, based on the flexural contraction of member (1) in each case being set equal to unity, i.e. ($e_1 = 1$), were computer-plotted in a bar-diagram against the member sequence numbers from 1 to 506 on the abscissa in Fig. 3.11.1, according to the three layers of the shell, (marked by vertical dotted lines outer, inner and middle layer).

The four mode bar-diagrams viewed from the top downwards, correspond to the function-ratios v_1, v_2, v_3 and v_4. It is seen from the bar-diagrams in Fig. 3.11.1, that the differences in the distribution of the flexural distortions in the buckling members throughout the structure, based on the four different bounding con-

Table 6
Summary of the characteristic numerical results in the determination of the most degrading dynamic buckling modes for **Model No. 1** subjected to hydrostatic pressure using four contours (v_1) to (v_4), computed by **Davister** (op.cit.)

Function ratio (v_i)	Unconstrained minimum value of (v_i)	Initial feasible value of (v_i)	Minimum constrained value of (v_i)	No. of iterations to obtain constrained minimum	No. of active constraints at minimum
v_1	−2.395	−2.395	−2.395	0	0
v_2	−63.286	−63.286	−63.286	0	0
v_3	−40.905	−16.465	−16.519	5	4
v_4	–	−40.861	−41.066	3	2

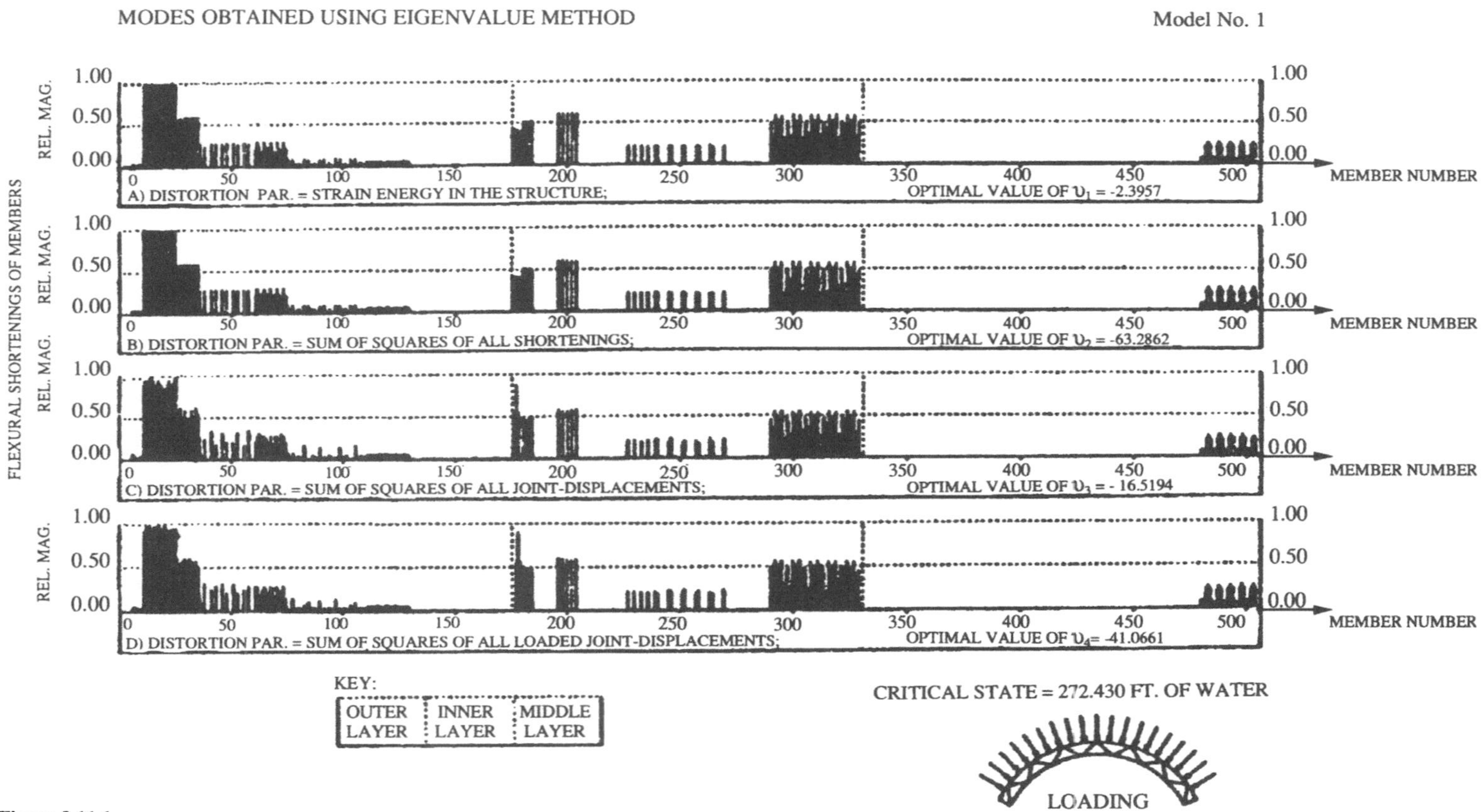

Figure 3.11.1
Comparison of the most degrading dynamic buckling modes for **Model No. 1** subjected to hydrostatic pressure computed on four different contours (v_1), (v_2), (v_3) and (v_4) by **Davister** (op.cit.)

tours, are *not* appreciable. The most distorted bars are the flexurally buckling members, situated in the outer layer at the beginning of the member sequence numbers around the top of the shell, further, in the inner layer near the sequence number 300, where the most heavily buckled members are situated close to and around the circumference of the shell. For a more detailed description of these regions of the lattice the reader is referred to the diagrams in Figs. 3.6.3a, 3.6.3b and 3.6.3c in Section 6 of this Chapter.

The interval of the bar-diagram corresponding to the flexural distortions in members 275 to 506, situated partly in the inner layer and partly in the whole of the middle layer, was found to be practically identical for all the modes obtained on the four different contours. Also, the first two modes from the top, corresponding to the contours described by the function-ratios (v_1) and (v_2) show a more symmetrically buckled shape of the reticulated shell, than the last two solutions corresponding to (v_3) and (v_4), although a certain circular symmetry in the distortion patterns is apparent in all four solutions.

Model No. 2

The computed characteristic results for the first optimized model or Model No. 2 are shown in Table 7, in this case there are five isostatic critical members, i.e. ($c = 5$), in the last isostatic subsystem before reaching the ultimate critical state at approximately 280 ft. of sea water. All four minimum values of the total potential energy function (v) on the four contours represented by the function-ratios (v_1), (v_2), (v_3) and (v_4), expressed in the same relative numbers as for Model No. 1, are now somewhat more negative than in the previous cases, displayed in Table 6, which indicates that the degree of instability (amount of total potential energy available for conversion into kinetic energy at the respective ultimate critical state) is somewhat greater than in the case of the unoptimized Model No. 1.

The computed bar-diagrams for Model No. 2 are given in Fig. 3.11.2. They correspond to the four contours defined by the four function-ratios (v_1) to (v_4),

Table 7
Summary of the characteristic numerical results in the determination of the most degrading dynamic buckling modes for **Model No. 2** subjected to hydrostatic pressure, using the four contours (v_1) to (v_4), computed by Davister (op.cit.)

Function ratio (v_i)	Unconstrained minimum value of (v_i)	Initial feasible value of (v_i)	Minimum constrained value of (v_i)	No. of iterations to obtain constrained minimum	No. of active constraints at minimum
v_1	−2.823	−2.823	−2.823	0	0
v_2	−74.349	−74.349	−74.349	0	0
v_3	−40.460	−18.108	−18.121	3	2
v_4	–	−48.253	−48.313	5	2

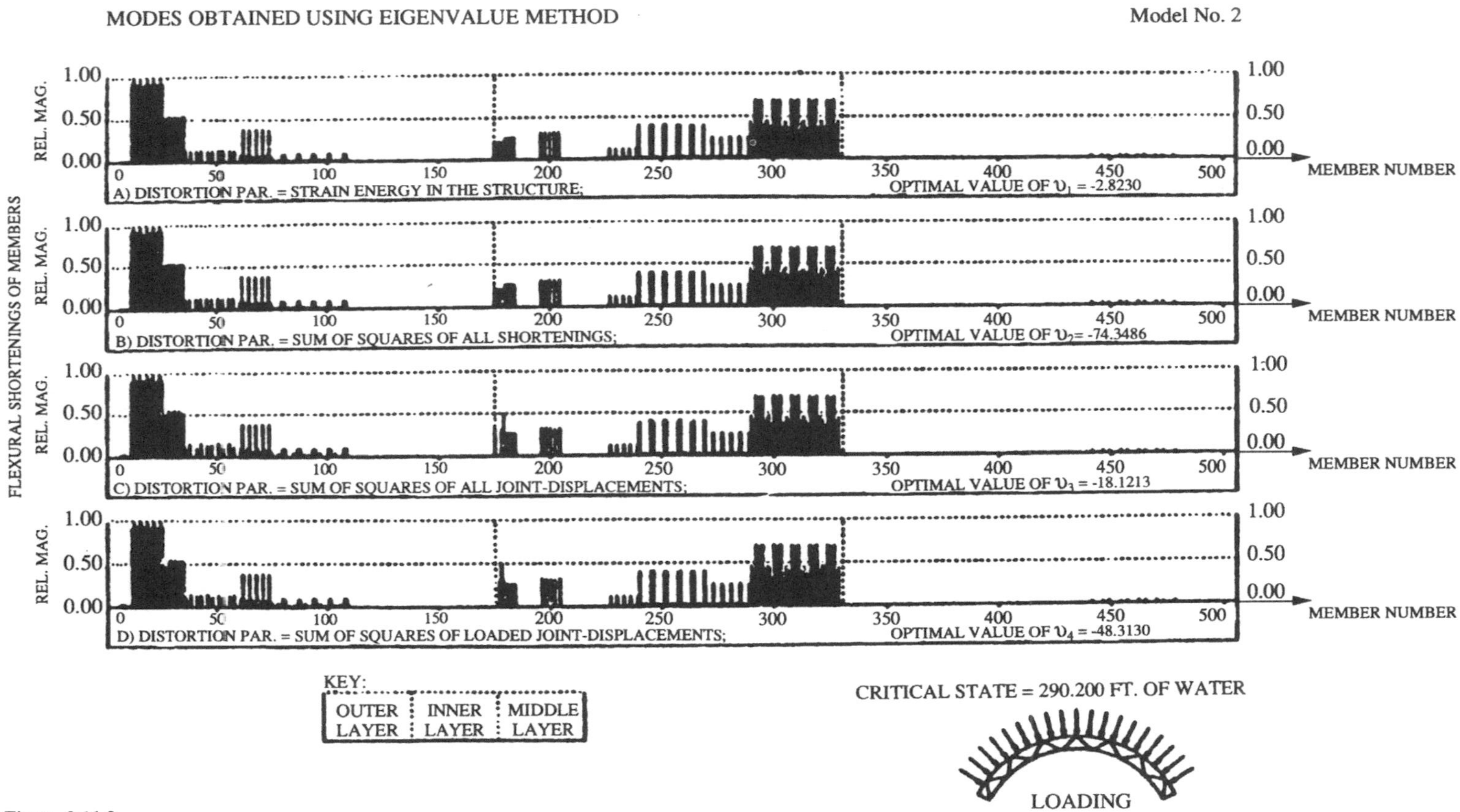

Figure 3.11.2
Comparison of the most degrading dynamic buckling modes for **Model No. 2** subjected to hydrostatic pressure computed on the contours (v_1), (v_2), (v_3) and (v_4) by **Davister** (op.cit.)

from the top downwards. The solutions associated with the ratios (v_1) and (v_2) are nearly the same and the solutions based on (v_3) and (v_4) are very similar. But, all four solutions show the same tendencies and, as before, there are only minor differences between them.

Model No. 3

The computed results for Model No. 3 are shown similarly as for the Models Nos. 1 and 2 in Table 8. In this case there are 10 simultaneously critically stressed isostatic compressive members in the last isostatic subsystem, $(c = 10)$, at the ultimate critical state corresponding to some 417 ft. of sea water. Again, the minimum values (v_i), expressed in relative numbers, are now even more negative than those for the Models Nos. 1 and 2, which indicates an even higher degree of instability for Model No. 3 at the ultimate critical state. This suggests that, if the initial geometrical imperfections and the imperfection-sensitivity of this system were considered, as this was done under statical equilibrium conditions in post-buckling, one would expect the actual ultimate peak hydrostatic head of the equivalent imperfect system, compared to the computed ultimate critical head in this case, to be lower for Model No. 3, than for the Models Nos. 1 and 2. In order to ascertain the exact influence of initial imperfections, dynamic tests of such models under controlled laboratory conditions are necessary. This influence strongly depends on the type of fabrication of the models. Such tests could not be conducted in the present study, because of the lack of funding, which is, therefore, confined only to the computed results based on the theory expounded in this Chapter.

The bar-diagrams depicting the distortion patterns of the buckled structure for Model No. 3 for the four function-ratios (v_1), (v_2), (v_3) and (v_4) are given in this sequence in Fig. 3.11.3 starting at the top, going downwards. In this, as in the case of Model No. 2, the patterns corresponding to the contours (v_1) and (v_2) are almost identical and so are the remaining two based on (v_3) and (v_4). Thus, based on this study, practically any of these function-ratios may be used to predict reasonably accurately the vector $(\mathbf{e}_b^*)$ (and the related vector $(\mathbf{e}_\gamma^*)$ defined by Eq.

Table 8
Summary of the characteristic numerical results in the determination of the most degrading dynamic buckling modes for **Model No. 3** subjected to hydrostatic pressure using the four contours (v_1) to (v_4), computed by **Davister** (op.cit.)

Function ratio (v_i)	Unconstrained minimum value of (v_i)	Initial feasible value of (v_i)	Minimum constrained value of (v_i)	No. of iterations to obtain constrained minimum	No. of active constraints at minimum
v_1	−5.559	−4.010	−4.048	6	5
v_2	−168.157	−96.710	−99.733	20	7
v_3	−176.539	−50.964	−52.226	8	7
v_4	–	−113.875	−116.370	33	9

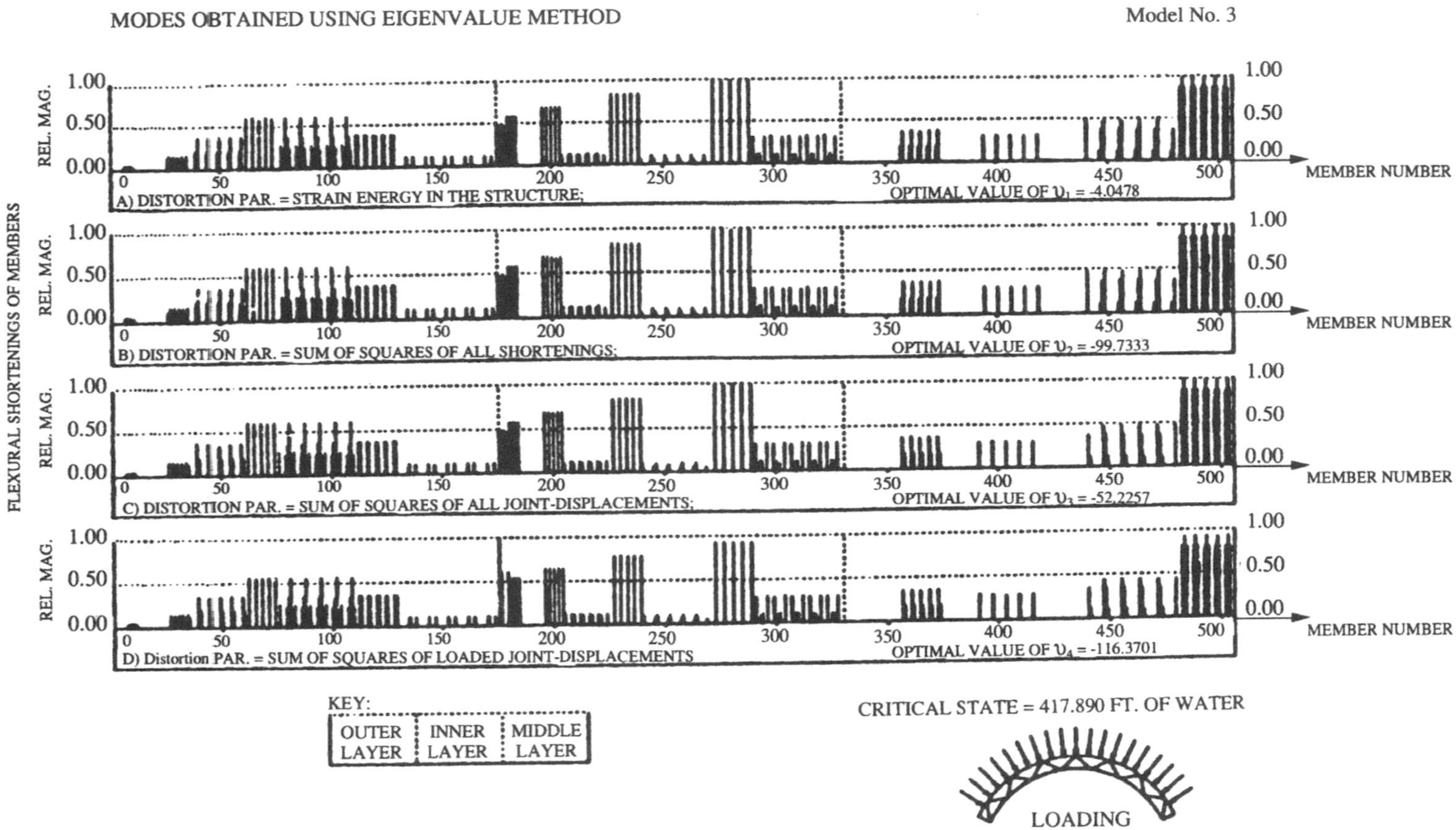

Figure 3.11.3
Comparison of the most degrading dynamic buckling modes for **Model No. 3** subjected to hydrostatic pressure computed on the contours (v_1), (v_2), (v_3) and (v_4) by **Davister** (op.cit.)

3.3.6(b)), corresponding to the constrained global minimum of the v-hypersurface and defining the most degrading dynamic buckling mode of a pin-jointed hyperstatic or isostatic elastic system.

If the evaluation of the most degrading buckling mode under statical conditions, subject to the usual kinematic constraints under the (proportionately) modified prescribed statical loading is desired, in that case the matrix (**B**) must be set equal to the matrix (**Q**), where (**Q**) is defined by Eq. (3.7.33).

It may be concluded from a comparison of the constrained global minima and of the most degrading dynamic post-buckling modes of the three models based on, say, the function-ratio (v_1), from the Tables 6, 7 and 8, that the modified or stiffened Model No. 3 is associated with the highest degree of dynamic instability in the respective ultimate critical *equilibrium* state, from which the dynamic collapse theoretically occurs. This is likely to be the case also in a corresponding *imperfect* real structure under hydrostatic pressure, when the ultimate critical state coincides with the peak-value on the imperfect equilibrium path, associated with either the external load parameter or the corresponding hydrostatic head of water, as explained in Chapter 2. The three minimum relative values of (v_1) in this comparison are then ($v_1 = -2.395$) for Model No. 1, ($v_1 = -2.823$) for Model No. 2 and ($v_1 = -4.048$) for Model No. 3.

A comparison of the results for the constrained global minimum values of the function-ratios (v_1) to (v_4) for the three models using the two methods of solution, referred to as: 1) The Eigenvector Search Direction Method or Eigenvector Descent Method and 2) The Conjugate Gradient Descent Method, which were explained before, are given in Table 9. The numerical values are given in the same relative numbers for all three models. A star (*) indicates that the unconstrained and the constrained minima, based on the usual kinematic constraints defined in (3.10.15) or in (3.3.6), coincide and are identical.

A comparison of the figures shows that the two independent methods of solution yield practically identical results.

Some Optimization Strategies in Hyperstatic Pin-Jointed or Quasi-Pin-Jointed Elastic Systems

The numerical evaluation of the most degrading static or dynamic post-buckling modes in pin-jointed or quasi-pin-jointed elastic hyperstatic lattices subjected to concentrated external loads applied centrally at the nodes, constitutes the *basic step* in any attempt to improve or optimize the performance of such a system. The next step in this optimization is to *eliminate* the computed dynamic or static most degrading mode. This can be accomplished by stiffening certain key critical members, responsible for the existence of such a mode, and preventing them from buckling. If one defines the (b) and (c) critically stressed compressive members in the *last* isolated isostatic subsystem, just before the ultimate critical

Table 9
Comparison of the results obtained for the constrained minima of the total potential energy hypersurfaces bounded by different contours (v_1), (v_2), (v_3) and (v_4) of the three models computed by the eigenvector and conjugate gradient descent methods by **Davister** (op.cit.)

Function ratio or parameter (v_i)	Model No.	(1) Constrained minimum based on the Eigenvector Descent Method	(2) Constrained minimum based on the Conjugate Gradient Descent Method
	1	−2.395*	−2.395
v_1	2	−2.823*	−2.823
	3	−4.048	−4.048
	1	−63.286*	−63.286
v_2	2	−74.349*	−74.349
	3	−99.733	−99.733
	1	−16.519	−16.519
v_3	2	−18.121	−18.121
	3	−52.226	−51.850
	1	−41.066	−41.017
v_4	2	−48.313	−48.313
	3	−116.370	−116.386

state is evaluated (as explained in Section 3.2), which *actually buckle* in this computed most degrading mode as the *key members*, then it suffices to *prevent one or several of these key members from buckling*, to eliminate this mode. This is accomplished most easily by increasing the flexural stiffness of the particular key bars. These key bars are usually few in number compared to the total number of compressive bars, the redundant ones included. It is often sufficient to stiffen *only one* of those key bars to accomplish the desired result. These key members are identified numerically by the vector ($\mathbf{e}_b^*$) defining the flexural contractions of the vector ($\mathbf{e}_b$) in the most degrading buckling mode, which is computed either from a dynamic collapse or in statical post-buckling unstable equilibrium according to the methods described in the preceding sections. Thus, the flexural shortenings of the key members are regarded as the independent variables in this minimization problem defined in (3.10.15).

The key members themselves may be considered as the *independent bars* in such an optimization step.

Alternatively, one or several *redundant* members which contract flexurally in the same mode may be stiffened to prevent this mode from developing. However, the choice of the stiffening or of the stiffened members is usually conditioned by the effect of this stiffening on the *subsequent* most degrading mode of the stiffened or modified structure. But, if a redundant buckling member is stiffened and the (b) key members are left unstiffened, it is a priori not certain that under the same proportionate loading the next most degrading mode of the stiffened structure is not located within an even more negative region of the total potential energy hypersurface of the *modified structure* and that the resulting subsequent mode does not turn out to be even more degrading than the eliminated one.

On the other hand, if one or several of the key isostatic critical members are prevented from buckling, which is equivalent to zeroing one or several of the independent variables (e_i^*; $1 \leq i \leq b$) in the vector ($\mathbf{e}_b^*$), then the reduced vector ($\mathbf{e}_b = \mathbf{e}_{b1}^*$), say, is located in the admissible feasible region of the v-hypersurface, where the *negative* value of (v) evaluated at ($\mathbf{e}_b = \mathbf{e}_{b1}^*$) is certainly greater than the most negative value of (v) in the same region. This is so, because the vector ($\mathbf{e}_{b1}^*$) is located on a new constraint-hyperplane ($e_i^* = 0$), or, if (i) takes several integer values, on the intersection of these hyperplanes, which are located in an admissible feasible region of (v). Thus, $v(\mathbf{e}_{b1}^*)$ may be regarded as another constrained minimum within the region mapped by the vector ($\mathbf{e}_b^*$) and, therefore, the constrained minimum $v(\mathbf{e}_{b1}^*)$ is certainly *greater* than or at worst equal to the unconstrained minimum in this region, as explained in the development following Eq. (3.10.27), i.e.,

$$v(\mathbf{e}_{b1}^*) \geq v(\mathbf{e}_b^*) \tag{3.11.15}$$

This follows from the analysis developed in the preceding sections and the solution of the minimization problem defined in (3.10.15). Thus, by stiffening one of the key critical members in the current most degrading post-buckling mode, defined by the vector ($\mathbf{e}_b^*$), one may, on repeating the minimization calculations, evaluate the next most degrading mode for the stiffened structure, which *results in a reduced degradation or in a reduced largest available amount of the total potential energy change of the system*, that may be converted into kinetic energy in a dynamic collapse.

Alternatively, in post-buckling equilibrium this results in the most degrading unstable or stable equilibrium paths or gradients, characterizing the relationship between external loads and the corresponding displacements of the loaded joints, such that the total potential energy change computed according to Eq. (1.2.6) in **Theorem 1** in Section 1.2, is less negative in the most degrading unstable *equilibrium* mode of the stiffened pin-jointed structure, than that in the unstiffened structure. Then, after each stiffening operation the most degrading modes must be re-calculated according to the procedure explained in the last section. Obviously, as shown in Chapter 1, this methodology applies only under *conservative* loading conditions, or when the non-conservative effects are so small that they may be neglected.

The step just described makes the system (structure and its loading) less unstable at the lowest ultimate critical state, characterized by the lowest eigen-value (w^0) of the load-proportionality parameter. If such an improved model structure were tested, it may reveal, on the whole, less steep theoretical negative gradients of the equilibrium paths, calculated on the assumption that the system is *perfect*. Under imperfect conditions, when the initial imperfections are present, as in a real model structure under test, this optimization step usually results in a higher peak-load at the knee of the imperfect path (see Chapter 2, Section 2.14), as this path now converges to a "flatter" negative gradient of the path for the improved equivalent perfect system, which is recalculated after each stiffening operation. Thus, a repeated recalculation of the most degrading theoretical modes, either in dynamic collapse or in unstable statical equilibrium, and the iterative stiffening of the key members in the pin- or quasi-pin-jointed system according to the procedure described herein, always results in an improved i.e. "less unstable" response of the so modified structure under the *same* loads, giving a better performance of the equivalent *real* structure under the influence of the intrinsic imperfections inherently present in *flexible* elastic members and in the assembled lattice.

The next step in an optimization procedure in lattices made of flexible members consists in eliminating the *entire* ultimate theoretical critical state. This is accomplished by stiffening *all* the key members, responsible for the generation of all buckling modes at the particular ultimate critical state, and zeroing all the flexural contractions of the members characterized by the vector $(\mathbf{e}_b^*)$, computed for the most degrading mode in the constrained minimization of the total potential energy hypersurface (v) for such a system. This is usually readily possible without any major increase in the dead weight of the lattice, as the key members in this case are usually few in number compared to the total number of bars in the lattice. (We saw that in the numerical examples of the model reticulated shells, Nos. 1, 2 and 3, the number (b) of such key members was for Models Nos. 1 and 2 equal to or less than 5 and for Model No. 3 equal to or less than 10, whereas the total number of members was 506. So, 5 to 10 members constitute only about 1% to 2% of the total number of bars and approximately the same amount of dead weight of the lattice.)

After the stiffening of all key critical bars, the lowest ultimate theoretical critical state of the original system is eliminated and the load parameter (w) may be increased further, until the next higher ultimate critical state is reached. This is accomplished following the numerical procedure described in Section 3.2. There, the optimization procedure is repeated, as described for the first or lowest ultimate critical state. The consecutive optimization is carried out until a practical limit of only small negative or positive total potential energy changes is obtained for the consecutively stiffened lattice, such that only drastically reduced amounts of the total potential energy may be available for conversion into kinetic energy in the case of a physical collapse of the system. It is sometimes possible to eliminate all unstable modes and stabilize the system completely by means of such selective stiffening of its members. This may not be possible under certain global geometrical configurations of the lattice, especially not in spherical or ellipsoidal or oval reticulated shells subjected to hydrostatic external pressures. However, certain improvements of the structural performance of such structures are usually possible, as we saw in the three model shells studied.

Numerical studies of this kind show that for lattices with a certain circular symmetry in the global geometrical configuration, subjected to circular symmetrical loadings, such as hydrostatic pressure, the critical states may *accumulate* on the loading path, so that the lowest state, characterized by the load parameter

(w^0), is followed closely by a number of other adjacent critical states as the structure is selectively and consecutively stiffened and the lowest critical states eliminated, so that the load parameter may be increased further. In fact one can speak of a "cluster" or of an accumulation of such adjacent critical states. We have shown in Ref. (15), Chapter 1, that in such a case "coupled" buckling modes become possible in which the primary branching equilibrium paths, such as we have evaluated for the models under consideration, may be coupled by new or *secondary* equilibrium paths which branch from the primary paths at secondary branching points. If this is the case, the possibility always exists that new unforeseen equilibrium states or paths of coupled buckling modes in a dynamic collapse or in unstable equilibrium may develop in the actual buckling process.

It is therefore necessary to isolate the ultimate critical state in an optimization process in order that the optimized equilibrium paths may develop in predictable *primary* post-buckling modes or loading paths at a predetermined critical value (w^0) of the load parameter. Similarly, dynamic collapse modes should be so conditioned that dynamic collapse modes become predictable and associated with a minimum of convertible total potential energy. To accomplish this and to eliminate or reduce the possibility of coupled buckling in such elastic systems, it is necessary to eliminate *a whole cluster* of adjacent ultimate critical states and to isolate a particular state, where a further elimination of the most degrading *primary* modes can be accomplished by a successive selective stiffening of the key sets of critical bars, as explained.

In practical design, when such lattices are made of flexible members, *specially designed connectors* should be used. Such special connectors and the relevent fittings for the assembly of a quasi-pin-jointed lattice are discussed in a separate subsequent section. In practice only the *potential* of the system to develop a predetermined *optimal mode* of collapse is required, as the loads are never increased to the catastrophic level, but are limited by a factor of safety to a subcritical level. If, under controllable laboratory conditions, such a collapse mechanism is allowed to develop in unstable equilibrium, or in an unstable accelerated motion away from the initial configuration, energy is dissipated in a localized controlled plastic deformation inside a restricted region of a specially designed part of the connector pins[10] in such a system. If the available convertible total elastic potential energy in this state is kept low, then the energy which may be absorbed in the totality of these localized plastic deformations may be greater than the available convertible energy and the system may *not* be launched into an onset of unstable motion, but collapse under gradually increasing localized plastic deformations in the restricted regions of the connector pins (at more or less constant or only slightly increasing or decreasing external loads), while the flexible members remain elastic in this process. Such localized plastic zones approximate then the plastic-hinges, which develop in specially designed restricted regions of the connector pin. Thus, a lattice pin-jointed by such

[10] e.g. The **Brishell System**, US. Patent 4,646,504, 1987.

connectors may be more efficiently optimized for global stability, as the remainder of the available total potential energy is not converted into kinetic energy, but dissipated or absorbed as heat in the localized plastic deformation of selected connector-pins. In this way dynamic or static collapse of an unstable *elastic* lattice may be mitigated or entirely prevented in the *limiting* or ultimate state under critical loading. The extent to which this is possible depends, in practice, on the degree to which unstable or degrading modes can be eliminated by a selective stiffening of the key members and, not least, on the global geometry of the lattice and its loading, as already described. Therefore, for this kind of optimization, some global geometries of the lattice are better in this respect than others. The elastic stability of the system depends strongly on the global geometrical configuration of the lattice, on its loading and on the way in which such a lattice is supported on a suitable foundation.

Structural properties of the connectors employed in connecting the assembled lattice are most important for the choice of the optimization strategy. Articulated joints in three dimensions in space are usually not feasible in practice and, if they are manufactured for special prefabricated deployable space structures, such articulated joints can only be implemented with large eccentricities. If special care is not taken to prevent this, a "local kinematic mechanism" caused by several such excentric articulations may develop in or near a critical equilibrium state, resulting in a collapse of the system, as such a mechanism can no longer be controlled. Therefore, quasi-pinned connectors, that is to say, such that approximate to an ideal hinge-action, must be carefully chosen and only a controlled hinge action, as that described, is practically useful. Practical optimization of lattices made of flexible light members for an improved elastic stability can be only achieved, if quasi-pinned connectors with predictable end-restraints in the members can be successfully physically implemented.

It should be mentioned in this regard that not all the members in a lattice connected by such connectors are affected by this optimization strategy. Many members in the less severely stressed parts of the lattice may be indifferent to this optimization and not be affected, especially if the collapse mechanism is local or partial. Such bars may be modified subject to *further optimization objectives*. For example, when the lattice is subjected to the working *subcritical* loads, which are limited by a factor of safety, the pinned connectors behave as *rigid* connectors and the whole structure, including the connector-pins is elastic and rigidly-jointed. Under those conditions the conventional elastic theories may apply, if the members are not too flexible. The members not needed for the global elastic stability optimization in the limiting range of loading may be partly modified to suit a further optimization objective under the working loads, such as an optimal local or global elastic stiffness, limited elastic deformation tolerances in certain parts of the lattice under the working loads or a reduced dead weight of the optimized lattice etc.

Often, it is not required to increase the limiting loads at a particular ultimate critical state, but only to eliminate the most degrading static or dynamic buckling modes at such a state and to *decrease* at the same time the dead weight of the structure. This is the case, for example, in many light space beams, reticulated dishes and other latticed support-structures for antennas, reactors and other instruments on space stations and on large orbiting bodies and satellites in outer space, where light reticulated structures of all types may be deployed or erected in a modular fashion. It is then of paramount importance to optimize and to make such systems as light as possible. In the first place, this reduces the cost of their transportation into outer space and secondly, safety and the maximum efficiency and functionality in their subsequent structural performance must be guaranteed by means of such optimization.

In such a case, the optimization problem becomes to reduce the dead weight of the respective lattice while keeping its load-carrying-capacity for a required loading scenario more or less constant. This type of optimization is relevant in all applications in outer space, but also in all cases on the earth, where the deformations caused by heavy masses in heavy trusses impair the efficiency of the system and of the instruments which such a structure may carry or support. An example of this kind are large radio-telescopes on earth. Rotation of the heavy telescope-dish during an operation may cause the local or global elastic deformations during an observation to become sufficiently large to interfere with and to distort the electromagnetic signals impinging on and being reflected from the telescope-dish, which such a heavy truss-type structure may be supporting. There, the problem becomes how to make the supporting lattice lighter, yet sufficiently stiff to carry the telescope without causing these undesirable distortions.

At the same time, such telescope supporting-structures must be designed to survive a catastrophic loading caused by a storm or by very strong winds, which the structure must overcome unimpaired. This becomes a problem in the minimization of the mass or dead weight of such a lattice and of a maximization of its *elastic* stability limit near the ultimate critical state of loading. Clearly, the problem is more complex than the one which we discussed and which concerns only the increased elastic stability of such systems under extreme loading. A full discussion of this and other related optimization problems is beyond the scope of this text. The preceeding theories, however, provide a basis and a starting point for the solution of such problems.

The main structural features of the connectors and the related fittings, which may be employed to build optimizable lightweight lattices in practice are summarized in the next section.

3.12 Structural and Material Features of Practical Optimizable Elastic Systems Pin-Jointed by Special Connectors—The Brishell Systems

Space lattices made from relatively slender quasi-pinned members can be shaped into arbitrary global geometries to fulfil an important role as high quality structural systems, capable of transmitting relatively very high useful loads to the foundation. They exhibit high useful load carrying capacity, compared to their own weight. The applications may be architectural, or in structural, industrial and especially in aero-space engineering, where coverage of and deployment over fairly large areas without intermediate supports is required. Up to now, mostly steel members have been used to form such lattices, supported on the ground, which consist of three-dimensional polyhedral elements, made from high-tensile tubular or solid steel members.

The present trends in the development of such systems are oriented towards composite structural materials. There are several advantages to be derived from light tubular members made from fibre-reinforced composite materials, such as fibre-glass, keflar (aramid fibre-) or carbon fibre-reinforced composites, which may be formed into tubular or solid members using winding or pultrusion manufacturing techniques resulting in superior mechanical properties, such as lightweight, high elastic modulus, a relatively very small thermal expansion coefficient in the longitudinal direction, while other physical characteristics of such structural composites can be maintained under control.

Structural lattices of this type may be enveloped by a cover-shell, whose function it is to transmit distributed loads through the peripheral nodes into the lattice. A relatively thin cover-shell has no other structural functions and it does not contribute appreciably to the global stability of the assembled system. A cover-shell or plate may be made thicker and it may be structurally integrated into the polyhedral lattice to form a new type of a *composite* reticulated shell.

As we saw in the preceding sections, a system of this type can be optimized for an unusually high global elastic stability limit, so that the global elastic stiffness of the system can be utilized under stable conditions, this resulting in a relatively high loading-capacity to dead weight ratio, which may be by an order of a magnitude higher (say 10 to 100 times) than that encountered in comparable heavy conventional structures of this type. There are definite economic advantages to be derived from new reticulated structures of this type, either made from steel or composite members, not only where the coverage of buildings is required, but also in special applications in maritime, aerospace and industrial engineering.

We refer to generic systems of this type, optimized and designed according to the principles explained earlier in this and the last Chapter, as the Brishell Systems[11]. In addition to its versatility of application, such a system is characterized by light weight, transportability, ease of assembly and erection, alignment to a predetermined geometry and if made from composite materials, by corrosion-resistance. Its chief advantage over other optically similar systems is that it can be optimized for a relatively high global elastic stability limit.

---

[11] S. J. Britvec, U.S. Patent, 4,646,504, 1987.

This optimization is made possible, by means of special connectors developed for this purpose. These are fabricated to make the system function as a quasi-pin-jointed system in the limiting range of loading, when the axial loads in the critical compressive members reach their critical or near-critical values. Therefore, essential for the high load-bearing capacity of such lattices are such special connectors, selectively fitted between the member ends and socketed nodal spheres, to which the members are attached. These connectors are rigid under the working loads, when the localized bending flexibilities in them are not required, thereby increasing the stiffness of the structure against elastic deformation. But, under the prescribed limiting loads a connector part may develop under a *prescribed* axial load a high local bending flexibility and rotation in a bending mode resembling a plastic hinge. Selectively fitted connectors of this type then permit the formation of safe constrained statically and kinematically admissable buckling mechanisms in the lattice, which develop under relatively stable and controllable conditions, so that a catastrophic failure beyond the ultimate or limiting loading range may be thereby mitigated or eliminated. This system then embodies a technical means for the implementation of the global optimization results based on the methodologies expounded in the last and the previous sections in this Chapter. Such connectors also permit the removal and insertion of members from and into the assembled lattice between any two fixed nodes. Moreover, they make it possible to design the members with precisely *predetermined* axial critical loads and therefore, enable a precise calculation of the stability limit of the entire structure.

A typical assembled connector of this type is shown in Fig. 3.12.1. Two opposite nodal spheres labelled by the number (30) are shown in the cross-sections. The tightening nuts (10), (11) and (19), (20) keep the oppositely threaded connector

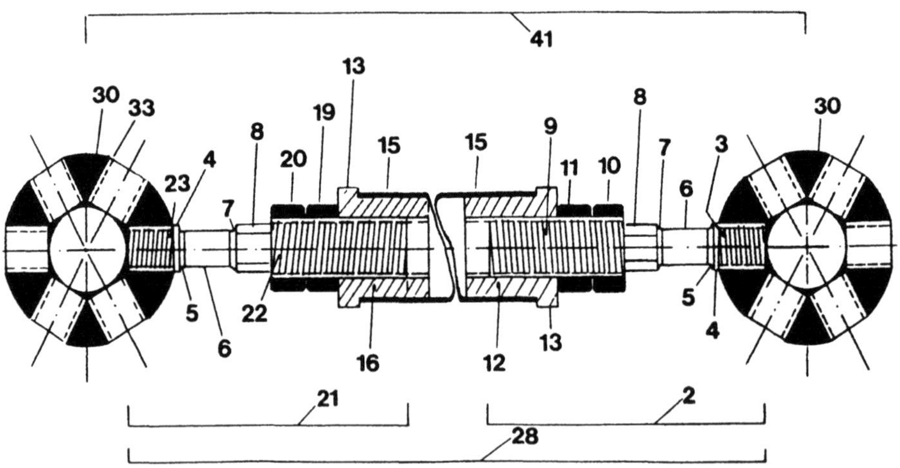

Figure 3.12.1
Typical connector in assembled configuration (U.S. Patent 4,646, 504, 1987)

pins (2) and (21) securely fastened in their working positions. These pins consist of several geometrically profiled parts and they are threaded in opposite directions, so that by rotating the loosened pins, these can be withdrawn from the nodal spheres. Once the pins have relinquished the nodal spheres their further withdrawal may be accomplished by rotating a typical member (1) itself (shown also in Fig. 3.12.2a), while the stationary pins are being withdrawn into the hollow end-pieces (12) and (16). Thereby, a member may be dismantled or inserted without disturbing the alignment of the nodal spheres. The portions marked (5), (6) and (7) have special geometrical characteristics necessary for the proper functioning of the pins during the localized plastic bending deformation in the formation of a post-buckling mechanism.

Fig. 3.12.2a shows a typical tetrahedron space element consisting of six typical members marked (1) jointed by means of the nodal spheres (30) and the oppositely threaded connectors (2) and (21). A typical nodal sphere (30) with two threaded holes (31), (32) is shown in Fig. 3.12.2b. In Fig. 3.12.2c an array of such tetrahedron–octahedron space elements is shown connected by nodal spheres at the joints (A) to (K).

A more complex array of cube–tetrahedron space elements in the shape of a hyperbolic paraboloid is displayed in Fig. 3.12.3 and an array of tetrahedron–octahedron space elements in the shape of another paraboloid reticulated shell in the form of a cooling tower in Fig. 3.12.4.

A typical nodal sphere and a matching connector of the Brishell type are shown in Fig. 3.12.5a. The connector is rigid under the working loads, as shown in Fig. 3.12.5a. Under the *predetermined axial critical load* a localized rotation develops in the connector-pin which makes the connection quasi pin-jointed in space. A localized hinge-like rotation developed by a specimen connector-pin of the

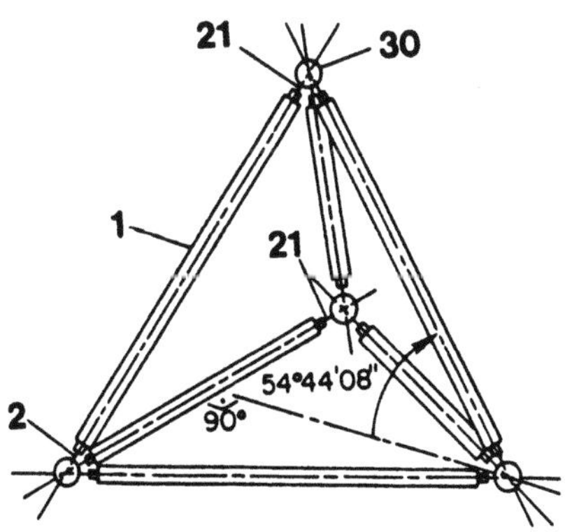

Figure 3.12.2a

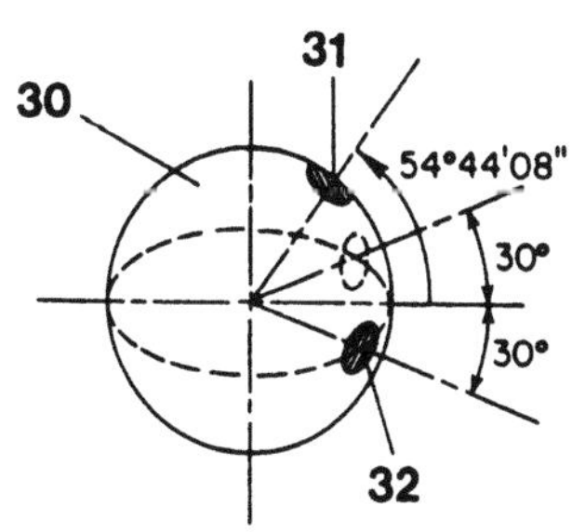

Figure 3.12.2b

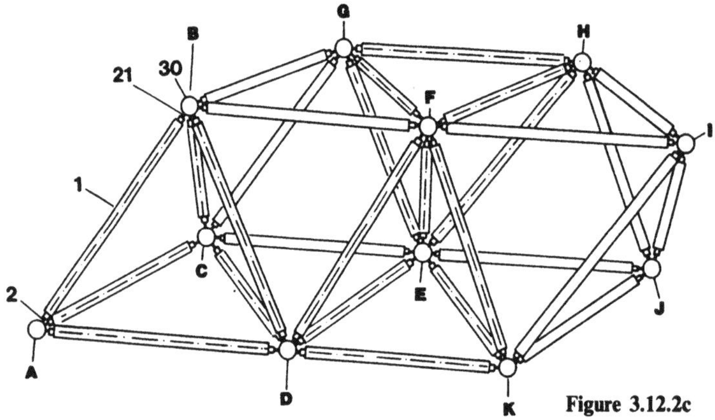

Figure 3.12.2c

Figure 3.12.2c
Typical arrays of polyhedral space elements in a lattice

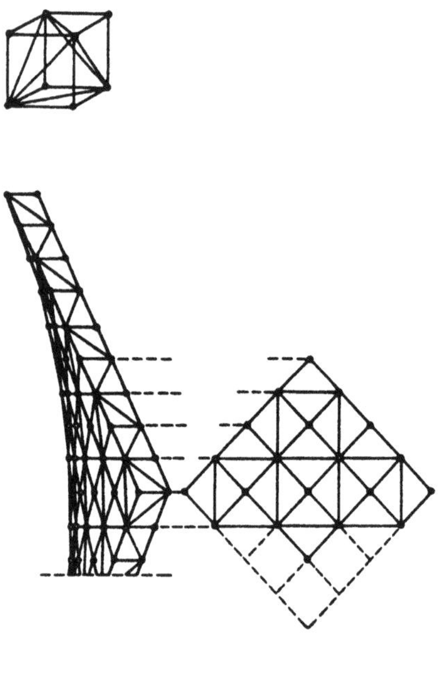

Figure 3.12.3

Brishell type, subjected to a predetermined critical axial load in a test, is shown in Fig. 3.12.5b. This makes it possible to optimize space lattices of this type in the ultimate critical state of loading, as explained before, to develop optimal collapse mechanisms at precisely predetermined levels of loading and, thereby, higher global elastic stability limits and stiffnesses under stable conditions, resulting in higher useful load-carrying capacity to weight ratios of lightweight space lattices.

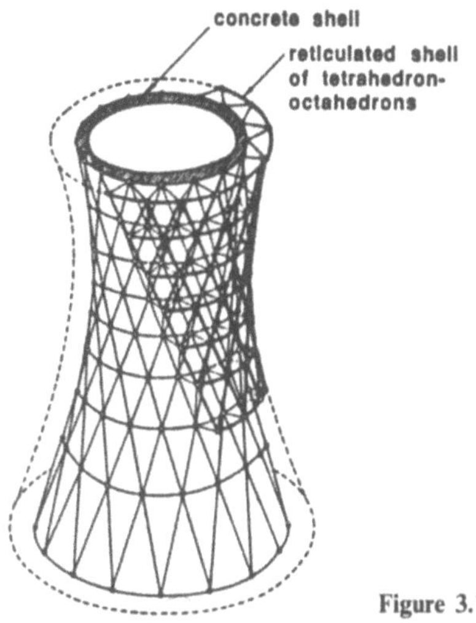

Figure 3.12.4

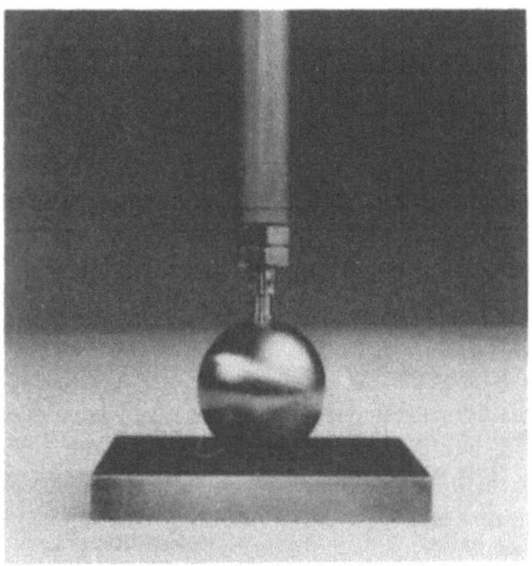

Figure 3.12.5a
Typical rigid connector and nodal sphere under the working axial loads

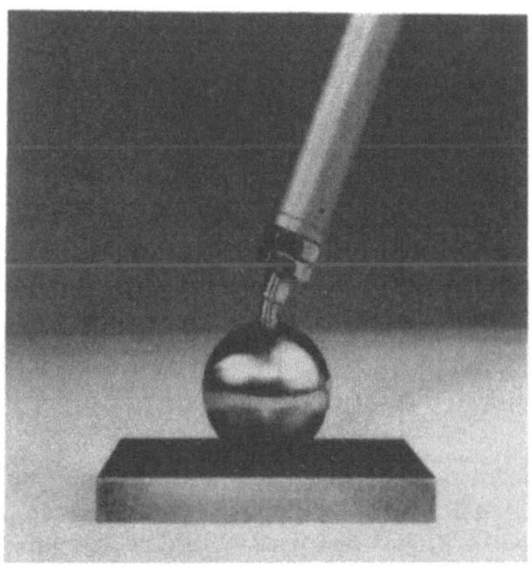

Figure 3.12.5b
A localized rotation develops in the Brishell connector under test subjected to a critical predetermined axial load, which makes the connection quasi pin-jointed in space

References and the Related Bibliography

1. Aitken, A.C., 1958. **Determinants and Matrices**, Oliver & Boyd, Edinburgh
2. Atluri, S.N. and Amos, A.K., 1987. **Large Space Structures: Dynamics and Control,** Springer Verlag
3. Bathe, K.J. and Wilson, E.L., 1976. **Numerical Methods in Finite Element Analysis,** Prentice Hall Inc. Englewood Cliffs, N.J.
4. Bellmann, R., 1960. **Introduction to Matrix Analysis,** MacGraw Hill, New York, p. 328
5. Bradbury, W.W. and Fletcher, R., 1966. **New Iteration Methods of Solution of the Eigenvalue Problem**, Numerical Math., Vol. 9
6. Britvec, S.J., 1960. **The Post-Buckling Behavior of Frames**, Dissertation, Cambridge University, Cambridge, England
7. Britvec, S.J., 1963. **Overall Stability of Pin-Jointed Frameworks after the Onset of Elastic Buckling**, Ingenieur-Archiv, Vol. 3, pp. 443–452
8. Britvec, S.J., 1963. **Elastic Buckling of Pin-Jointed Frames**, Int. J. Mech. Sci., Vol. 5, pp. 447–461
9. Britvec, S.J. and Chilver, A.H., 1963. **Elastic Buckling of Rigidly-Jointed Plane Frames**, Jour. Eng. Mech. Div. ASCE, Vol. 89, pp. 217–255
10. Britvec, S.J. and Chilver, A.H., 1963. **Elastic Buckling of Rigidly Jointed Braced Frames** (A discussion of axial extensibility in the members), Jour. Eng. Mech. Div. ASCE, Vol. 89, pp. 257–263
11. Britvec, S.J., 1966. **Some Aspects of the Elastic Stability of Pin-Jointed Space Frameworks and Reticulated Surfaces**, International Conference on Space Structures, Proceedings, University of Surrey, England and "Space Structures", pp. 417–484, John Wiley & Sons, Inc. New York
12. Britvec, S.J., 1966. **Sur le flambage thermique des anneaux et coques cylindriques précontraints**, Journal de Mechanique, Vol. 5, No. 4, Decembre 1966
13. Britvec, S.J., 1972. **A Treatise on the Stability of Elastic Systems**, Habilitationsschrift, University of Stuttgart, Also, National Science Foundation Report No. GK2879, University of Pittsburgh, Pennsylvania, August 1972
14. Britvec, S.J. and Yu, M.T., 1972. **Buckling of a Hyperstatic Segment of a Reticulated Cylindrical Shell**, Report No. GK20 National Science Foundation, Washington D.C. and University of Pittsburgh, PA., January 1972
15. Britvec, S.J., 1973. **The Stability of Elastic Systems**, Pergamon Press, Oxford, New York, (460 pp.)
16. Britvec, S.J. and Nardini, D., 1976. **Some Aspects of the Nonlinear Elastic Behavior and Instability of Reticulated Shell-Type Systems**, Developments in Theoretical and Applied Mechanics, 8th SECTAM, Blacksburg, Va.
17. Britvec, S.J., 1977. **Einige Bemerkungen über das Verhalten und die Berechnung von hochbelastbaren Flächentragwerken**, Proceedings GAMM, Graz, 1976, published in ZAMM 57, T59–T60, 1977
18. Britvec, S.J., Manacorda, T., Wesolowski, Z., Wozniak, C. and Seth, B.R., 1979. **Nonlinear Dynamics of Elastic Bodies—On the Nonlinear Behavior and the Stability of Reticulated Elastic Systems**, Edit. Z. Wesolowski, Springer Verlag, Wien, Austria and New York, N.Y.
19. Britvec, S.J., 1980. **High-Pressure Shells in Off-Shore Engineering I**—Existing Approaches, Their Limitations and Alternatives in the Design of High-Pressure Shells for Subsea Applications—**The Post-Buckling Analysis of Reticulated Shells.**—Paper presented in the Lyapunov Session on Modern Problems in Off-Shore Engineering, The International Centre for Mechanical Sciences (CISM), Udine, Italy
20. Britvec, S.J. and Davister, M.D., 1980. High-Pressure Shells in Off-Shore Engineering II.—**An Investigation of a Model Reticulated Shell Under Large Hydrostatic Pressures**, Paper presented in the Lyapunov Session on Modern Problems in Off-Shore Engineering at the International Centre for Mechanical Science (CISM), Udine, Italy
21. Britvec, S.J., 1982. **Finite Element Analysis of the Post-Buckling Equilibrium and the Dynamic Collapse of Rigidly-Jointed Large Space Structure and Reticulated Shells**, Memorandum No. 296, Analytical Mechanics Division, Martin-Marietta Corp., Denver, Colorado
22. Britvec, S.J. and Davister, M.D. **Post-Buckling Equilibrium of Hyperstatic Lattices**, Paper presented at the ASCE/EMD Specialty Conference Purdue University, West Lafayette, Indiana, May 23–25, 1983
23. Britvec, S.J. and Davister, M.D. **Post-Buckling Equilibrium of Hyperstatic Lattices**, Jour. of Eng. Mech. ASCE, Vol. 111, No. 3, March 1985, presented at the ASME/ASCE Mechanics Conference in Albuquerque, N.M., June 24–26, 1985

24. Britvec, S.J., 1987. **The Response of Slender Flexible Space Structures and The Control of Their Stability and Shape**, Final Report, Naval Research Laboratory (NRL) Project No. N00014-86-K-2014, Department of Mechanical Engineering, University of Delaware, Newark, Delaware 19716, July 1987
25. Britvec, S.J., 1987. **Nonlinear Response and Stability of Flexible Space Structures,** Original Manuscript, Research Project No. N00014-86-K-2014, Supported by the Naval Research Laboratory (NRL), Department of Mechanical Engineering, University of Delaware, Newark, Delaware 19716, July 1987
26. Chilver, A.H. and Britvec, S.J., 1963. **The Plastic Buckling of Aluminium Columns**, Proceedings, Symposium on Aluminium in Structural Engineering, paper 4, pp. 1–12; also, **Aluminium Federation Symposium** (pp. 47–58), Aluminium Federation, London 1964
27. Davister, M.D., 1983. **The Post-Buckling Equilibrium and the Dynamic Collapse of Complex Hyperstatic Pin-Jointed Lattices and Reticulated Shells**, Ph.D. Thesis, University of Colorado at Boulder, Dept. of Civil Engineering, (approved in 1983). (University Microfilms Int., 300 N. Zeeb Road, Ann Arbor, Mi. 48106, U.S.A.)
28. Davister, M.D. and Britvec, S.J., 1984. **Evaluation of The Most Unstable Dynamic-Buckling Mode of Pin-Jointed Space Lattices**, Proceedings of the 3rd International Conference on Space Structures, University of Surrey, Guildford, UK, 11–14 September, (pp. 468–473)
29. Fletcher, R. and Reeves, C.M., 1963. **Function Minimization by Conjugate Gradients**, Computer Journal, Vol. 6 (2)
30. Fox, A.L., 1971. **Optimization Methods for Engineering Design**, Addison-Wesley Publishing Company 1971
31. Fox, R.L. and Kapoor, M.P., 1968. **A Minimization Method for The Solution of Eigenproblems Arising in Structural Dynamics**, Proceedings, 2nd Conference on Matrix Methods in Structural Mechanics, Wright-Patterson AFB, Ohio
32. Grill P.E. and Murray, W., Editors, 1974. **Numerical Methods for Constrained Optimization**, Academic Press, New York
33. Huseyin, K., 1986. **Multiple Parameter Stability Theory and its Applications**, Oxford Engineering Science Series 18, Clarendore Press, Oxford
34. Klöppel, K. and Schardt, R. 1962. **Zur Berechnung von Netzkuppeln**, Der Stahlbau, Vol. 31, pp. 129–136
35. Kondoh, K. and Atluri, S.N., 1985. **Influence of Local Buckling on Global Instability: Simplified, Large Deformations, Post-Buckling Analyses of Plane Trusses**, Computers and Structures Vol. 21, No. 4, pp. 613–627
36. Kondoh, K. and Atluri, S.N., 1986. **A Simplified Finite Element Method for Large Deformation, Post-Buckling Analyses of Large Frame Structures, Using Explicitely Derived Tangent Stiffness Matrices**, Int. Jour. for Numerical Methods in Eng., Vol. 23, pp. 69–90
37. Livesley, R.K., 1975. **Matrix Methods of Structural Analysis,** 2nd Edition, Pergamon Press, New York
38. Murray, W., 1972. **Numerical Methods for Unconstrained Optimization**, Academic Press, New York, N.Y.
39. Noor, A.K., Anderson, M.S. and Greene, W.H., 1978. **Continuum Models for Beam- and Platelike Lattice Structures**, AIAA Jour., Vol. 16, December 1978, pp. 1219–1228
40. Papadrakakis, M., 1981. **Post-Buckling Analysis of Spatial Structures by Vector Iteration Methods**, Computers and Structures, Vol. 14, No. 5–6, pp. 393–402
41. Pflüger, A., 1964. **Stabilitätsprobleme der Elastostatik**, 2. Auflage, Springer Verlag
42. Pirlet, J., 1921. **Kompendium der Statik der Baukonstruktionen**, Zweiter Band, Erster Teil, Springer Verlag
43. Roorda, J., 1965. **The Buckling Behavior of Imperfect Structural Systems**, Jour. Mech. Phys. Solids, Vol. 13, pp. 267–280
44. Roorda, J., 1965. **Stability of Structures with Small Imperfections**, Jour. Eng. Mech. Div. ASCE, Vol. 91, pp. 87–106
45. Supple, W.J., 1966. **Coupled Buckling Modes of Structures**, Thesis, Dept. of Civil and Municipal Engineering, University College, London
46. Timoshenko, S.P. and Gere, J.M., 1961. **Theory of Elastic Stability**, McGraw Hill Co.
47. Britvec, S.J., 1990. **The Post-Buckling Analysis of Rigidly-Jointed Lattices in Statical Equilibrium**, Report No. B-TF V-324, Industrieanlagen-Betriebsgesellschaft m.b.H. (IABG), Ottobrunn, Germany

Index

The Figure Source Index

Chapter 1.

The Figs. 1.1.1, 1.1.2, 1.1.3, 1.1.4, 1.1.5, 1.1.6, 1.2.1, 1.2.2, 1.2.3, 1.2.4, 1.2.5, 1.2.6, 1.2.7, and 1.2.8 are reprinted from "The Stability of Elastic Systems" by S. J. Britvec, Copyright 1973, 460 pages, with the permission from Pergamon Press Ltd., Headington Hill Hall, Oxford OX3 0BW, UK. The Figs. 1.1.3, 1.1.4, 1.1.5, 1.1.6, 1.2.1, 1.2.2, 1.2.3, 1.2.4, 1.2.5, 1.2.7 and 1.2.8 appear also in the article entitled "Elastic Buckling of Pin-Jointed Frames" by S. J. Britvec, Int. Jour. of Mechanical Sciences, Vol. 5, pp. 447–452, Copyright 1963, and are also reprinted by permission from Pergamon Press Ltd., Headington Hill Hall, Oxford OX3 0BW, UK.

Chapter 2.

The Fig. 2.2.1 is reprinted from "The Stability of Elastic Systems" by S. J. Britvec, Copyright 1973, 460 pages, with permission from Pergamon Press Ltd., Headington Hill Hall, Oxford OX3 0BW, UK.

The Figs. 2.13.1 and 2.13.2 are reprinted from the original photographs supplied by the NASA Headquarters in Washington D.C., which are gratefully acknowledged.

The Figs. 2.1.1, 2.2.1, 2.3.1, 2.6.1, 2.7.1, and the Figs. 2.14.1, 2.14.2, 2.14.3, 2.14.4a and 2.14.4b are reproduced from Ref. 25 and the Figs. 2.13.1b, 2.13.1c, 2.13.2, 2.13.3, 2.13.4, 2.13.5, 2.13.6, 2.13.7, 2.13.8, 2.13.9, 2.13.10, 2.13.11, 2.13.12, 2.13.13, 2.13.14, 2.13.15, and Figs. 2.14.5, 2.16.1, 2.16.2, 2.16.3, 2.16.4, 2.16.5, 2.16.6, 2.16.7, 2.16.8, 2.16.9, 2.16.10, 2.16.11 and 2.16.12 from the Refs. 24 and 25 reporting on the results obtained in the Sole Source Research Project No. N00014-86-K-2014 entitled "The Response of Slender Flexible Space Structures and the Control of Their Stability and Shape" conducted by S. J. Britvec as the Principal Investigator and Visiting Professor at the Department of Mechanical and Aerospace Engineering at the University of Delaware, Newark, Delaware 19716 in 1986 and 1987 and supported by the Naval Research Laboratory (NRL) in Washington D.C. The Author would like to use this opportunity to acknowledge this support, as well as, the continued interest in this research by Dr. Robert Lindberg, formerly of the Naval Research Laboratory in Washington D.C., who made this investigation possible.

Chapter 3.

The Figs. 3.1.1, Table 1, Figs. 3.9.1(a) and (b) and Table 4 are reproduced by permission of the American Society of Civil Engineers (ASCE) from the article entitled "The Post-Buckling Equilibrium of Hyperstatic Lattices" by S. J. Britvec and M. D. Davister, published in the Jour. of Engineering Mechanics, Vol. 111, No. 3, March 1985.

The Figs. 3.6.1, 3.6.2(a), 3.6.2(b), 3.6.2(c), 3.6.3(a), 3.6.3(b), 3.6.3(c), 3.6.4(a), 3.6.4(b), 3.6.4(c) 3.6.5, 3.6.6, 3.6.7, Table 2, Figs. 3.8.1, 3.8.1(a), 3.8.2, 3.8.3, 3.8.4, 3.8.5, 3.8.6, 3.8.6(a), 3.8.7, 3.8.8, 3.8.9, 3.8.10, 3.8.11, and the Tables 3(a), 3(b), 3(c) are reproduced by the permission of the authors from the papers entitled: "The Post-Buckling Analysis of Reticulated Shells" and "An Investigation of a Model Reticulated Shell Under Large Hydrostatic Pressures", by S. J. Britvec and M. D. Davister, which appeared in the series entitled: "High-Pressure Shells in Off-Shore Engineering I and II". These papers were presented in the Lyapunov-Session on Modern Problems in Off-shore Engineering held at the International Centre for Mechanical Sciences (CISM) in Udine, Italy in 1980.

The Figs. 3.10.1, 3.10.2, 3.10.3, 3.10.4, Tables 5 and 6, Figs. 3.11.1, Table 7, Fig. 3.11.2, Table 8, Fig. 3.11.3 and Table 9 are reproduced with permission of M. D. Davister from "The Post-Buckling Equilibrium and the Dynamic Collapse of Complex Hyperstatic Pin-Jointed Lattices and Reticulated Shells", Ph.D. Thesis by M. D. Davister, approved by the University of Colorado at Boulder in 1983. The Figs. 3.10.1, 3.10.2 and 3.11.1 and the Tables 6, 7, 8 and 9 appear also in the article entitled "Evaluation of the Most Unstable Dynamic-Buckling Modes of Pin-Jointed Space Lattices" by M. D. Davister and S. J. Britvec published in the Proceedings of the 3rd. International Conference on Space Structures, University of Surrey, Guildford, U.K., September 1984 and are reproduced here by permission from the Elsevier Science Publishers Ltd.

The Figs. 3.12.1, 3.12.2, 3.12.3 and 3.12.4 are reproduced from the U.S. Patent Document, Patent Number 4,646,504 of March 3, 1987 by permission of the Inventor (S. J. Britvec). The Figs. 3.12.5(a) and 3.12.5(b) show the load-responsive means of the connector-fastener covered by this patent, subjected to the **working** load axially applied to the member under test, Fig. 3.12.5(a), and this load-responsive means in action under the **predetermined critical** load, in excess of the working axial load, Fig. 3.12.5(b), when the connector-pin becomes locally plastically deformable.

BIRKHÄUSER CIVIL ENGINEERING

Vibration Problems in Structures – Practical Guidelines

Authors:

H. Bachmann,
W. Ammann,
ETH Zürich,
Switzerland,
F. Deischl,
J. Eisenmann,
I. Floegl,
G.K. Klein,
G.J. Lande,
O. Mahrenholtz,
H.G. Natke,
H. Nussbaumer,
A.J. Pretlove,
J.H. Rainer,
E-U. Saemann,
L. Steinbesser

1994. 256 pages. Hardcover
120 illustrations b/w
ISBN 3-7643-5148-9

Large structures such as factories, gymnasia, concert halls, bridges, towers, masts and chimneys can be detrimentally affected by vibrations. These vibrations can cause either serviceability problems, severely hampering the user's comfort, or safety problems.
The aim of this book is to provide structural and civil engineers working in construction and environmental engineering with practical guidelines for counteracting vibration problems.
Dynamic actions are considered from the following sources of vibration:
- human body motions,
- rotating, oscillating and impacting machines,
- wind flow,
- road traffic, railway traffic and construction work.

The main section of the book presents tools that aid in decision-making and in deriving simple solutions to cases of frequently occurring "normal" vibration problems. Complexer problems and more advanced solutions are also considered. In all cases these guidelines should enable the engineer to decide on appropriate solutions expeditiously. The appendices of the book contain fundamentals essential to the main chapters.

Please order through your bookseller or write to:
Birkhäuser Verlag AG
P.O. Box 133
CH-4010 Basel / Switzerland
FAX: ++41 / 61 / 271 76 66
e-mail:
100010.2310@compuserve.com

For orders originating in the USA or Canada:
Birkhäuser
333 Meadowlands Parkway
Secaucus, NJ 07094-2491 / USA

Please contact us for being regularly informed about our publications in this field.

Birkhäuser

Birkhäuser Verlag AG
Basel · Boston · Berlin